24.95
58

Advances in Circuits and Systems

SELECTED PAPERS ON

STATISTICAL DESIGN OF INTEGRATED CIRCUITS

EDITED BY

ANDRZEJ J. STROJWAS

Associate Professor of Electrical and Computer Engineering
Carnegie Mellon University

A SERIES PUBLISHED FOR THE
IEEE CIRCUITS AND SYSTEMS SOCIETY

IEEE PRESS

The Institute of Electrical and Electronics Engineers, Inc., New York

PRINTED IN THE UNITED STATES OF AMERICA

IEEE Order Number: PP0222-0

Library of Congress Cataloging-in-Publication Data

Statistical design of integrated circuits.

(Advances in circuits and systems)
Includes index.
"IEEE order number: PP 0222-0"—T.p. verso.
1. Integrated circuits—Design and construction—
Data processing—Statistical methods. I. Strojwas, Andrzej. II. Series.
TK7874.S776 1987 621.381′73 87-3510

ISBN 0-87942-226-2

Contents

Introduction

DUE to the statistical fluctuations inherent in the manufacturing process of any electronic circuit, there exist significant variations in the performance of these circuits. Yield is defined as the ratio of the number of circuits that meet the desired specifications to the total number of manufactured circuits. Achieving an acceptable yield is a key objective for a circuit designer. This goal can be accomplished by choosing a nominal design which maximizes the yield. We refer to such an activity as *design centering*.

Statistical analysis tools have been a salient part of CAD for a number of years. These tools began to evolve after advances in circuit simulation techniques made even complex analyses quite feasible. Computations of such quantities as design margins or element tolerances became a part of design practice. Worst-case analyses were performed for a majority of designs prior to the manufacturing process. However, since the worst-case analysis produces pessimistic results and thus overly conservative designs, statistical analysis techniques started to emerge. The first paper in this selection, by Balaban and Golembeski, presents an interesting approach to statistical analysis that uses the Monte Carlo method. The authors stress the importance of statistical modeling and characterization of circuit elements. They present relatively complex examples in which the statistical models, including correlation coefficients among the model parameters, were derived from extensive measurements made in the manufacturing process.

The next group of papers in this selection deals with the design centering problem. In order to solve this problem, it is necessary to determine the so-called feasible region, or region of acceptability, in the design parameter space. This region contains all points in the design space which result in circuits that meet the desired specifications. The paper by Director and Hachtel proposes a computationally efficient technique for feasible region approximation. Their method, called *simplicial approximation*, approximates the boundary of a feasible region in an n-dimensional design space with a polyhedron of bounding $(n - 1)$ simplices. The design centering problem is solved by determining the location of the center of the maximal hyperellipsoid inscribed within this polyhedron. The yield can be efficiently estimated by performing an inexpensive Monte Carlo analysis in the design parameter space. An alternative approach to feasible region approximation is presented in the paper by Bandler and Abdel-Malek. Their approach involves the approximation of the design constraints by low-order multidimensional polynomials which are continually updated during the design centering procedure. The paper presents algorithms for both worst-case design and design centering in terms of statistically independent design parameters. Efficient yield estimation techniques are included as well. Solid theoretical foundations for the design centering problem are developed in a paper by Polak and Sangiovanni-Vincentelli [1] in which an interested reader can find a very thorough treatment of the statistical optimization problem. The next paper, by Brayton *et al.*, proposes an interesting generalization of the yield maximization and worst-case design problems for arbitrary statistical distributions of the design parameters. This paper introduces the concept of *norm body* associated with the probability distribution of the design parameters. This concept enables the use of a linear program to inscribe a maximal norm body into a simplicial approximation of the feasible region. The simplicial approximation algorithm described above was developed under the assumption that the feasible region in the design parameter space is convex. However, the convexity assumption is not always valid and the solutions obtained using the simplicial approximation can be deceptive. In the paper by Vidigal and Director, a computationally efficient method based on sequential solution of subproblems is proposed. In each of the subproblems, the convexity asumption is valid.

An integrated approach to yield estimation and design centering is presented in the next paper by Tahim and Spence. The boundaries of the feasible region for a given nominal design and a set of element tolerances are determined via linear searches along radial directions in the design space. The set of boundary points enables the yield to be estimated. These boundary points are also used in the design centering algorithm. Efficient yield prediction using the Monte Carlo method is also the subject of the paper by Rankin and Soin. The authors propose a variance reduction technique based on the control variate method. A significant reduction in the sample size, i.e, the number of full circuit analyses, is demonstrated in this paper. The paper by Singhal and Pinel presents another approach aimed at increasing the efficiency of estimating yield and yield gradients. The authors advocate first the construction of a database containing the results of statistical simulation, and then the evaluation of yield and its gradients within the statistical optimization loop by a *parametric sampling* method. Cost-effective design centering and tolerancing algorithms are also proposed in this paper. Algorithms for yield estimation and optimization are also the subject of a paper by Styblinski *et al.* [2]. Antreich and Koblitz describe in their paper a design centering method based on a yield prediction formula which also reduces the sample size of the necessary Monte Carlo analyses. The yield prediction formula is used to aid in the iterative yield maximization algorithm. In addition, a variance prediction formula is employed to speed up the convergence.

The design centering and tolerance assignment problems described above make simplified assumptions about circuit element parameters. Therefore, these approaches are useful primarily for discrete circuits with a relatively small number of designable parameters. As indicated in the paper by Balaban and Golembeski, circuit element parameters are very often strongly correlated, and their variance has both global and local components. A paper by Maly *et al.* elaborates on the yield maximization problem formulation in the case of monolithic integrated circuits (IC's). An extension of the previous methods for yield maximization is proposed in which only fabrication process controls and layout dimensions are the designable parameters. This paper indicates the need for employing a statistical process simulator which relates a circuit element electrical behavior to the designable parameters. An early implementation of such a simulator was developed by Kennedy *et al.* in 1964 [3]. The paper by Maly and Strojwas introduces the first statistical simulator of a complete IC fabrication process, FABRICS. This simulator contains a set of fabrication step models and device models which allows for the simulation of typical bipolar IC's. The models have been designed to achieve a trade-off between the accuracy and efficiency of the simulation. The paper contains illustrations of FABRICS applications in a variety of statistical design problems. Since then, a number of statistical process simulators have been developed, including CASTAM [4] and a new generation of FABRICS [5], that are targeted at MOS IC's. Another approach to statistical modeling of MOS transistors is presented in the paper by Cox *et al.* A simplified statistical model of a MOSFET has been proposed in terms of the principal factors responsible for the statistical variations in device characteristics. The model parameters are automatically extracted from device measurements. This model has been successfully implemented in a parametric yield estimator—SPYE.

The next paper by Hocevar *et al.* deals with a yield extrapolation technique which is very effective in maximizing the yield along a search direction. Yield estimates are computed based upon a quadratic model of the circuit performance functions and correlated sampling is used to extrapolate along the search direction. This approach can handle the case of variant parameter statistics, i.e., variances changing with the mean values. An integrated approach to IC yield maximization is presented in the paper by Styblinski and Opalski. The authors describe the methodology, a set of algorithms and software tools for statistical design centering with respect to fabrication process parameters. The yield gradient estimation and yield optimization algorithms are based on the *stochastic approximation* approach and the *method of random perturbations*. The software system is presented and illustrated by yield optimization examples.

All the above papers have been devoted to performance failures and consequently to *parametric yield* prediction and maximization. However, in presently manufactured VLSI circuits catastrophic failures play a significant role in yield losses. The last paper in this selection by Maly *et al.* establishes a unified framework for yield prediction that aims at bridging the gap between the traditional concepts of parametric and catastrophic yield. A thorough classification of causes of yield loss is presented and efficient techniques for yield prediction are discussed. Finally, the yield is related to manufacturing costs which provides a common denominator for the discussion of the manufacturing process efficiency.

In conclusion, this selection of papers attempts to demonstrate the evolution of research in the statistical design of IC's. Due to space limitations, only a partial picture of this research area has been presented. We believe that the relevance of this field is increasing and that new methods must be developed to address the extremely challenging problems that exist in VLSI design and manufacturing. We hope that this selection can trigger an interest in this challenging area and provide a good starting point for researchers embarking on research in this field. Much work remains to be done before we can say that statistical design of VLSI circuits is a routine activity.

Andrzej J. Strojwas
Carnegie Mellon University
Pittsburgh, Pennsylvania

February 16, 1987

References

[1] E. Polak and A. Sangiovanni-Vincentelli, "Theoretical and Computational Aspects of the Optimal Design Centering, Tolerancing, and Tuning Problem," *IEEE Trans. Circuits and Systems*, vol. CAS-26, no. 9, pp. 795–813, September, 1979.

[2] M. A. Styblinski, J. T. Ogrodzki, L. J. Opalski and W. Strasz, "New Methods of Yield Estimation and Optimization and Their Application to Practical Problems," in *Proc. of ISCAS*, pp. 131–134. Chicago, IL, April, 1981.

[3] D. P. Kennedy, P. C. Murley and R. R. O'Brien, "A Statistical Approach to the Design of Diffused Junction Transistors," *IBM J. Res. Dev.*, vol. 8, no. 5, pp. 482–495, November, 1964.

[4] Y. Aoki, T. Toyabe, S. Asai, and T. Hagiwara, "CASTAM: A Process Variation Analysis Simulator for MOS LSI's," *IEEE Trans. Electron Devices*, vol. ED-31, pp. 1462–1467, 1984.

[5] S. R. Nassif, A. J. Strojwas and S. W. Director, "FABRICS II: A Statistically Based IC Fabrication Process Simulator," *IEEE Trans. Computer-Aided Design of Integrated Circuits and Systems*, vol. CAD-3, no. 1, pp. 40–46, January, 1984.

Statistical Analysis for Practical Circuit Design

PHILIP BALABAN, SENIOR MEMBER, IEEE, AND JOHN J. GOLEMBESKI, MEMBER, IEEE

Abstract—**Statistical analysis using the Monte Carlo method is shown to have significant value in the design of practical circuits. Three integrated circuit examples are described; a logic circuit with critical dc properties, a gate with stringent transient requirements, and a hybrid circuit which combines precise linear high-frequency and transient performance.**

Characterization and modeling problems are described in detail. A particular device family is described by deriving statistics for the parameters of a single nonlinear model, which is then used directly for dc and transient cases and can be linearized for ac applications. Parasitic elements are included for fast high-frequency applications.

Complex performance criteria are derived in a post-processor in order that the results be expressed by figures of merit which are practically meaningful, such as noise immunity of a logic gate and operating margin of a digital repeater.

I. Introduction

THE EARLY USE of CAD involved computation of nominal or worst-case circuit performance in the frequency or time domain. More recent developments in modeling and analysis programs permit relatively complex simulation and computation of such quantities as the design margin of a digital repeater (to evaluate error rate) or noise margin of a logic circuit. Worst-case analysis has been widely used in the past. It is known to be unnecessarily conservative in setting test limits which, in turn, can result in increased production costs. Statistical analysis, on the other hand, helps assure both proper performance in the field and minimum production costs.

Monte Carlo analysis is a relatively straightforward technique which is capable of dealing with difficult practical problems, and has become practical with fast computers, efficient network analysis programs, and statistical modeling and characterization of devices [1], [2], [5]. Monte Carlo analysis is also shown to be useful in identifying critical combinations of network elements, establishing their relative importance when more than one such combination exists (tolerance assignment), and in evaluating ranges of internal voltages and currents.

A key requirement of any statistical analysis is the modeling and characterization of semiconductor and other components of the circuits. New approaches and maximum accuracy are required, and some results of the considerable efforts in this area are described.

The application of the Monte Carlo method is demonstrated by three distinct analyses of two relatively complex circuits. The first is a nanosecond emitter-coupled logic (ECL) integrated logic gate which is useful in a variety of communication and switching systems. The second case is a signal equalizer for a high-speed digital transmission line. These applications are representative practical problems which require dc, ac, and transient analysis. The programs described below were written specifically for

Manuscript received January 14, 1974; revised May 6, 1974.
The authors are with Bell Laboratories, Holmdel, N.J.

Reprinted from *IEEE Trans. Circuit Syst.*, vol. CAS-22, no. 2, pp. 100–108, February 1975.

these applications, but have adequate generality to be useful in other applications as well.

A. Analysis Procedure

The Monte Carlo analysis of a circuit consists of three distinct activities. A large number of circuits are generated, each analyzed in turn, and the resulting output saved. The performance index is then computed from the statistical data.

The input contains the particular circuit description, a set of nominal input parameters and their statistical descriptions, and a list of frequencies or time steps at which the analysis is to be performed. A random number generator is employed to select individual elements from within their possible ranges. The random number generator is given a starting number which permits subsequent runs to be made with the same sets of parameters. The output of the input processor consists of as many circuits as are stipulated for a statistically significant population.

The analysis portion accepts the circuits generated sequentially by the preprocessor and analyzes each in turn. The output variables themselves are recorded on magnetic tape and saved for use by the output processor. DC and transient analysis is done with the NICAP program [3], and ac analysis with CAPECOD [4].

The output processor performs the task of assembling the results of all the analyses. It performs additional computation from the analysis results if necessary, and presents the output in the desired form (e.g., histogram of gain, medians, scatter plots, design margin, etc.).

Note that the input processor is relatively general since it is required to generate the population of components required for the given family of circuits. The output processor, however, must be tailored to the particular problem and depends to a much greater extent on the details of the particular problem.

In analyzing a large network, it is sometimes desirable to partition it into subnetworks and express statistics of the larger network in terms of the subnetwork statistics. This is done in a 2-step process. A large number of the subnetworks are generated, analyzed, and stored. These, in turn, are used as components when the larger networks are generated. This partitioning operation was done in the case of the linear signal equalizer described in Section III where the subnetworks are described by a set of frequency-dependent transfer functions.

The number of analyses performed in each Monte Carlo run was determined by an empirical rule used to gauge uncertainty in the statistics. The analysis continued until such time that a 10-percent increase in the number of cases analyzed produces no significant change in the computed statistical quantities, such as histograms. Experience to date has resulted in 200–400 cases for a typical problem.

II. Transistor Modeling for Statistical Analysis

The critical ingredient in getting correct answers to a statistical circuit analysis lies in the device modeling, the objective of which is the modeling of an entire device population. Modeling for Monte Carlo analysis requires a statistical model capable of generating individual transistor models which are then used in an analysis program. The problem was approached successfully for the cases to be described by a combination of theoretical considerations and the maximum practical use of experimental data [5], [6], [8]. It was required that the model parameters of the discrete transistors derived from the statistical model correspond to a physically realizable transistor with the proper frequency of occurrence. This requires a certain degree of accuracy[1] in the discrete and statistical models and that the sample selected for measurement be an accurate representation of the total population.

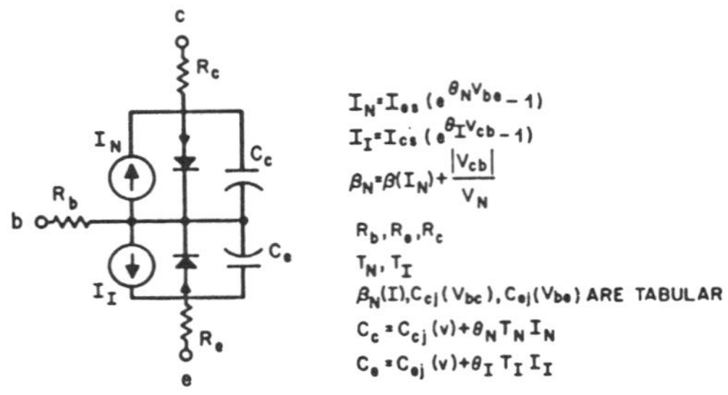

Fig. 1. Nonlinear NICAP model.

A. Nominal Model

The nonlinear NICAP model given in Fig. 1 is an extension of the Ebers–Moll model, which is useful directly in dc or transient applications, and can be linearized conveniently for linear ac analysis [9]. It is idealized in the sense that high frequency parasitic effects are ignored, although the extrinsic elements R_b and R_c are included. The capacitances are the sum of depletion and diffusion components, the relative size of each dependent on the value of the junction voltage and diffusion current. The depletion capacitances and the current gain $\beta_N(I_N)$ are made tabular to achieve good accuracy.

B. Statistical Model

Although statistical models are expected to vary between transistor types, several common problems can be identified. Scalars and functions must be characterized statistically. It must be determined which parameters are independent of the others, which are correlated, and which of the correlated parameters should be taken as independent when deriving the complete model.

A single independent parameter is described by giving its range and distribution (nominal value and tolerance). Single or multiple correlations, e.g., tolerance of resistors either within a slice or globally, can be described as

$$R = R_0[1 + T(\rho X_g + (1 - \rho)X_l)] \qquad (1)$$

where

R_0 nominal value;
T fractional tolerance;

[1] Note that the accuracy of a given transistor model is a function of the circuit being analyzed and the performance index to be evaluated.

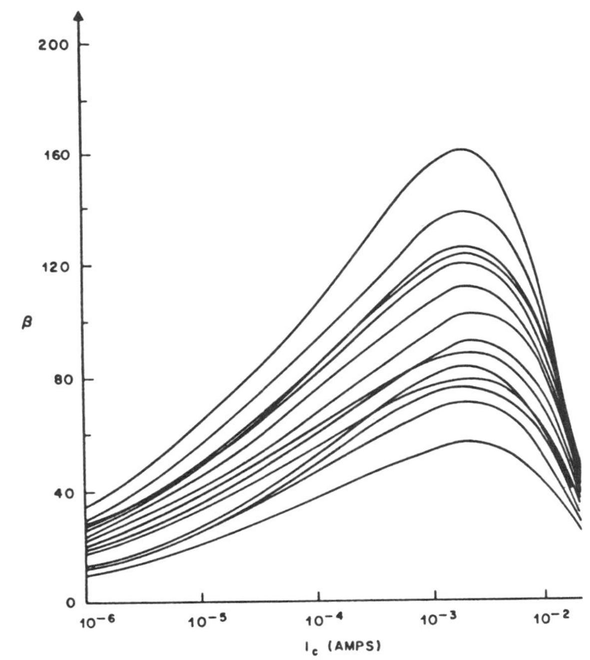

(a)

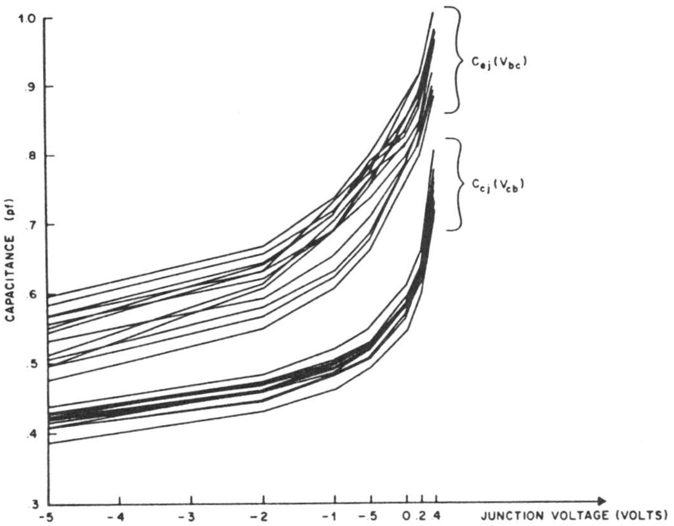

(b)

Fig. 2. (a) $\beta(I)$ for transistors tested. (b) $C_J(V)$ for transistors tested.

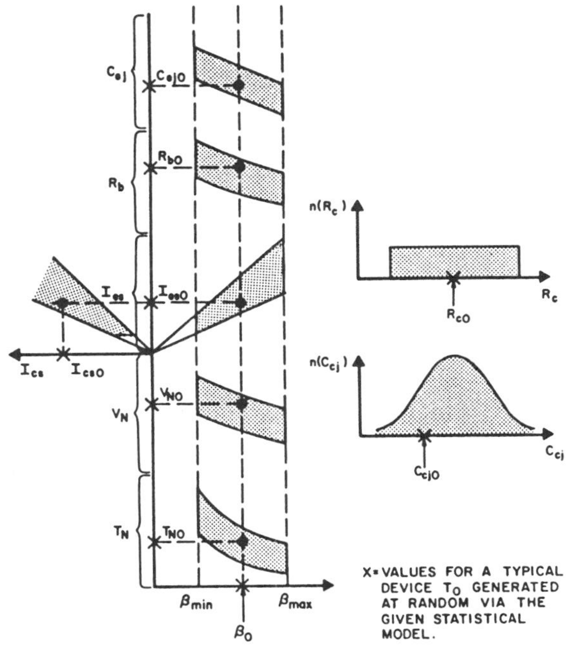

Fig. 3. Statistical model dependencies.

X_g, X_l independent random numbers $-1 \le [X_g, X_l] \le 1$;
ρ correlation between local (slice) and global variations, $0 \le \rho \le 1$.

Functionally dependent quantities such as $\beta(I)$ or $C(v)$ can be characterized in a manner suggested by inspection of measured data. Fig. 2(a) and (b) show a comparison of $\beta(I)$, $C_{cj}(v)$ and C_{ej} functions for representative sample of 40 transistors of the same type. A nominal function can be estimated for each family of curves by identifying the device nearest the median of the envelope. The nominal function is then varied about its nominal value over a range which corresponds to the width of the envelope and with a distribution determined by a histogram of β_{max} [10] or $C_{CJ}(0)$. Correlations between model parameters are determined by inspecting their scatter plots, and the ranges from the scatter plots or the corresponding histograms.

A useful statistical analysis requires accurate modeling analysis and interpretation. Modeling errors can arise when the quantity of transistor data is insufficient. Limitations of the discrete transistor model will carry over to its statistical extension. The modeling and characterization described in this paper was based on two sources of model parameters.

Extensive data measured as part of the normal manufacturing process was used to extract most of the model parameters for statistically significant population size.[2] This established the distributions for each model parameter. Correlations between model parameters were established by assembling a collection of 50 transistors thought to be representative of the population at large, modeling each completely (dc and reactive) and taking the correlations thus found as representative of the entire device population. The following example comprises the statistical characterization of a single type of planar epitaxial Si transistor used in one of the following examples. Measurements were made and the correlation among parameters of the statistical model is determined by inspection of scatter plots of measured model parameters. The results are the correlations shown in Fig. 3, in which the correlated quantities are taken as β, T_N, $C_{ej}(0)$, V_N, I_{es}, I_{cs}, and R_b. β is taken as the independent variable[3] to which others are correlated, with

[2] See Butler [10] for an approach in which large quantities of data are measured specifically for modeling purposes and then used to generate dc statistical models.

[3] These results are in agreement with those presented by Fox *et al.* [5], since their independent variable I_{DB} is strongly correlated with β.

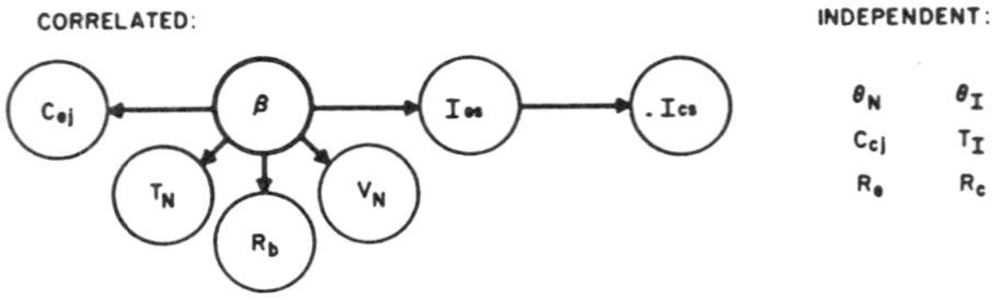

Fig. 4. Composite model statistics.

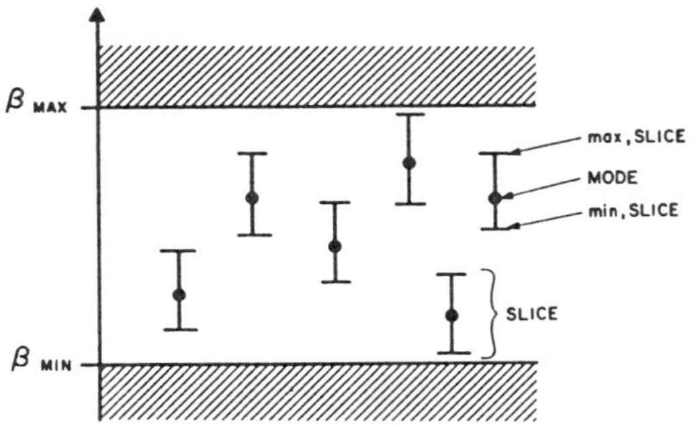

Fig. 5. Typical device population.

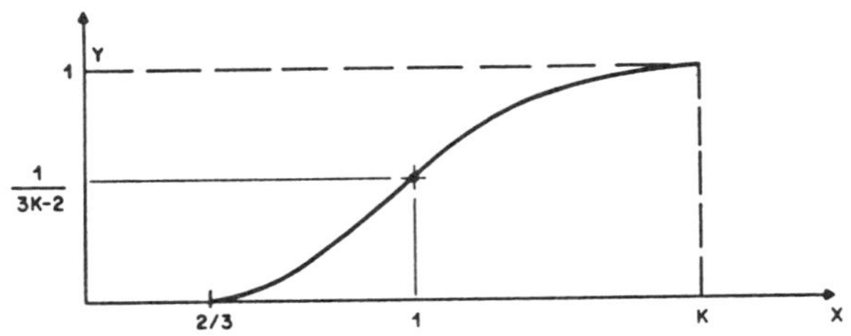

Fig. 6. Cumulative density function for slice β.

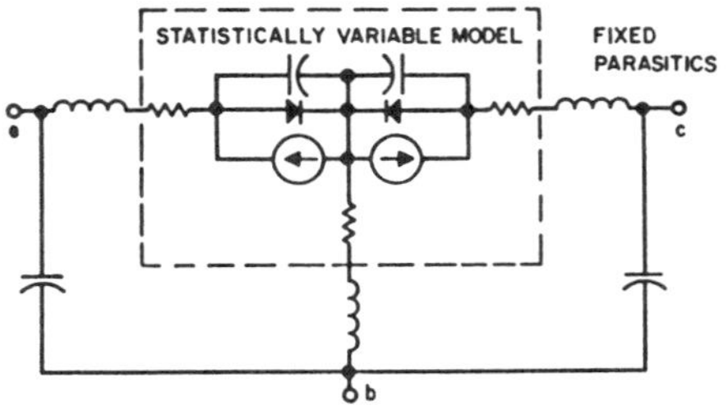

Fig. 7. Parasitic elements in statistical model.

the exception of I_{cs} which is made dependent on I_{es}. The parameters T_I, θ_N, θ_I, C_{cj}, and R_c showed no significant correlation with other model parameters and are taken as independent.

The results are summarized in the composite scatter plot illustrated in Fig. 3, which illustrate the dependencies described in Fig. 4, the range covered by each variable and the variation of range with independent variable β. The parameters for transistors selected by the input processor are generated in the following way:

$$I_{es} = 0.0448\beta_0(1 + 0.3516X_{r1})10^{-18} \text{ [A]} \quad (2)$$

$$I_{cs} = 3.45I_{es}(1 + 0.463X_{r2}) \text{ [A]} \quad (3)$$

$$R_B = (400 - 1.17\beta_0)\left(1 + \frac{1}{(8.89 - 0.026\beta_0)} X_{r3}\right) [\Omega] \quad (4)$$

$$C_{ej} = (0.833 - 0.0014\beta_0)(1 + 0.2X_{r4}) \text{ [pF]} \quad (5)$$

$$T_N = \left(2.67 + \frac{0.382}{\sqrt{\beta_0}}\right) 10^{-10}(1 + 0.2X_{r5}) \text{ [s]} \quad (6)$$

$$V_N = \left(\frac{32}{\beta_0} - 1.33\right)(1 + 0.3X_{r6}) \text{ [V]} \quad (7)$$

where X_{ri} are independent random numbers selected from a uniform distribution over the range $-1 \leq X_{ri} \leq 1$.

A typical device population is shown in Fig. 5 in which β_p, the peak value of $\beta(I)$ in Fig. 2, is the parameter illustrated. Each line represents the range of β_p for a single slice of devices; gain variations within a slice range from a minimum value $\beta_{\min}$ through the mode value β_M to a maximum value $K\beta_M$. The distribution of gain within a slice has been determined empirically for this population to be given by

$$Y = 0, \qquad \text{for } X \leq \tfrac{2}{3} \quad (8)$$

$$Y = \frac{1}{3K - 2}(9X^2 - 12X + 4), \qquad \text{for } \tfrac{2}{3} \leq X \leq 1 \quad (9)$$

$$Y = \frac{1}{(3K - 2)(1 - K)}(3X^2 - 6KX + 5K - 2), \qquad \text{for } 1 \leq X \leq K \quad (10)$$

$$Y = 1, \qquad \text{for } X > K \quad (11)$$

where Y is the empirical cumulative distribution function for $X = \beta_p/\beta_M$. β_M and K are uncorrelated, and their values are selected randomly from flat distributions with the range $40 \leq \beta_M \leq 100$ and $1.2 \leq K \leq 2.0$. An individual transistor is then identified from the distribution for Y given in (8)–(11) and Fig. 6. The table which represents the nominal curve $\beta(I)$ in Fig. 2 is scaled to achieve the value of β_p. The scatter plots of Fig. 3 are then employed, together with the required additional random numbers X_{ri}, to determine values of the remaining elements (correlated and uncorrelated) which account for local variations. This specifies the complete statistical nonlinear transistor model.

Parasitic Effects: Problems which arise from parasitic effects on device performance become important when high-frequency or fast transient simulations are required. These are associated with extrinsic reactances such as lead inductance, can capacitance, etc., and are modeled apart from the device statistical model. The parasitic effects are first determined through wide-band measurements. An arrangement of lumped reactive elements is then determined via optimization which simulates their effects [11]. Since the parasitic effects are attributed to causes extrinsic to the transistor, they are held constant during the Monte Carlo analysis. The parasitic arrangement shown in Fig. 7 permits

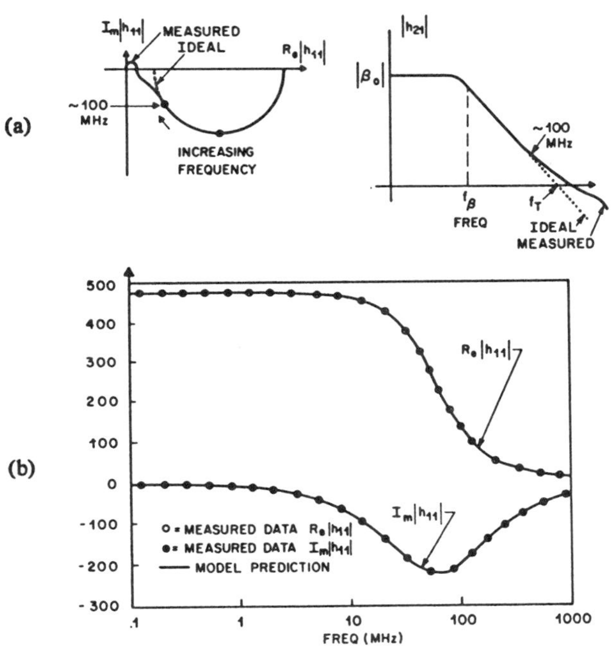

Fig. 8. Model validation.

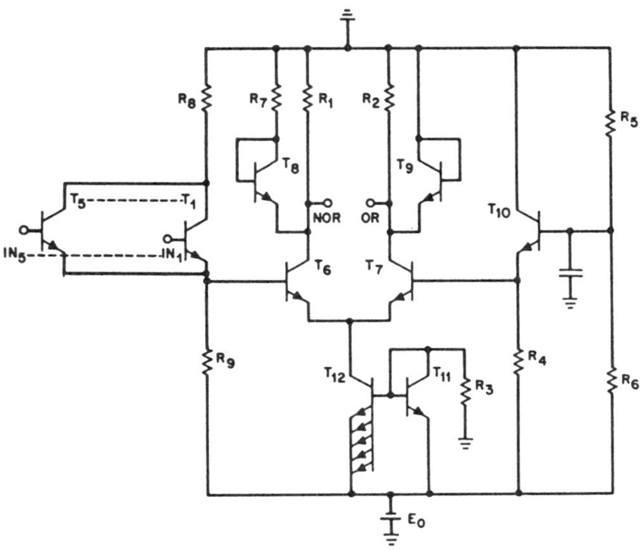

Fig. 9. Integrated ECL gate.

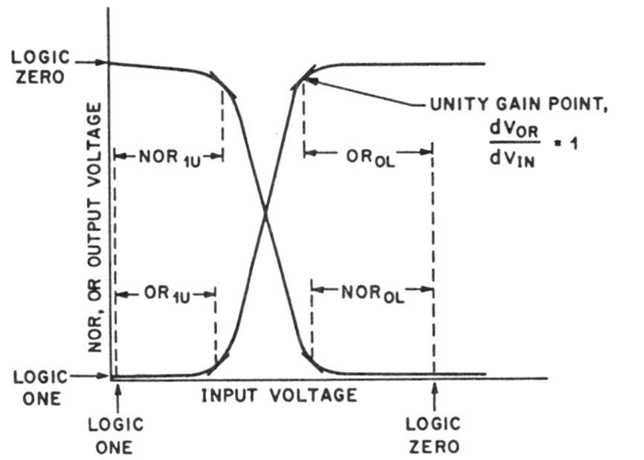

Fig. 10. DC transfer function and noise immunity definition.

the necessary exchange of transistors during the Monte Carlo analysis while the parasitic elements remain intact.

Parasitic element values are determined in a relatively straightforward manner as follows. A single transistor is measured over a wide frequency range. Experience has shown that transistors of this type exhibit little or no parasitic effects at frequencies below 100 MHz (see Fig. 8), so data below that frequency are used in optimizing element values in the intrinsic linear transistor model. These elements are then fixed at their optimum values. Parasitic elements are then allowed to vary and their values adjusted by optimization to match data at frequencies over 100 MHz. The final values of the parasitic elements are assumed to be constant for all devices of that type and held fixed during all cases in the subsequent Monte Carlo analysis. Typical results are illustrated in Fig. 8 which show close agreement at frequencies up to 1 GHz for $h_{11}(\omega)$. Similar agreement was obtained for $h_{12}(\omega)$, $h_{21}(\omega)$, and $h_{22}(\omega)$ but is not presented here.

Subsequent linearization of the NICAP model to simulate linear high-frequency performance is done by linearizing the intrinsic transistor to produce a hybrid-pi model, leaving the extrinsic ring of parasitic elements intact.

III. Transient and DC Statistical Analysis of an Integrated NOR-OR ECL Gate

The circuit to be analyzed is an integrated fast ECL gate which was designed for use in a variety of communication applications [12]. A statistical analysis was done to evaluate and compare design alternatives and to determine whether the circuit meets the design criteria for the anticipated component tolerances. The semiconductors used in the circuit were planar, epitaxial high-frequency transistors. The device modeling was done as described in Section II. Since the analysis preceded the production of the gate, the statistical transistor model was extrapolated from measurements on devices which bore a known relation to those which were to be used in the actual circuit.

The schematic of the gate is shown in Fig. 9. The gate has five inputs and OR and NOR outputs. One input was used during the analysis and the other four inputs were biased at cutoff and therefore ignored.

DC and transient analyses were performed for the evaluation of gate performance.

A. DC Analysis

The dc transfer function of the gate is shown in Fig. 10. The unity gain points (indicated by 45° tangents) are taken as a measure of noise immunity because noise above this level will be amplified. This may cause an error in a logic circuit which usually contains a cascade of such gates. The difference between the unity gain point and the appropriate logic level is therefore considered a measure of noise immunity of the gate. A calculation of absolute worst-case noise immunity is unduly pessimistic since the probability of occurrence of such a case is generally much smaller than the failure rate of the components. A statistical calculation of the distribution of noise immunity margin is a more meaningful criterion for such an assessment [12].

The analysis was performed by the method described in Section I, using the transistor model described in Section II. A composite histogram of the outputs is shown in Fig. 11 in which the logic zero level and the unity gain points are plotted. Variations in the unity gain points and logic zero voltage are attributed primarily to differences in V_{BE} and resistor tracking on the chip. The logic "1" level is determined by the input transistor base currents of the loading gates and, since it is small and proportional to fanout, it is not shown in the figure. Two numbers are given for each noise margin. The upper number applies to the median of the distribution (nominal case), and the lower number to the end points of the distribution. Thus the two values of noise margin can be classified as an expected value and an estimated worst-case value, respectively. The analysis was performed on the original circuit at 25, 60, and 95°C. The initial results led to a redesign in order that the minimum noise margins for the "1" and "0" states be equal.

The results shown in Fig. 11 apply to the final circuit design at 95°C. The estimated worst-case values were 50 mV larger than the actual peak noise anticipated in a working circuit, and the circuit design performance is considered adequate by the standard. By comparison, the absolute worst case was calculated by combining the appropriate worst-case parameters. The resulting noise margin was smaller than the noise in a working circuit, and would have erroneously predicted that the circuit would not operate correctly.

B. Transient Analysis

The gate of Fig. 9 was designed to have a propagation delay of 2 ns. Numerous discrete analyses indicated that nominal performance would be adequate, and statistical analysis was used to predict the variation of propagation delay with the device variations expected in manufacture. Accurate simulations of high-performance circuits of this type require that parasitic effects be included in the device model, as indicated in Section II, and in the circuit description by the appropriate substrate and interelectrode capacitances [13].

The nominal transient response of the gate is shown in Fig. 12. The distribution of the propagation delay was evaluated for both 0–1 and 1–0 level transfers, and the resulting histogram for the 0–1 transfer is given in Fig. 13.

The nominal analysis is confirmed, since $T = 2$ ns is located virtually at the median of the distribution. Furthermore, 90 percent of the circuits are seen to be within 2 ± 0.3 ns for this design.

An additional benefit of the Monte Carlo analysis was its use in assigning tolerances for an acceptance test. The predicted parameter distributions were considered accurate and produced acceptable range of circuit performance. This permitted the end points of the statistical distributions to be taken directly as the appropriate tolerances on the circuit specifications [14].

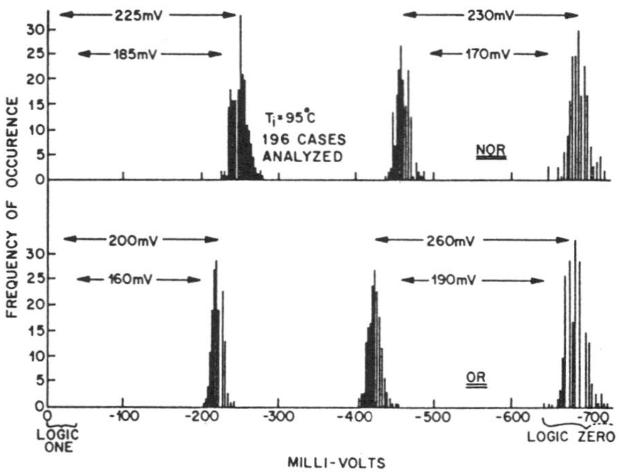

Fig. 11. Unity gain and logic levels histogram.

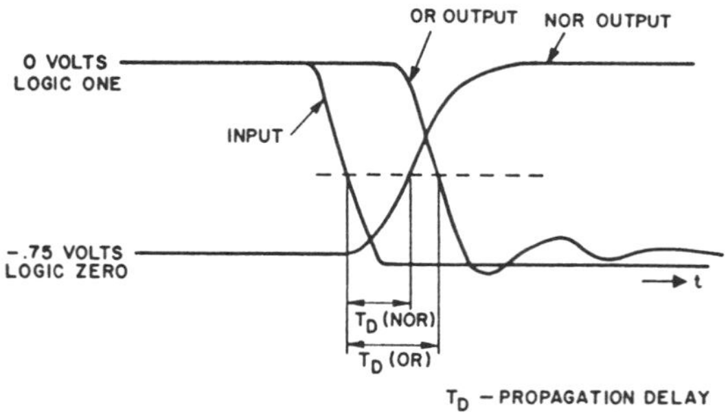

Fig. 12. Transient response of the gate, one to zero transition.

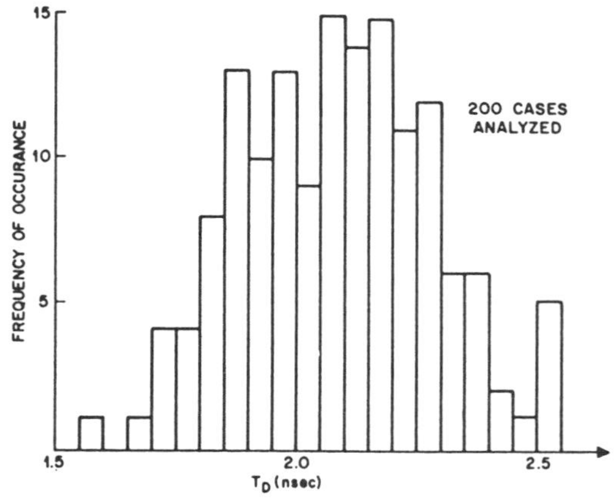

Fig. 13. Propagation delay T_D (OR) of an ECL gate

IV. Statistical Frequency Domain Analysis

A digital repeater consists of two parts: a linear equalizer and a nonlinear regenerator. The equalizer compensates for the linear distortion of the transmission line, while the regenerator decides if the equalized signal is a "0" or a "1" and generates an appropriately shaped and timed output pulse. Statistical frequency domain analysis was used to evaluate the performance of the linear equalizer of a high-speed digital repeater [15]. The primary objective of the

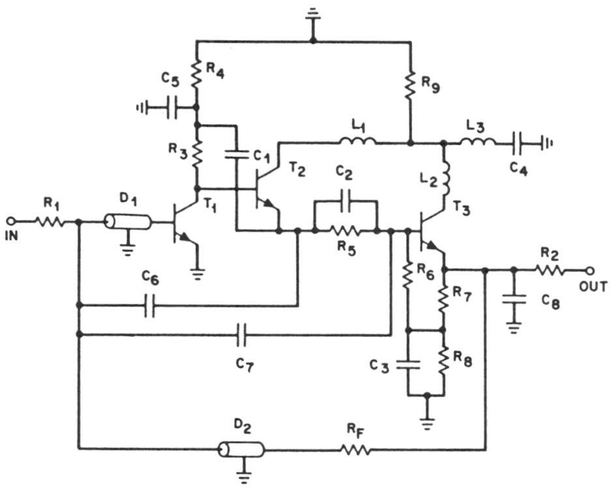

Fig. 14. Broad-band amplifier including layout parasitics.

analysis was to relate variations in the equalizer transistors to the corresponding variations in repeater performance. The transistors which affect the performance of the linear equalizer are contained in seven similar high-frequency amplifiers. The amplifiers were analyzed first by a Monte Carlo run which generated 300 circuits. Each amplifier run was saved to make up a population of candidates for assembly into repeaters, and the complete repeater was then analyzed statistically using the characterized amplifiers as building blocks. The statistical analysis was partitioned into this 2-step procedure for three reasons. First, repeater performance can conveniently be expressed in terms of the amplifier subnetwork properties; second, the study of the individual amplifiers was a necessary part of the amplifier design process; and finally, partitioning made the entire process computationally efficient.

A. Monte Carlo Analysis of the Amplifier

The amplifier of Fig. 14 is a broad-band amplifier with a nominal closed loop gain of 14 dB and 700-MHz bandwidth [16]. The primary objective placed on the amplifier can be expressed as a constraint on its frequency response, and the analysis was used to determine the effect of statistical transistor variations on amplifier response. The distribution of gain at 300 MHz was the quantity to be studied.

The statistical transistor model is based on the NICAP model which is converted to the appropriate linear small signal model in the preprocessor [9]. The dc operating points for the transistors are held essentially constant by the dc feedback and were not varied in the linearization process and the subsequent analysis.

The amplifiers were analyzed at 25 frequencies between 10 MHz and 1 GHz. Histograms of gain at each frequency and composite envelopes of the expected worst cases were produced by the output processor. A composite envelope consists of response extremes at each frequency. The envelope boundaries do not usually represent any single amplifier response. Fig. 15(a) illustrates a histogram of gain at 300 MHz, while Fig. 15(b) illustrates a composite envelope of the expected worst cases and also shows the

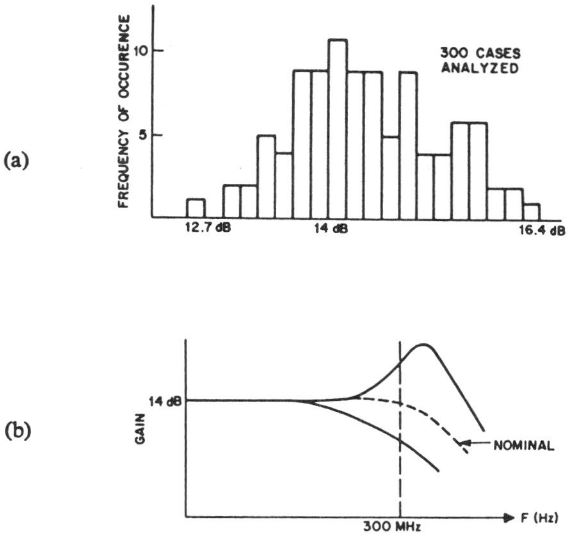

Fig. 15. Statistical response of amplifier. (a) Typical histogram at 300 MHz. (b) Composite envelopes.

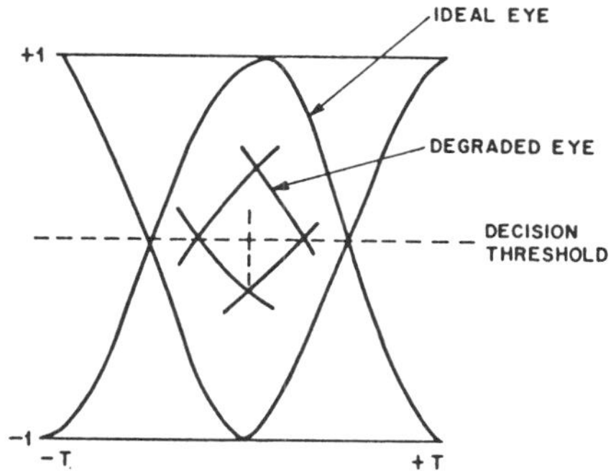

Fig. 16. Eye diagram for binary transmission

performance of the nominal circuit. Note here that the nominal performance is approximately centered within the envelope and histogram.

Analysis of the scatter plots confirmed that the collector junction capacitance C_{cj} of T_1 produced most of the variation bounded by the envelope. The Monte Carlo analysis was then used to determine the range of C_{cj} which corresponds to an acceptable range of amplifier gain characteristic. The analysis also identified the specific worst-case circuits. These amplifiers were later studied in detail to resolve questions concerned with tuning adjustments.

B. Monte Carlo Analysis of the Complete Repeater

Although the analysis of the equalizer portion of the repeater was done in the frequency domain, the scalar performance index of the repeater is defined in the time domain by the eye diagram [17] illustrated in Fig. 16. This diagram is the result of superimposing all-possible pulse sequences at two signaling intervals. The eye must be "open" to regenerate the pulse without introducing error.

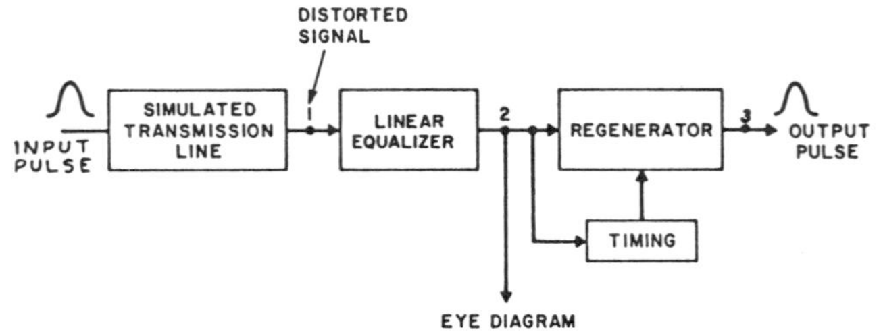

Fig. 17. Simplified block diagram of digital repeater simulator.

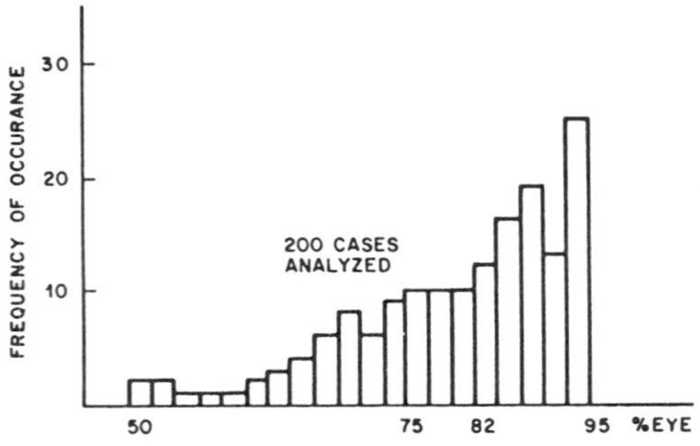

Fig. 18. Histogram of the eye for a digital repeater

In practice, pulse degradation reduces the size of the ideal eye and a suitable scalar index is the percentage of "eye opening."

In this case, the postprocessor performs the entire second step of the analysis procedure. The amplifier responses stored previously are called in groups of seven. The "eye" of the repeater, consisting of the seven randomly selected amplifiers is computed, and the histogram of the eye opening for a prescribed number of repeaters is generated.

The simulation of the repeater is done by a special purpose program (PTP) [18]. The block diagram of this simulation is shown in Fig. 17. The distorted pulse appears at the end of a simulated transmission line at point 1. This signal is shaped by the linear equalizer, which ideally would have a frequency response reciprocal to that of the transmission line. The eye diagram is generated at point 2, since this is the point at which the decision is made whether a "0" or "1" was transmitted. The regenerated pulse appears at point 3. The histogram of the eye diagram is shown in Fig. 18. The ideal eye, which corresponds to having a set of nominal amplifiers, was 95 percent.

Statistical analysis has also proven useful in assessing different strategies concerning tuning adjustments during manufacture. In the following example, two tuning philosophies are compared. The choice involves tuning individual substrates which are then assembled without further adjustment, or assembling the substrates first and tuning the resulting circuit. The decision will be based on the total number of tuning operations required. The first case is simulated by selection of seven amplifiers at random, assembling them into an equalizer and evaluating its eye diagram. (It is assumed that one adjustment per equalizer is adequate if the eye opening is unsatisfactory.) For the second case, an amplifier substrate is assumed to require

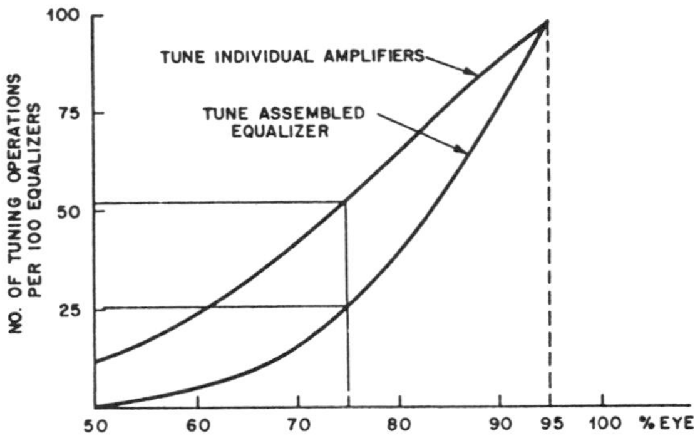

Fig. 19. Comparison of tuning methods for equalizer.

tuning if, when assembled with six nominal amplifiers, the resulting equalizer exhibits an unsatisfactory eye opening. The comparison of the two cases is summarized in Fig. 19, in which it is seen that the use of pre-tuned amplifier substrates required a greater number of tuning operations for any given eye opening. It is expected that 100 equalizers with an eye opening of 75 percent or better will require 25 tuning operations on the assembled equalizer, or 53 tuning operations on individual substrates. This is attributed to the fact that for randomly selected amplifiers some have a wide and some a narrow frequency response and therefore a certain amount of cancellation is observed in the equalizer. It is also in agreement with the law of computation of variances for a sum of random variables. Based on the above results, the adjustment and tuning is performed on the assembled equalizer rather than on each individual subnetwork. The latter would have produced an unnecessarily conservative design.

V. Conclusion

Statistical analysis has been shown to be an effective computer aid to the design of practical circuits. Through Monte Carlo analysis, factors can be evaluated which have significant manufacturing implications but are not available through other techniques. This is equivalent to the simulation, not of the circuit, but of the manufacturing process itself from the standpoint of the performance of the final product.

The cases studied are practical applications involving nonlinear dc, transient, and ac circuits. Accurate results are shown to be obtainable, but powerful analysis programs on large computers and extensive transistor characterization and modeling are required.

Acknowledgment

The authors wish to thank D. B. Brown, C. A. von Roesgen, F. D. Waldhäuer, and W. I. H. Chen, the designers of the circuits, who introduced the authors to the problems and provided circuit expertise and helpful suggestions; D. B. Brown, E. M. Butler, N. J. Elias, M. M. Hower, and W. H. Becker, who provided characterization assistance and much useful discussion on statistical characterization; N. K. Schellenberger and R. W. Kolar, authors of the PTP program, who were helpful in the problems of the eye diagram computation; and J. C. Selbo, who wrote the programs used in the analyses.

REFERENCES

[1] Special issue on "Statistical Circuit Design" *Bell Syst. Tech. J.*, vol. 50, Apr. 1971.

[2] D. M. Bohling and L. A. O'Neill, "An interactive computer approach to tolerance analysis," *IEEE Trans. Comput.*, vol. C-19, pp. 10–16, Jan. 1970.

[3] I. A. Cermak and D. B. Kirby, "Nonlinear circuits and statistical design," *Bell Syst. Tech. J.*, vol. 50, pp. 1173–1195, Apr. 1971.

[4] E. D. Walsh, unpublished material.

[5] P. E. Fox, F. C. Wernicke, and R. Narayanasamy, "Statistical analysis of digital integrated circuits," *Digest Tech. Papers, ISSCC*, vol. XV, pp. 66–67, Feb. 16–18, 1972.

[6] C. L. Wilson and R. I. Dowell, "Modeling for computation of IC-gate yield from Processing data," *Digest Tech. Papers, ISSCC*, vol. XV, pp. 68–69, Feb. 16–18, 1972.

[7] H. M. Ghosh *et al.*, "Computer-aided transistor design, characterization and optimization," *Proc. IEEE*, vol. 55, Nov. 1967.

[8] R. Berry, "Correlation of diffusion process variations with variations in electrical parameters of bipolar transistors," *Proc. IEEE*, vol. 57, Sept. 1969.

[9] J. Logan, "Characterization and modeling for statistical design," *Bell Syst. Tech. J.*, vol. 50, Apr. 1971.

[10] E. M. Butler, "Techniques for dc statistical modeling of bipolar transistors," *Proc. Int. Symp. Circuit Theory*, San Francisco, Calif., Apr. 21–24, 1974.

[11] J. J. Golembeski, "Linear circuit models derived via computer optimization," *IEEE Trans. Educ., Special Issue on CAD*, vol.E-12, pp. 152–222, Sept. 1969.

[12] D. B. Brown, unpublished material.

[13] P. Balaban, "Calculation of the capacitance coefficients of planar conductors on a dielectric surface," *IEEE Trans. Circuit Theory*, vol. CT-20, pp. 725–731, Nov. 1973.

[14] C. A. von Roesgen, personal communication.

[15] F. D. Waldhauer, *High Speed Digital Repeater*, to be published.

[16] ——, "Feedback conceptual or physical?" *IEEE Proc. Int. Symp. Circuit Theory*, Toronto, Ont., Canada, 1973.

[17] *Transmission Systems for Communication*, Bell Laboratories, 1970, p. 626–676.

[18] R. W. Kolor and N. K. Schellenberger, unpublished material.

The Simplicial Approximation Approach to Design Centering

STEPHEN W. DIRECTOR, SENIOR MEMBER, IEEE, AND GARY D. HACHTEL, MEMBER, IEEE

Abstract—The basis of a method for designing circuits in the face of parameter uncertainties is described. This method is computationally cheaper than those methods which employ Monte Carlo analysis and nonlinear programming techniques, gives more useful information, and more directly addresses the central problem of design centering. The method, called simplicial approximation, locates and approximates the boundary of the feasible region of an n-dimensional design space with a polyhedron of bounding $(n-1)$-simplices. The design centering problem is solved by determining the location of the center of the maximal hyperellipsoid inscribed within this polyhedron. The axis lengths of this ellipsoid can be used to solve the tolerance assignment problem. In addition, this approximation can be used to estimate the yield by performing an inexpensive Monte Carlo analysis in the parameter space without any need for the usual multitude of circuit simulations.

I. Introduction

Designing in the face of statistical uncertainty is a pervasive engineering problem, particularly in the field of semiconductor integrated circuits [1]–[6]. Here the values of all circuit components are subject to statistical fluctuations inherent to the thermal, chemical, and optical processes used in this technology. Due to these fluctuations it is likely that some percentage of the total number of circuits manufactured will not meet the desired design specifications. In such a situation design emphasis is placed primarily on obtaining a satisfactory yield, i.e., choosing a nominal design which maximizes the number of circuits which meet specification. We refer to this type of design as *design centering*. In this paper we present a new, simple, efficient method for design centering which is particularly well suited for integrated circuits.

Although the design centering problem is of great importance, relatively little work has been reported on its solution. For the most part, Monte Carlo analysis [4],[5], albeit crude and expensive, has been widely used as an approach to design centering. More recently, Bandler *et al.* [1] have proposed a design centering procedure which is directly related to classical nonlinear programming methods [7]. This approach is based upon selecting, for each individual problem, a scalar objective functional which when minimized, inscribes a hypercube in the "feasible region" R of the n-dimensional parameter space.[1] However, the nonlinear programming approach does not attempt, as does the new procedure described below, to approximate the boundary of the feasible region but only determines points on this boundary.[2] Although the simplicial approximation method has only been used on a few problems, the results obtained suggest that this method will require far fewer circuit simulations (i.e., solutions of the differential-algebraic equations) than the nonlinear programming and Monte Carlo approaches in typical applications. This point is important because the transient simulation of an integrated circuit usually requires a significant amount of computer time. In fact on some of the problems solved using this method, the number of circuit simulations required was about the same as that required for a single unconstrained optimization [7]. On the other hand, the constrained optimization procedure required by the nonlinear programming approach inherently involves a *sequence* of unconstrained optimizations. The constrained nonlinear program is solved by means of a Fiacco–McCormick penalty function method. While this technique is less dependent on the convexity assumption made in the method to be described below, its utility seems to be more suitable for linear circuit applications than the time domain, nonlinear circuit applications which arise in integrated circuit designs.

The proposed method, which we call "simplicial approximation," is based on explicitly approximating the boundary, ∂R, of the feasible region, R, of an n-parameter design space by a polyhedron made up of n-dimensional simplices.[3] Approximation is necessary since ∂R is generally known only in terms of nonlinear inequality constraints which express acceptable circuit performance in terms of voltages and currents which depend implicitly

Manuscript received November 19, 1976; revised February 10, 1977. This work was supported in part by the National Science Foundation under Grant ENG 75-23078.

S. W. Director is with the Department of Electrical Engineering, University of Florida, Gainesville, FL 32611. This work began while he was a Visiting Scientist at IBM T. J. Watson Research Center.

G. D. Hachtel was a Regent's Lecturer at the University of California, Los Angeles, CA. He is now with the Department of Mathematics, University of Denver, Denver, CO 80210.

[1] The feasible region, or region of acceptability, R, is an n-dimensional subspace of the parameter space P which contains all points $p \in P$ which result in circuits that meet the desired specifications.

[2] As will be pointed out later, an explicit approximation of the boundary of the feasible region is useful for solving a variety of other statistical design problems [2], [3], [6] as well as allowing one to obtain a Monte Carlo estimate of the yield at a reduced computational cost.

[3] Butler, in [6], was one of the first to try and approximate the feasible region, however he was interested in graphical displays of the region and therefore approximated only projections of this region onto two dimensional subspaces..

Reprinted from *IEEE Trans. Circuit Syst.*, vol. CAS-24, no. 7, pp. 363–372, July 1977.

upon the design parameters.[4] We assume that the constraint functions are locally convex, i.e., that the sequence of points generated on ∂R are extreme points of a convex set. The case where this assumption is violated involves mathematical representation of a nonconvex polyhedron, but can still be handled, as we shall discuss in a subsequent report.

We begin our formal discussion in Section II by defining more precisely some of the terms we use, such as yield and design center. A conceptual framework for the simplicial approximation approach to design centering is outlined in Section III. In this section a systematic procedure for initializing and refining the simplicial approximation of ∂R is also given. Furthermore, a theorem is given which describes conditions under which improvement of the simplicial approximation to the boundary ∂R of the feasible region may be assured. In Section IV we describe a scaling strategy which converts the inscription of a hypersphere into that of a hyperellipsoid. We then give the results (Section V) of applying an implementation of the simplicial approximation procedure for determining the boundary ∂R of the feasible region R which arises from a practical integrated circuit design problem. In order to provide some perspective to our work, in Section VI we relate what has been done here to some related work which has been reported in the operations research literature. Finally, in Section VII we discuss some areas of future work.

II. Definitions of Yield and Design Center

Let the network under consideration obey the set of differential-algebraic equations

$$\mathfrak{N}(x,\dot{x},p,t)=\mathbf{0}, \qquad 0 \leqslant t \leqslant T \tag{2-1}$$

where $\boldsymbol{x}$ is a vector of node voltages, branch voltages, and branch currents, and $\boldsymbol{p}$ is an n-vector of statistically varying parameters with a joint probability density function $g(\boldsymbol{p})$. Further let P denote the n-dimensional parameter space so that $\boldsymbol{p} \in P$. Thus a network is defined by a nominal set of parameter values $\boldsymbol{p}^0 \in P$. In addition a network will be considered acceptable if the solution $\boldsymbol{x}(t,\boldsymbol{p})$ of (2-1) obeys the constraints

$$\Phi_i(\boldsymbol{p}) = \int_0^T \phi_i(x,\dot{x},p,t)\,dt \leqslant 0, \qquad i=1,2,\cdots,n_c. \tag{2-2}$$

Note that the constraints Φ_i depend on $\boldsymbol{p}$ implicitly through the solutions $\boldsymbol{x}(t,\boldsymbol{p})$ of the differential-algebraic equations, as well as explicitly through their role as arguments of the $\phi_i(x,\dot{x},p,t)$. That is, the specifications $\Phi_i \leqslant 0$ are given in terms of output variables, $\boldsymbol{x}$, i.e., the output space X (e.g., currents, voltages, delay times, etc.) whereas the statistical design takes place in terms of the designable parameters, $\boldsymbol{p}$, i.e., the parameter space, P (e.g., in integrated applications these are diffusion depths, sheet resistivities, and line widths).

Given a set of constraints (2-2), the parameter space can be separated into two regions: the region of acceptability, R, and the failure region, F. These regions are defined as follows:

$$R \equiv \{\, p \mid \Phi_i(p) \leqslant 0, \quad \text{for all} \quad i \in \{1,2,\cdots,n_c\}\} \tag{2-3}$$

$$F \equiv \{\, p \mid \Phi_i(p) > 0, \quad \text{for at least one } i \in \{1,2,\cdots,n_c\}\}. \tag{2-4}$$

In what follows we assume that R is convex. Of particular interest will be the boundary of R denoted by ∂R and defined by

$$\partial R \equiv \{\, p \mid \Phi_i(p) \leqslant 0, \quad \text{for all } i \in \{1,2,\cdots,n_c\}$$

and

$$\Phi_k(p) = 0, \qquad \text{for at least one } k \in \{1,2,\cdots,n_c\}\}. \tag{2-5}$$

We can now formally define *yield* in terms of the probability density function g and the feasible region, R, by the relation

$$Y = \text{prob}\,(p \in R) = \int\int_R \cdots \int g(\boldsymbol{p})\,d\boldsymbol{p} \tag{2-6}$$

where the integration in (2-6) is over the entire feasible region R.

The purpose of design centering can now be stated as that of choosing the nominal values $\boldsymbol{p}^0$ of the designable parameters so that yield Y is maximum for a given distribution $g(\cdot)$. Since in general the distributions are known and fixed, the design centering problem can be defined to have two clearly delineated phases, namely,

a) determining the feasible region R and the design center, and
b) evaluating the yield integral (2-6).

III. Design Centering by Simplicial Approximation

The method of simplicial approximation is based on the approximation of the boundary ∂R of the feasible region R by a polyhedron, i.e., by the union of those portions of a set of n-dimensional hyperplanes which lie inside ∂R or on it. The procedure begins by determining any $m \geqslant n+1$ points $p_1, p_2, \cdots, p_m$ on the boundary ∂R. Usually m is taken to be either $n+1$ or $2n$. (One way to find $2n$ points is by first finding a feasible set of designable parameters inside R and then performing one-dimensional line searches in the positive and negative coordinate directions. See Appendix A for a description of the line search.) The feasible point can be obtained either by designer insight or through the use of an optimization program. Given a set of m points on R, the convex hull (poly-

[4]It is important to recognize that determination of the voltages or currents in a circuit for a given set of design parameters requires a complete circuit simulation, i.e., the solution of a set of nonlinear differential-algebraic equations.

hedron) of these points is then constructed.[5] The convex hull is characterized by the set of m_H inequalities

$$c_k^T p \leqslant b_k, \qquad k=1,2,\cdots,m_H \tag{3-1}$$

where c_k is a unit "outward pointing," vector normal to the kth hyperplane defined by

$$c_k^T p = b_k, \qquad k=1,2,\cdots,m_H \tag{3-2}$$

and b_k is a measure of the distance of the kth hyperplane from the origin. Moreover, the convex hull of the set of points p_j, $j=1,2,\cdots,m_H$ approximates ∂R in an *interior* sense. That is, under our assumption of convexity of R, every point interior to these boundary of hyperplanes is also inside ∂R.

Given the first approximation to ∂R we can find a first estimate of the design center by determining the center of the largest hypersphere which can be inscribed inside the polyhedron of m_H hyperplanes described by (3-2). The center of the largest hypersphere which can be inscribed within this polyhedron can be found quite easily using a linear programming approach. First recognize that the distance from a point p^* inside the polytope to the kth hyperplane is

$$d_k = c_k^T p^* - b_k. \tag{3-3}$$

Therefore the center and radius of the largest hypersphere can be found by determining the maximum value of r and the point p^* for which

$$c_k^T p^* + r \leqslant b_k, \qquad k=1,2,\cdots,m_H. \tag{3-4}$$

This linear program is the dual of the following standard (primal) linear programming problem.

Minimize the objective function

$$\Phi = \sum_{k=1}^{m_H} b_k \lambda_k \tag{3-5}$$

subject to the constraints

$$\begin{bmatrix} c_1 & c_2 & \cdots & c_{m_H} \\ \downarrow & \downarrow & & \downarrow \\ \hline 1 & 1 & \cdots & 1 \end{bmatrix} \begin{bmatrix} \lambda_1 \\ \lambda_2 \\ \vdots \\ \lambda_{m_H} \end{bmatrix} = \begin{bmatrix} 0 \\ 0 \\ \vdots \\ 1 \end{bmatrix} \tag{3-6}$$

and $\lambda_k \geqslant 0$ for all k. The primal form of this linear programming problem is more desirable because the number of unknowns is less than the number of constraints. Any standard linear programming computer subroutine can be used to solve (3-5) and (3-6) and as a byproduct yield the results of the dual problem, i.e., the radius r and center p^* of the largest inscribed hypersphere.

[5]The convex hull can easily be constructed by considering the hyperplanes defined by combinations of the m boundary points taken n at a time. All hyperplanes which have all boundary points to one side are part of the convex hull. There are more efficient procedures to determine the convex hull, however since this step is performed only once in the entire procedure, efficiency is not of prime importance here.

It is important to recognize the fact that the computational effort required to determine the convex hull of a set of boundary points and then inscribe a hypersphere in this convex hull is usually considerably less than the computational effort required to perform a nonlinear transient circuit simulation. In fact an assumption that is made throughout this paper (and experience shows it to be a reasonable assumption) is that the computational cost of performing a circuit simulation far exceeds the computational cost of solving any linear program or of performing any of the matrix-vector manipulations required by our method.

The next step in the design centering procedure is to improve the simplicial approximation to ∂R and then update the design center. This step is accomplished by determining which of the m_H "faces" of the polyhedron (i.e., which of the m_H truncated hyperplanes (3-2)), that are tangent to the largest inscribed hypersphere is the "largest." The largest tangential face is of particular interest since it is the one which is most likely the poorest approximation to ∂R. The faces of the polyhedron which are tangent to the largest inscribed hypersphere are those associated with the "zero slack variables" used in solving the linear program (3-5), (3-6). The largest tangential face is defined as the face in which the largest $(n-1)$-dimensional hypersphere may be inscribed.

The procedure for finding the largest hypersphere which can be inscribed in a face of the approximating polyhedron is only a slight variation of the linear programming procedure outlined above for finding the largest hypersphere inscribed in the polyhedron itself. Suppose we wish to find the largest hypersphere inscribed in the jth face of the polyhedron. The center of this hypersphere, denoted by p_j^*, must be on the jth hyperplane so that

$$c_j^T p_j^* = b_j. \tag{3-7}$$

Now let $c_j^\perp$ denote a unit vector perpendicular to c_j, i.e.,

$$(c_j^\perp)^T c_j = 0. \tag{3-8}$$

Thus $c_j^\perp$ lies in the jth hyperplane and the surface of a hypersphere of radius r_j centered at p_j^* which lies in the jth hyperplane is described by

$$p_j^* + r_j c_j^\perp. \tag{3-9}$$

The largest such hypersphere is the one with as large an r_j as possible subject to the constraint that all points on the surface of this hypersphere lie within the approximating polyhedron, i.e., they satisfy the constraints

$$c_k^T(p_j^* + r_j c_j^\perp) \leqslant b_k, \quad \text{for } k=1,2,\cdots,m_H; \quad k \neq j \tag{3-10}$$

which can be shown to be equivalent to the constraints

$$c_k^T p_j^* + r_j \sin\phi_{jk} \leqslant b_k \tag{3-11}$$

where ϕ_{jk} is the angle between c_k and c_j. Thus the linear program to be solved is to determine p^* and maximize r subject to (3-8) and (3-11). Note that (3-8) can be expressed as

$$c_j^T p^* \leq b_j \tag{3-12}$$

and

$$-c_j^T p^* \leq -b_j \tag{3-13}$$

so that this linear program is the dual form of the following standard (primal) linear programming problem.

Minimize the objective function

$$\Phi = \sum_{k=1}^{m_H+1} b_k \lambda_k \tag{3-14}$$

subject to the constraints

$$\left[\begin{array}{cccccc} c_1 & c_2 & \cdots & c_j & \cdots \; c_{m_H} & -c_j \\ \downarrow & \downarrow & & \downarrow & \downarrow & \downarrow \\ \hline \sin\theta_{1j} & \sin\theta_{2j} & \cdots & 0 & \cdots \; \sin\theta_{m_H j} & 0 \end{array}\right] \begin{bmatrix} \lambda_1 \\ \lambda_2 \\ \cdot \\ \cdot \\ \cdot \\ \lambda_{m_H+1} \end{bmatrix} = \begin{bmatrix} 0 \\ 0 \\ \cdot \\ \cdot \\ \cdot \\ 1 \end{bmatrix} \tag{3-15}$$

and $\lambda_k \geq 0$ for all k. Note that in (3-14) $b_{m_H+1} = -b_j$.

The above procedure is repeated for each tangential face, the largest face being the one with the largest inscribed hypersphere.[6] Once the largest face is determined, a one-dimensional search is made in the outward normal direction, starting from the center of the largest inscribed hypersphere in that face, for a new point on the boundary. This new point is then added to the set of boundary points previously found and a new convex hull is found.[7] Thus our approximation to ∂R has been improved and a new design center can be found. The entire procedure can then be repeated.

In summary, the proposed design centering procedure, illustrated for two dimensions in Fig. 1, is as follows.

[6]In general, for an n-dimensional parameter space, there will be $n+1$ tangential faces. Thus there are about $n+1$ linear programs that must be solved, which is much fewer than if the size of all the faces of the polyhedron had to be found.

[7]Determining the new convex hull at this point requires two steps. First all faces of the original polytope for which the defining inequality (3-1) no longer holds when the new boundary point is considered must be deleted. Next hyperplanes formed by combinations of $(n-1)$ points taken from the set of points which define the deleted faces and the new boundary point are tested to see if all boundary points are on one side. If this test is passed, the hyperplane is part of the convex hull, if not it is ignored.

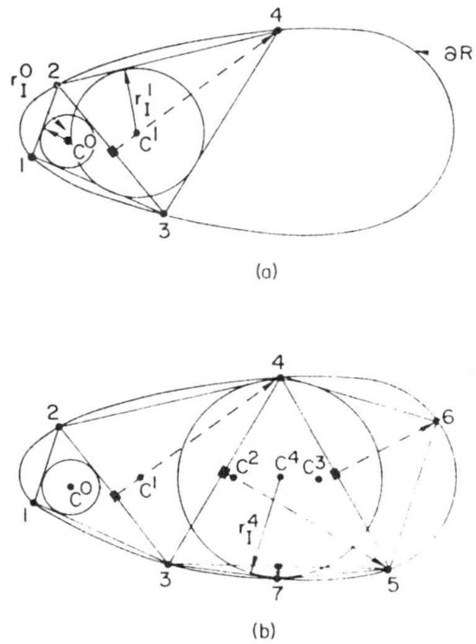

Fig. 1. Illustration of simplicial approximation. (a) Initial polyhedron defined by points (1,2,3) and breaking largest face (2-3). (b) Resulting polyhedron after four inflation iterations.

Design Centering Procedure

a) Determine a set of $m \geq n+1$ points on the boundary ∂R.

b) Find the convex hull of these points and use this polyhedron as the initial approximation to ∂R. Set $k=0$.

c) Inscribe (using the linear program (3-5)) the largest hypersphere in this approximating polyhedron and take as the first estimate of the design center the center of this hypersphere.

d) Find (using the linear program (3-14)) the "midpoint" of the "largest" face in the polyhedron which is tangent to the inscribed hypersphere.

e) Find a new boundary point[8] on ∂R by searching (using, for example, the search strategy outlined in Appendix A) along the outward normal of the "largest" face found in step d) extending from the "midpoint" of this face.

f) Inflate the polyhedron by forming the convex hull of all previous points plus the new point generated in step e).

g) Find (using the linear program (3-5)) the center p_c^k and radius r_I^k of the largest hypersphere inscribed in the new polyhedron found in step f). Set $k=k+1$. Go to step e).

A key step in the above is step e) since it involves a line search along a given direction. Each step in this search involves first solving the circuit equations (2-1) and then evaluating the constraints (2-2) to see which, if any, are violated. The search ends when point p_B is located for which

$$\Phi_j(p_B)=0, \quad \Phi_i(p_B) \leq 0, \qquad \forall i \neq j. \tag{3-16}$$

The above procedure will rapidly generate a sequence of points whose convex hull forms an interior polyhedron,

[8]It can occur that the largest face actually lies in ∂R. We test for this case, and if it occurs, the face is flagged, and never again considered as a candidate for step d). In this case, step d) is repeated using the next largest face.

∂R_I, which approximates ∂R. Step g), the process of inscribing the largest hypersphere in ∂R_I, can be used as a means of monitoring the convergence of $\partial R_I \rightarrow \partial R$. If R is convex, as per our assumption, the sequence of centers $\{p_c^k\}$ will converge to the center of the largest hypersphere that can be inscribed in ∂R, and the associated sequence of radii, $\{r_I^k\}$, will converge to the radius of the largest inscribed hypersphere. This sequence is considered to be converged when

$$|r_I^{k+1} - r_I^k| \leq \epsilon_R r^k + \epsilon_A \tag{3-17}$$

where ϵ_R and ϵ_A are given relative and absolute convergence parameters.

Convergence is expected on the basis of the following.

Theorem (3-18)

1) Let r_I^k be the radius of the largest hypersphere inscribed in a polyhedron ∂R_I^k formed from $n+1+k$ points on a convex surface ∂R in n-space.

2) Let r^{k+1} be the radius of the largest hypersphere inscribed in a polyhedron ∂R_I^{k+1} formed by applying the inflation step of the design centering procedure to ∂R_I^k, then

$$r_I^{k+1} \geqslant r_I^k. \tag{3-19a}$$

3) Further, if the outward normals to the faces of ∂R_I^k are not parallel and if the inscribed hypersphere is tangent to only $n+1$ faces, then

$$r_I^{k+1} > r_I^k. \tag{3-19b}$$

Note that condition 3) of Theorem (3-18) is not restrictive in practice because if it is violated, a finite number of passes through the design centering procedure will cause condition 3) to be satisfied. For example, if two tangential faces are parallel, breaking the largest face may not allow the sphere to grow. However, upon reapplication of the procedure often enough (no more than n reapplications are necessary), then one of the parallel faces will eventually be broken. Thus we have as a corollary the following.

Corollary

$$r_I^{k+1+l} > r_I^k, \qquad 0 < l \leqslant n. \tag{3-19c}$$

Assume that there are no unsuspected surface pathologies which would cause the simplicial approximation process to converge to some limit *inside* ∂R.[9] Then, it would follow from the above theorem and corollary that the approximating simplex ∂R_I^k would, as $k \rightarrow \infty$, fill ∂R like a ball being inflated by an air pump. Theorem (3-18) can be proven by observing that the new polyhedron contains the old polyhedron. The corollary follows from the fact that if two faces are parallel, breaking one of them destroys this relationship.

A word about the computational effort required to use the above procedure is in order. Our basic premise is that this procedure is to be used in situations where the number of statistically varying design parameters is small compared to the number of the algebraic-differential equations that need to be solved in order to determine circuit behavior. This situation is common in the design of integrated circuits where the variations in circuit parameters are related to the variations in a much smaller number of processing parameters. For these cases the dimensionality of the linear programs that are solved in the above procedure is relatively small when compared to the dimensionality of the circuit simulation problem. Thus the time spent in generating and inflating the convex hull is small when compared to the time involved in performing a transient simulation of a large nonlinear network. Based upon the above reasoning the most expensive operation in the above procedure is the search step e) to find a new boundary point. Therefore the search step must be efficient. At this juncture we do not have sufficient data to determine the relationship between the total number of circuit simulations required to perform design centering and the dimensionality of the parameter space. However, since there is no nonlinear programming involved here, we feel that the total number of circuit simulations required would be considerably fewer than that required by some other design centering procedures.

Before leaving this section we note that with each interior approximation ∂R_I^k may be associated a yield

$$Y_I^k = \int \int_{R_I^k} \cdots \int g(p, p^0, \sigma_i)\, dp \tag{3-20}$$

where R_I^k is the volume in n-space bounded by ∂R_I^k, i.e., the kth-interior approximation to R. Note also that since ∂R_I^k consists of interior bounding hyperplanes

$$Y_I^{k-1} < Y_I^k \leqslant Y$$

where Y is computed over the true feasible region R (cf. (3)). That is, the interior simplicial approximation gives a lower bound on the yield. Moreover calculation of this yield requires only generation of random sets of parameter values and determination of the number which fall inside ∂R_I, i.e., only a Monte Carlo analysis on the input space is required; no multitude of circuit simulations is necessary. (For further discussion of these ideas see [20].)

IV. Scaling–Inscribing a Hyperellipsoid

If the design centering procedure outlined in Section III is employed when the region of acceptability is rectangular in shape, the best possible results may not be obtained. For elongated regions of acceptability, it would be more appropriate to determine the design center by inscribing a hyperellipsoid rather than a hypersphere. Fortunately this

[9]Theorem 32 of [11] shows that the above procedure will produce a convergent subsequence as $k \rightarrow \infty$. Theorem 33 of [11] shows that ∂R can be approximated by a polyhedron to arbitrary accuracy. However, we have as yet no proof that $\partial R^k \rightarrow \partial R$ as $k \rightarrow \infty$. This has not been a problem in practice.

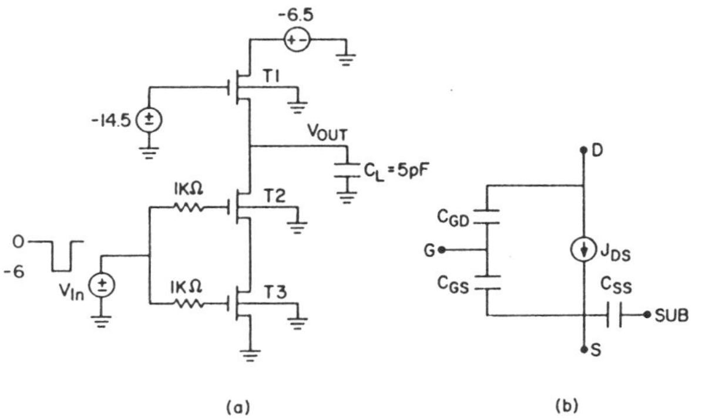

Fig. 2. (a) Two input NAND gate used in the sample. (b) FET model used: $J_{DS} = G_m(V_{GS} - V_T)^2 W/L$ above pinch-off and $J_{DS} = G_m V_{DS}(V_{GS} - V_T - V_{DS}/2) W/L$ below pinch-off, where W is width and L is length of device. Capacitances are functions of W and L: $C_{GS} = (L+\alpha)W\beta$; $C_{GO} = \alpha W\beta$; and $C_{SS} = (W/\gamma + \delta)\epsilon$; where $\alpha, \beta, \gamma, \delta$, and ϵ are constants.

result is easily obtained by merely scaling the parameters prior to employing the design centering procedure. One scheme for scaling, which has proven to be successful, is based upon using an estimate of the spreads in possible parameter values. This technique is employed in our computer implementation and is described below.

Assume that the mth step of the design centering procedure has been carried out and that $n+m+1$ boundary points $\{p_k\}$, $k=1,2,\cdots,n+m+1$ have been found. The lower and upper bounds of each parameter are now determined

$$p_i^L = \min_k \{p_{ik}\}$$

and

$$p_i^U = \max_k \{p_{ik}\}$$

for $i=1,2,\cdots,n$ where p_{ik} is the ith coordinate of the kth-boundary point. The scale factor for the ith component of the parameter vector is given by

$$S_i = p_i^U - p_i^L, \qquad i=1,2,\cdots,n$$

and a scale matrix is defined

$$\boldsymbol{S} = \text{diag}\{S_1, S_2, \cdots, S_n\}.$$

A scaled set of boundary points are now defined

$$\hat{p}_k = \boldsymbol{S}^{-1} p_k, \qquad k=1,2,\cdots,n$$

and steps c) and d) of the design centering procedure can be carried out as before using the set of points $\{\hat{p}_k\}$. Observe though that the radius of the largest hypersphere, $\hat{r}$, found in step c) can be unscaled to give the "radii," or axis lengths, of a hyperellipsoid

$$r_i = S_i \hat{r}, \qquad i=1,2,\cdots,n$$

which can be thought of as being inscribed in the unscaled polyhedron defined by the points $\{p_k\}$. Similarly, if $\hat{C}$ denotes the design center of the scaled problem found in step c), the design center of the unscaled problem is

$$C = S\hat{C}.$$

This interpretation of C as the center and the r_i as the radii of a hyperellipsoid comes about by recognizing that in the scaled problem the largest hypersphere of radius r and center $\hat{C}$ is characterized by the locus of points $\hat{p}$ for which

$$\sum_{i=1}^{n} (\hat{p}_i - \hat{C}_i)^2 = r^2 \tag{4-1}$$

or

$$\sum_{i=1}^{n} \left(\frac{p_i - C_i}{S_i r} \right)^2 = 1. \tag{4-2}$$

In terms of the unscaled space (4-2) is the equation of a hyperellipsoid. Note that r_i is *not* the distance between the design center C and the ith face of the polyhedron (as it would have been without scaling) but the distance from the design center to the surface of the hyperellipsoid along the ith axis.

Clearly the results of the above procedure are dependent upon the scaling. It is suggested that once the design centering procedure has converged, the lower and upper bounds on the parameters be redetermined, considering now all the boundary points of ∂R_I^{m+1}, and the scaling adjusted. The design centering procedure should then be repeated to ensure that a good nominal point had been found.[10]

V. Example

To illustrate the design centering algorithm, consider the three transistor NAND gate circuit of Fig. 2. In order that we may graphically illustrate the steps in the procedure, we allow only three designable parameters: the width W_1 of the load device $T1$, the width W_{23} of the input devices $T2$ and $T3$ (both $T2$ and $T3$ have the same width), and the threshold voltage, V_T, of the three devices (the same for all three). The lengths of all three devices

[10]Note that all boundary points previously found as well as the approximating polyhedron can be rescaled and used as the starting approximation here. Thus, in general, little additional work will be required to repeat the design centering procedure.

TABLE I
INITIAL BOUNDARY POINTS

W_1	W_{23}	V_T	A	T_D	V_o
12.85	230.0	1.2	2500	68.5	-.6
8.28	230.0	1.2	2442	110.0	-.395
12.0	231.1	1.2	2500	73.8	-.555
12.0	180.7	1.2	1989	70.0	-.7
12.0	230.0	2.	2490	78.7	-.631
12.0	230.0	1.	2390	72.8	-.543

are held fixed. Therefore we have two geometric parameters and one processing related variable to adjust.

We have three constraints which must be satisfied in order for the circuit to be considered acceptable: the total area A, occupied by the three devices must be less than 2500 mils^2; the delay time, T_D (i.e., the length of time between which the input passes through -3 V and the output falls through -3.25 V) must be less than 110 ns, and the output voltage V_0, be greater than -0.7 V when the gate is on. In addition there are *box* constraints on the parameters

$$5.0 \leqslant W_1 \leqslant 50.0 \quad \text{(microns)}$$
$$50.0 \leqslant W_{23} \leqslant 250.0 \quad \text{(microns)}$$

and

$$1.0 \leqslant V_T \leqslant 2.0 \quad \text{(volts)}.$$

The feasible starting point was $W_1 = 12$, $W_{23} = 230$, and $V_T = 1.2$. For this set of parameter values, $A = 2490$, $T_p = 73.8$, and $V_0 = -0.558$.

The first step in the procedure is to find 6 points on the boundary of the feasible region by searching in the positive and negative coordinate directions. The six points found and the associated constraint values are given in Table I. The constraints which are satisfied exactly are framed. In two cases the box constraints on V_T were the limiting constraints.

A sketch of the initial approximating simplex defined by these boundary points is shown in Fig. 3. The x axis corresponds to $W_1/4.6$, the y axis corresponds to V_T, and the z axis corresponds to $W_{23}/50.3$. The scaling was determined by the initial spread in parameter values as determined by the coordinate searches above. The axes emanate from the "center" of the body. (Note that this figure represents a perspective drawing and does not necessarily display the actual relative size of each face.) The first estimate of the design center, based upon the first approximating simplex, is $W_1 = 11.4$, $W_{23} = 219$, and $V_T = 1.32$. The radii (i.e., axis) of the largest inscribed ellipsoid (i.e., unscaled sphere radius) is 1.1 for W_1, 11.7 for W_{23}, and 0.23 for V_T. In other words, given this design center, acceptable circuits would be obtained for a deviation of 9.6 percent in W_1 if W_{23} and V_T were kept fixed, a 5.3-percent deviation in W_{23} if W_1 and V_T were kept fixed and a 17.4 percent deviation in V_T if W_1 and W_{23} were kept fixed. Note that since we are inscribing a hyperellipsoid in the region of acceptability, and not a hypercube whose sides are twice the maximum deviation, we are not guaranteed of acceptable circuits when two or more parameters deviate maximally from the nominal.

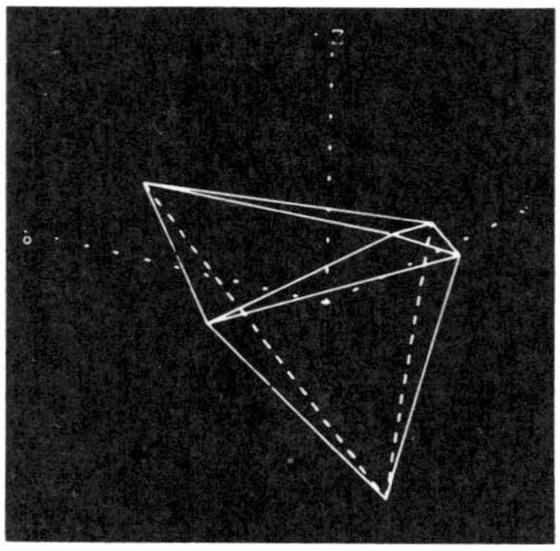

Fig. 3. Initial (scaled) approximating polyhedron to region of acceptability for NAND gate example. Axes eminate from estimate of design center.

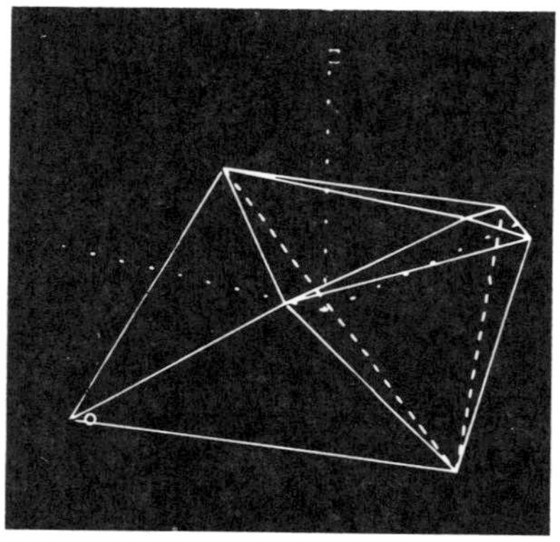

Fig. 4. Resulting polyhedron when largest face of polyhedron in Fig. 3 was broken.

The next step involves searching for a new point on the boundary starting from the center of the largest face and proceeding in an outward normal direction. A new polyhedron is then constructed using this new point and a new estimate of the design center made. Fig. 4 shows the new polyhedral approximation. This procedure is repeated until no increase in the size of the inscribed hypersphere is possible. The final approximating polyhedron is shown in Fig. 5. The final design center was $W_1 = 10.6$, $W_{23} = 209$, and $V_T = 1.47$; and the radius of the inscribed hyperellipsoid was 2.2 for W_1, 24 for W_{23} and 0.47 for V_T. Thus acceptable circuits would be formed for a 21 percent change in W_1 if W_{23} and V_T were kept fixed, an 11.5 percent change in W_{23} if W_1 and V_T were kept fixed, and a 32 percent change in V_T if W_1 and W_{23} were kept fixed.

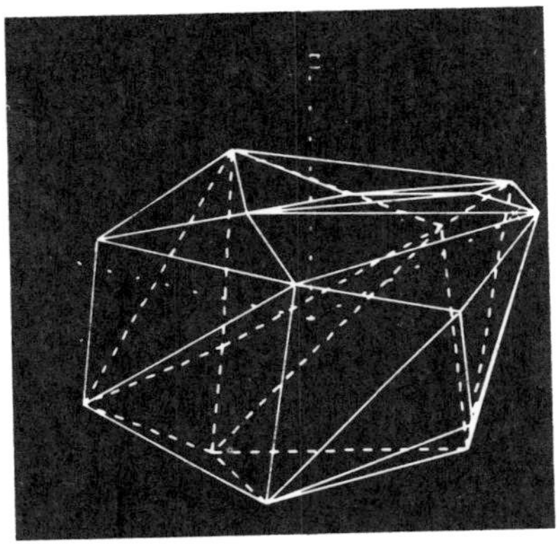

Fig. 5. Final polyhedral approximation.

Therefore we see that the design centering procedure increased the allowable tolerances of the designable parameters by a factor of two over the value quoted above for the first computed design center (which was itself an improvement over the initial feasible point).

VI. Relationship of Simplicial Approximation to Existing Operations Research Literature

In two obvious respects, the design centering problem itself, and the simplicial approximation method proposed for solving it, are generally related to various problems and algorithms [11]–[19] in operations research because

a) the problem requires design *optimization* as well as simulation, and
b) the basic approach does, as do most methods in operations research, employ the pervasively useful properties of the simplex method [16].

Because some of this pertinent literature may not be familiar to those who wish to apply the above method and because this literature helps put what we have done in perspective, we briefly review it here.

In its statistical aspect the design centering problem is tangentially related to the so-called chance constrained programming problem [12], [18] which has, in some special cases, been successfully attacked. Two major distinctions exist between previous work in this area, and the problem and methods of this paper, namely,

1) the feasible region, R, is not known in advance but must be implicitly determined, and
2) the constraints implicitly determining R are not only nonlinear, but are functionals of the solutions of a (usually large) set of differential-algebraic equations.

Because of these distinctions, what integrated circuit designers consider a large, realistic problem differs from the conventional definitions of operations research. Whereas in operations research a large scale problem may contain hundreds of thousands of (linear inequality) constraints and variables, in integrated circuit design situations we are typically concerned with only a few design variables and a small number of constraints. In the latter case, it is mainly the size of the underlying system of nonlinear differential-algebraic equations, often of very high (10^3–10^4) order, which describes the circuit that determines the practical realism of the problem.

It is, partially, this dichotomy which persuaded us to attempt what was known to be possible (cf. [11, theorems 32 and 33]) but has, apparently, heretofore been considered too large (cf. [13, p. 35, ¶3]) to consider, that is, the approximation in n-space of the boundary ∂R of an arbitrary convex set R by an inductively constructed polyhedron (cf. [15, ch. 5] for undeveloped ideas which might lead to alternative constructions.) The procedure given above for such "simplicial approximation" contains various features which appear to be novel. However, the operations research literature does contain many possibilities which may speed up the proposed algorithm. These alternate implementations need further investigation. Of particular relevance in this respect is the so-called "decomposition" principle of Dantzig and Wolfe [17]. Much work in this apparently fruitful direction will be necessary before the full economic impact of the proposed method can be properly evaluated. However, we emphasize again that the major portion of the computational cost associated with the above procedure is due to the circuit simulations required to find new boundary points and not in the solution of the associated linear programs.

VII. Conclusions

To summarize, we have described a method, called the simplicial approximation method, which locates with repeated line searches, an increasingly dense set of points on the boundary ∂R of the feasible region R. These points are used to define a polyhedron of interior bounding hyperplanes for the region R. At each stage of approximation, the design center is (after appropriate scaling) the center of the maximal hypersphere inscribed in the polyhedron. The approximation is refined by searching for the intersection, with the true boundary ∂R, of the outward normal line emanating from the "midpoint" of the "largest" of the faces of the polyhedron which are tangent to the maximal inscribed hypersphere.

We have constructed, and verified by experiment, an algorithm based on the process described above. We have also been able to show that the radius of the maximal hypersphere is monotone increasing as the approximation is refined. This has enabled us to compute a monotone increasing lower bound on the yield, i.e.,

$$Y^k < Y^{k+1} < Y$$

where Y is the true maximum available yield and Y^k is the available yield for the kth approximation to ∂R.

It has been suggested by Wolfe [8] that an exterior set of supporting hyperplanes could be constructed with a procedure essentially dual to the procedure presented in Section III. In the corresponding "exterior" approximating polyhedron, ∂R_E^k, the location of a new boundary point on ∂R would be followed by a gradient evaluation, yielding a supporting hyperplane containing the new point. All extreme points of ∂R_E^k which are outside the new supporting hyperplane would then be deleted from

∂R_E^k, and the convex hull of the new set of extreme points would constitute ∂R_E^{k+1}. Ideas related to "exterior" simplicial approximation have been discussed in [13]–[15]. The cutting plane method of Kelley, [14], is quite similar in method, although different in purpose.

Also significant in the list of future research possibilities are the establishment of conditions for, and rates of, convergence and of procedures for handling nonconvex, locally convex, and/or quasi-convex cases.

In the meantime, it is important to note that the described implementation is based on a straightforward interfacing of code for the new algorithm with standard programs for (small scale) LP problems and for network simulation. More specifically, the method of "interior" simplicial approximation (unlike the nonlinear programming method of [1]) requires *no* gradient evaluations. Thus any readily available conventional network simulator, e.g., SPICE, [9], or ASTAP-II, [10], can be employed to provide prospective users with immediate access to design centering via interior simplicial approximation.

Appendix A–The One-Dimensional Search Strategy

Consider the n_c constraint functions $\Phi_i(p)$, $i=1,2,\cdots,n_c$ of the n-parameter vector p. For any feasible point p^0

$$\Phi_i(p^0)\leqslant 0, \qquad i=1,2,\cdots,n_c.$$

A point on the boundary of the feasible region, p^*, is defined as a point for which

$$\Phi_j(p^*)=0$$

and

$$\Phi_i(p^*)\leqslant 0, \qquad i=1,2,\cdots,n_c; \qquad i\neq j$$

for some $j\in[1,m]$. During the design centering procedure we wish to find a point on the boundary of the feasible region starting from some feasible point p^0 and proceeding in a given direction s. More specifically, we wish to determine the step size α for which

$$\Phi_j(p^0+\alpha s)=0$$

and

$$\Phi_i(p^0+\alpha s)\leqslant 0, \qquad i=1,2,\cdots,n_c; \qquad i\neq j$$

for some j. The one-dimensional search procedure which has been implemented is based on the secant method and proceeds as follows.

1) Evaluate $\Phi_i(p^0)$, $i=1,2,\cdots,m$. Define $C_i^0=C_i^l=\Phi_i$ and $\alpha^0=\alpha^l=0$.

2) Determine the smallest α such that the point $(p^0+\alpha s)$ is on one of the box constraints and evaluate $\Phi_i(p^0+\alpha s)$. If $\Phi_i(p^0+\alpha s)\leqslant 0$ for all i then the box constraint is active and the point $p^0+\alpha s$ is on the boundary, otherwise define $\alpha^N=\alpha^u=\alpha$ and $C_i^N=C_i^u=\Phi_i$ for all $\Phi_i>0$. Let I denote the set of indices $\{i\}$ for which the constraints are violated.

3) If $(\alpha^u-\alpha^l)s_i>\Delta p_k$ (where Δp_k is the minimum allowable change in p_k) for any $k=1,2,\cdots,n$, then proceed, if not then stop and take current value of $p^0+\alpha s$ as boundary point.

4) Calculate the value of α given by

$$\alpha\rightarrow\min_{i\in I}\left\{\alpha^N-C_i^N(\alpha^N-\alpha^0)/(C_i^N-C_i^0)\right\}$$

If $\alpha<\alpha^l$ then set α to α^l, or if $\alpha>\alpha^u$ then set α to α^u. Then set α^0 to α^N, C_i^0 to C_i^N, $i\in I$, and α^N to α.

5) Evaluate $\Phi_i(p^0+\alpha^N s)$, $i\in I$, and set C_i^N to Φ_i, $i\in I$. If $\Phi_i<0$ for all $i\in I$ set α^l to α^N and C_i^l to Φ_i and repeat from step 3). If there is at least one $i\in I$ for which $\Phi_i>0$, than set α^u to α^N and $C_i^u<\Phi_i$ for all $i\in I$ and repeat from step 3).

Observe that step 4) is essentially the secant method for determining α. Moreover, at every step in the procedure there is an upper and lower bound on the step size α. When the interval, $[\alpha^l,\alpha^u]$ gets small enough, the procedure terminates.

Acknowledgment

We wish to thank R. Brayton, H. Crowder, F. Gustavson, A. Wilson, and P. Wolfe for many helpful discussions. A special debt is owed to G. Geortzel for sharing outstanding unpublished work in related areas.

References

[1] J. W. Bandler, P. C. Liu, and H. Tromp, "Practical design centering, tolerancing and tuning," in *Proceedings of the 1975 IEEE International Symposium on Circuits and Systems*, Apr. 1975, pp. 206–209.

[2] J. F. Pinel and K. A. Roberts, "Tolerance assignment in linear networks using nonlinear programming," *IEEE Trans. on Circuit Theory*, vol. CT-19, pp. 475–479, Sept. 1972.

[3] A. R. Thorbjornsen and S. W. Director, "Computer-aided tolerance assignment for linear circuits with correlated elements," *IEEE Trans. on Circuit Theory*, vol. CT-20, pp. 518–524, Sept. 1973.

[4] B. J. Karafin, "An efficient computational procedure for estimating circuit yield," Ph.D. dissertation, University of Penn., 1974.

[5] N. J. Elias, "New statistical methods for assigning device tolerances," in *Proceedings of the 1975 IEEE Symposium on Circuits and Systems*, pp. 329–332, Apr. 1975.

[6] E. M. Butler, "Realistic design using large change sensitivities and performance contours," *IEEE Trans. on Circuit Theory*, vol. CT-18, pp. 58–59, Jan. 1971.

[7] R. Fletcher and M. J. D. Powell, "A rapidly convergent descent method for minimization," *Computer Journal*, vol. 6, pp. 163–168, 1963.

[8] P. Wolfe, private communication.

[9] L. W. Nagel, "SPICE 2: A computer program to simulate semiconductor circuits," Ph.D. dissertation, Univ. of Calif., May 1975.

[10] *Advanced Statistical Circuit Analysis Program (ASTAP) Program Reference Manual*, SH20-1118-0, IBM Corporation.

[11] H. Eggleston, *Convexity*. London, England: 1958, Cambridge University Press.

[12] A. Charnes and W. Cooper, "Chance constrained programming," *Management Science*, vol. 6 pp. 73–80, 1959.

[13] A. M. Geoffrion, Ed., *Perspectives on Optimization*. Reading, MA: Addison-Wesley, 1972.

[14] J. E. Kelley, "The cutting plane method for solving convex programs," *J. Soc. Indus. Appl. Math*, vol. 8, pp. 703–712, 1960.

[15] B. Grunbaum, *Convex Polytopes*. New York: Wiley-Interscience, 1967.

[16] G. Dantzig, *Linear Programming and Extensions*. Princeton, NJ: Princeton University Press, 1963.

[17] G. Dantzig and P. Wolfe, "Decomposition principle for linear programs," *Operations Research*, vol. 8, no. 1, pp. 101–111, 1960.

[18] L. Cooper and D. Steinberg, *Methods and Applications of Linear Programming*. Philadelphia, PA: W. B. Saunders, 1974.

[19] D. Luenberger, *Introduction to Linear and Nonlinear Programming*. Reading, MA: Addison-Wesley, 1973.

[20] S. W. Director and G. D. Hachtel, "Yield estimation using simplicial approximation," in *Proceedings of the 1977 IEEE Symposium on Circuits and Systems*.

Optimal Centering, Tolerancing, and Yield Determination via Updated Approximations and Cuts

JOHN W. BANDLER, FELLOW, IEEE, AND HANY L. ABDEL-MALEK, STUDENT MEMBER, IEEE

Abstract—**This paper presents a new approach to optimal design centering, the optimal assignment of parameter tolerances and the determination and optimization of production yield. Based upon multidimensional linear cuts of the tolerance orthotope and uniform distributions of outcomes between tolerance extremes in the orthotope, exact formulas for yield and yield sensitivities, with respect to design parameters, are derived. The formulas employ the intersections of the cuts with the orthotope edges, the cuts themselves being functions of the original design constraints. Our computational approach involves the approximation of all the constraints by low-order multidimensional polynomials. These approximations are continually updated during optimization. Inherent advantages of the approximations which we have exploited are that explicit sensitivities of the design performance are not required, available simulation programs can be used, inexpensive function and gradient evaluations can be made, inexpensive calculations at vertices of the tolerance orthotope are facilitated during optimization and, subsequently, inexpensive Monte Carlo verification is possible. Simple circuit examples illustrate worst case design and design with yields of less then 100 percent. The examples also provide verification of the formulas and algorithms.**

I. Introduction

OPTIMAL tolerance assignment is the process of associating the largest tolerances with design parameters to minimize cost. Design centering is the process of defining a set of nominal parameter values to maximize the tolerances or to maximize the yield for known but unavoidable statistical fluctuations. This paper integrates the concepts of design centering, the optimal assignment of parameter tolerances and the determination and optimization of production yield into an overall optimal design process.

Our computational approach should be viewed in the context of the following important work in this area: the nonlinear programming approach of Bandler *et al.* [1], [2] and by Pinel and Roberts [3], the branch and bound method of Karafin [4], the Monte Carlo approach of Elias [5], and the Director and Hachtel technique involving approximations of the feasible region [6]. It makes use of approximations of all the constraints by low-order multidimensional polynomials. These approximations are continually updated in critical regions identified during optimization and integrated with the nonlinear program which inscribes an orthotope in the constraint region by minimizing a suitable scalar objective function. This orthotope will actually be the optimum tolerance region for a worst case design problem with independent variables.

The readily differentiable approximations permit efficient gradient methods of minimization to be employed as well as inexpensive calculations at vertices of the tolerance orthotope, which tend to locate the critical regions. The yield problem commences when the orthotope is allowed to expand beyond the boundary of the constraint region. Attention is then directed to the critical regions which contribute to the yield calculation.

Section II describes the nature of the tolerance problem and discusses the implications of the assumption of one-dimensional convexity [7], [8]. Section III formally introduces the multidimensional polynomial. Our approach to choosing suitable interpolation base points is given. The section includes an efficient algorithm for evaluating the approximations and their derivatives at different vertices in different well-chosen interpolation regions. Section IV presents algorithms for worst case design: Phase 1 deals with a single interpolation region, and Phase 2 involves two or more interpolation regions. These interpolation regions are updated according to desired accuracy for the approximate constraints in critical regions.

Based upon multidimensional linear cuts of the tolerance orthotope and uniform distributions of outcomes between tolerance extremes in the orthotope, Section V presents exact formulas for yield and yield sensitivities with respect to design parameters. The formulas employ the intersections of the cuts with the orthotope edges, the cuts themselves being functions of the original design constraints. Ways of treating linear and quadratic constraints (actual or approximate) are discussed so that results obtained by implementing the material of the previous sections can be followed up.

Section VI details an algorithm embodying all the ideas and results of Sections II to V. It deals with optimization involving yield less than 100 percent. Appropriate approximations to the boundary based on a single function of least pth type [9] within each critical region are utilized.

Manuscript received October 13, 1976; revised June 15, 1977 and January 6, 1978. This work was supported by the National Research Council of Canada under Grant A7239. This paper is based on material presented at the 1977 IEEE International Symposium on Circuits and Systems, Phoenix, AZ, April 25–27, 1977.

J. W. Bandler is with the Group on Simulation, Optimization and Control and the Department of Electrical Engineering, McMaster University, Hamilton, Ont. L8S 4L7, Canada.

H. L. Abdel-Malek was with the Group on Simulation, Optimization and Control and the Department of Electrical Engineering, McMaster University, Hamilton, Ont. L8S 4L7, Canada. He is now with the Department of Engineering Physics and Mathematics, Faculty of Engineering, Cairo University, Giza, Egypt.

Reprinted from *IEEE Trans. Circuit Syst.*, vol. CAS-25, no. 10, pp. 853–871, October 1978.

Some illustrative examples are also included. A two-section quarter-wave transmission-line transformer is used to explain how a worst case design is obtained and, further, is used for yield determination and optimization. A worst case design and a well-centered design for yield less than 100 percent for a three-section low-pass *LC* filter are included. Practical examples of nonideal two-section and three-section waveguide transformers are described.

II. Nonlinear Programming Formulation of the Tolerance Problem

Introductory Concepts

An engineering design can be described by a vector of nominal parameters $\boldsymbol{\phi}^0$ and an associated vector of manufacturing tolerances $\boldsymbol{\epsilon}$, where

$$\boldsymbol{\phi}^0 \triangleq \begin{bmatrix} \phi_1^0 \\ \phi_2^0 \\ \cdot \\ \cdot \\ \cdot \\ \phi_k^0 \end{bmatrix} \geqslant \mathbf{0}, \quad \boldsymbol{\epsilon} \triangleq \begin{bmatrix} \epsilon_1 \\ \epsilon_2 \\ \cdot \\ \cdot \\ \epsilon_k \end{bmatrix} \geqslant \mathbf{0} \tag{1}$$

and where k is the number of designable parameters. Accordingly, any design outcome is represented by a point which lies inside a tolerance region R_ϵ as shown in Fig. 1. For simplicity as well as the implications of a uniform distribution of outcomes between tolerance extremes $\phi_i^0 \pm \epsilon_i$, we define

$$R_\epsilon \triangleq \{\boldsymbol{\phi} | \boldsymbol{\phi} = \boldsymbol{\phi}^0 + \boldsymbol{E}\boldsymbol{\mu}, \ \boldsymbol{\mu} \in R_\mu\} \tag{2}$$

where

$$R_\mu \triangleq \{\boldsymbol{\mu} | -1 \leqslant \mu_i \leqslant 1, \ i = 1, 2, \cdots, k\} \tag{3}$$

and where $\boldsymbol{E}$ is a $k \times k$ matrix with diagonal elements set to ϵ_i and $\boldsymbol{\mu}$ is a random vector distributed according to the joint probability distribution function of the outcomes. Any value for $\boldsymbol{\mu}$ identifies a point in R_ϵ. The tolerance region R_ϵ as defined in (2) is an orthotope in the k-dimensional space (see Coxeter [10]). Consequently, the tolerance region will often be referred to as the tolerance orthotope. The vertices of this orthotope are the points for which all parameters are at extreme values (positive or negative extremes), i.e., $\mu_i \in \{-1, 1\}$, $i = 1, 2, \cdots, k$. See Fig. 1. The number of these vertices is 2^k and they are, for convenience, uniquely indexed by $\boldsymbol{\phi}^r$, $r \in I_v$, where

$$I_v \triangleq \{1, 2, \cdots, 2^k\}. \tag{4}$$

Thus the set of vertices is given by

$$R_v = \{\boldsymbol{\phi}^r | r \in I_v\}. \tag{5}$$

This numbering scheme will allow us to identify a vertex by the number r only.

Let R_c be the constraint region, illustrated in Fig. 1, defined by m_c functions $g_i(\boldsymbol{\phi})$ and given by

$$R_c \triangleq \{\boldsymbol{\phi} | g_i(\boldsymbol{\phi}) \geqslant 0, \text{ for all } i \in I_c\} \tag{6}$$

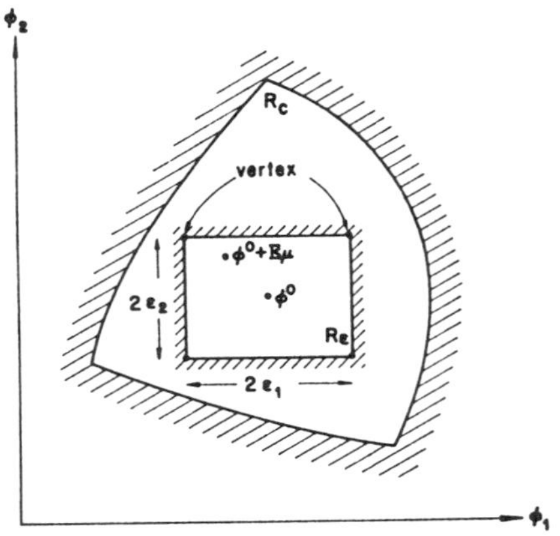

Fig. 1. Illustration of the constraint region R_c and the tolerance region R_ϵ. Also, the nominal point $\boldsymbol{\phi}^0$, an outcome $\boldsymbol{\phi}^0 + \boldsymbol{E}\boldsymbol{\mu}$ and vertices are indicated.

where

$$I_c \triangleq \{1, 2, \cdots, m_c\}. \tag{7}$$

Worst Case Design

For a worst case design [1], [2], sometimes called a design with 100-percent yield, it is required that all design outcomes satisfy the specifications, i.e.,

$$R_\epsilon \subset R_c. \tag{8}$$

If the constraint region R_c is one-dimensionally convex [7], it is sufficient that all vertices of R_ϵ belong to R_c to guarantee that (8) is satisfied, i.e., it is sufficient to have

$$R_v \subset R_c \tag{9}$$

where, formally

$$R_v \triangleq \{\boldsymbol{\phi} | \boldsymbol{\phi} = \boldsymbol{\phi}^0 + \boldsymbol{E}\boldsymbol{\mu}, \ \mu_i \in \{-1, 1\}, \ i = 1, 2, \cdots, k\}. \tag{10}$$

Bandler and Liu [8] and Brayton *et al.* [11] have considered the implications of one-dimensional convexity for certain classes of circuits.

The foregoing discussion leads to the following nonlinear programming problem for worst case design involving, in general, both centering of $\boldsymbol{\phi}^0$ and optimal assignment of $\boldsymbol{\epsilon}$:

$$\text{WCD} \begin{cases} & \underset{\boldsymbol{\phi}^0, \boldsymbol{\epsilon}}{\text{minimize}} \ C(\boldsymbol{\phi}^0, \boldsymbol{\epsilon}) \\ \text{subject to} & \\ & g_i(\boldsymbol{\phi}^r) \geqslant 0, \quad \text{for all } i \in I_c, \text{ and all } r \in I_v \end{cases} \tag{11}$$

where C is a suitable cost function and the constraints (11) are an explicit formulation of the constraint (9). The total number of constraints involved in WCD is $m_c \times 2^k$. The one-dimensional convexity assumption allowed us to have this finite number of constraints rather than the infinite number of constraints implied by (8).

Methods for solving nonlinear programs are well developed in the literature. We simply note here that efficient evaluation of the constraints, rapid determination of active constraints as well as the use of gradient techniques in

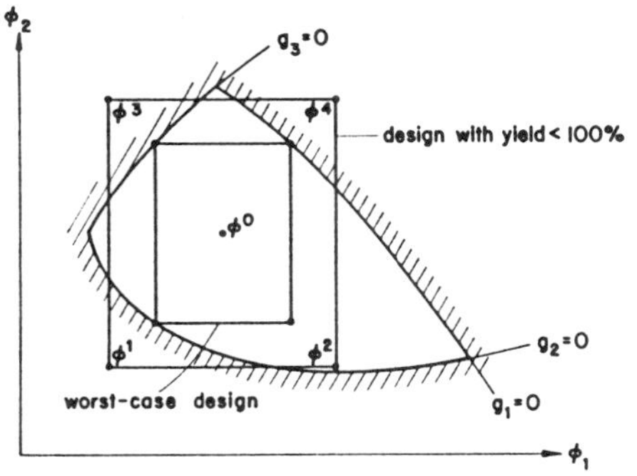

Fig. 2. Example of a worst case design and a design with yield <100 percent. For the worst case design the set of active vertices is $S_{av}=\{1,3,4\}$. These vertices indicate critical regions where constraint violations are most likely to occur for a design with yield <100 percent.

the search for the optimum values of ϕ^0 and ϵ are computationally highly desirable.

The active vertices at the worst case optimum, i.e., at the solution of WCD, are those which lie on the boundary of R_c. The set of active vertices is given by

$$S_{av} \triangleq \{r \mid g_i(\phi^r)=0, r\in I_v, i\in I_c\}. \qquad (12)$$

See Fig. 2 for an illustration of a worst case design.

Yield Less Than 100 Percent

When the yield is allowed to drop below 100 percent we have $R_\epsilon \not\subset R_c$. An appropriate nonlinear program in this case is

$$\text{YNP}\begin{cases}\underset{\phi^0,\epsilon}{\text{minimize}}\ C(\phi^0,\epsilon,\mu)\\ \text{subject, for example, to}\\ \quad Y(\phi^0,\epsilon,\mu)\geqslant Y_L\end{cases} \qquad (13)$$

where Y_L is a yield specification. A design with yield <100 percent is depicted in Fig. 2.

Again, this nonlinear program is to be solved for optimum values of ϕ^0 and ϵ. It is not necessary that all components of ϕ^0 and ϵ be allowed to vary. Some of them might be fixed. The constraint on yield might be removed if the yield is represented in the cost. This case might arise, for example, if the distribution of outcomes is fixed and ϕ^0 is allowed to vary in order to meet maximum yield. Although design constraints do not seem to appear explicitly in YNP they are all implicitly accounted for in the consideration of yield.

Approximations to R_c

Unlike optimization problems in which a single point is of interest, tolerances and uncertainties create a region of interest. The solution is usually characterized by several critical points or regions so that more information about the constraint region is required. Under the foregoing assumptions it seems reasonable to assume that for a high but less than 100-percent yield the active vertices determined by a worst case design will indicate regions where constraint violations are most likely (see Fig. 2). Accordingly, our interest must be directed to the active vertices as locations for centering reliable approximations to the boundary, which is the subject of the following section.

III. Interpolation by Quadratic Polynomials

Worst case design, yield analysis, and optimization involve a mass of calculations. Inadequate information on cost functions, component distributions, model uncertainties, etc., already hinders precise design solutions. Consequently, multidimensional approximations to design constraints appear to be a computational necessity without, it is felt, any significant sacrifice in design accuracy.

An approximate representation of a function $g(\phi)$, typically a constraint, using its values at a finite set of points ϕ is possible. These points are called nodes or base points. Interpolation is adopted since it is not only a simple approach to approximation but also because it requires relatively few actual function evaluations. In general, interpolation can be done by means of a linear combination of the set of all possible monomials [12], [13]. A monomial in ϕ of the order m is given by

$$\Phi(\phi)=(\phi_1)^{\gamma_1}(\phi_2)^{\gamma_2}\cdots(\phi_k)^{\gamma_k}, \qquad \sum_{i=1}^{m}\gamma_i=m \qquad (14)$$

where the integers $\gamma_i \geqslant 0$, $i=1,2,\cdots,k$.

Since the accuracy of the approximation depends upon the size of the interpolation region, the critical parts of R_ϵ may not be covered by a single interpolation region. Thus the use of more than one interpolation region will be discussed.

The Quadratic Polynomial

A quadratic polynomial in k variables can be written as

$$\begin{aligned}P(\phi)=&a_1(\phi_1)^2+a_2(\phi_2)^2+\cdots+a_k(\phi_k)^2\\ &+a_{k+1}\phi_1\phi_2+a_{k+2}\phi_1\phi_3+\cdots+a_{N-k-1}\phi_{k-1}\phi_k\\ &+a_{N-k}\phi_1+a_{N-k+1}\phi_2+\cdots+a_{N-1}\phi_k+a_N\end{aligned} \qquad (15)$$

where

$$N=(k+1)(k+2)/2 \qquad (16)$$

is the number of monomials and at the same time the number of the unknown coefficients $a_1,a_2,\cdots,a_N$. In order to find these coefficients, the values of $P(\phi)$ at N base points ϕ_b are required. By setting

$$P(\phi_b^i)=g(\phi_b^i), \qquad i=1,2,\cdots,N \qquad (17)$$

a set of N simultaneous linear equations is constructed. A solution for this system exists if the base points are degree-2 independent [14]. A set of N points is said to be degree-m independent if there exist no constants β_j, except $\beta_j=0$, $j=1,2,\cdots,N$, such that

$$\sum_{j=1}^{N}\beta_j\Phi(\phi^j)=0 \qquad (18)$$

where Φ is the monomial given in (14).

Selection of Base Points

Suppose that the function $g(\phi)$ is to be approximated at a particular region in the parameter space. We identify this interpolation region $\bar{R}$ through a "center" of interpolation $\bar{\phi}$ and a step size δ. We define, accordingly

$$\bar{R} \triangleq \{\phi \,|\, |\phi_i - \bar{\phi}_i| \leqslant \delta_i,\ i=1,2,\cdots,k\} \tag{19}$$

and require that the base points should satisfy

$$\phi_b^i \in \bar{R}, \qquad i=1,2,\cdots,N. \tag{20}$$

This requirement is satisfied if the set of base points is given by

$$[\phi_b^1 \quad \phi_b^2 \quad \cdots \quad \phi_b^N] = D[\mathbf{0} \quad \mathbf{1}_k \quad -\mathbf{1}_k \quad B] + [\bar{\phi} \quad \bar{\phi} \quad \cdots \quad \bar{\phi}] \tag{21}$$

where D is a $k \times k$ diagonal matrix with elements δ_i, $\mathbf{0}$ is the zero vector of dimension k, $\mathbf{1}_k$ is a $k \times k$ unit matrix, B is a $k \times (k(k-1)/2)$ matrix defined by

$$B = [\mu_b^1 \quad \mu_b^2 \quad \cdots \quad \mu_b^L] \tag{22}$$

in which

$$L = k(k-1)/2 \tag{23}$$

where μ_b^j are randomly selected such that

$$\mu_b^j \in R_\mu, \qquad j=1,2,\cdots,L. \tag{24}$$

See, for example, Fig. 3.

This choice of base points preserves one-dimensional convexity/concavity of the approximated function, since there are three base points along each axis (see Appendix).

Polynomial Evaluation at Vertices

In solving the nonlinear program WCD, the values of the constraints and their derivatives at the vertices are required. Here, we develop an efficient technique for evaluating approximations to the constraints along with their derivatives for subsequent use in conjunction with gradient optimization methods.

The technique exploits simple properties of a quadratic approximation. The following two equations are used to obtain the polynomial value and its gradients at any vertex ϕ^r using values at another vertex ϕ^s.

$$P(\phi^r) = P(\phi^s) + (\phi^r - \phi^s)^T \nabla P(\phi^s) + \tfrac{1}{2}(\phi^r - \phi^s)^T H (\phi^r - \phi^s) \tag{25}$$

$$\nabla P(\phi^r) = \nabla P(\phi^s) + H(\phi^r - \phi^s) \tag{26}$$

where

$$\nabla \triangleq \begin{bmatrix} \partial/\partial\phi_1 \\ \partial/\partial\phi_2 \\ \vdots \\ \partial/\partial\phi_k \end{bmatrix} \tag{27}$$

and

$$H \triangleq \nabla\nabla^T P \tag{28}$$

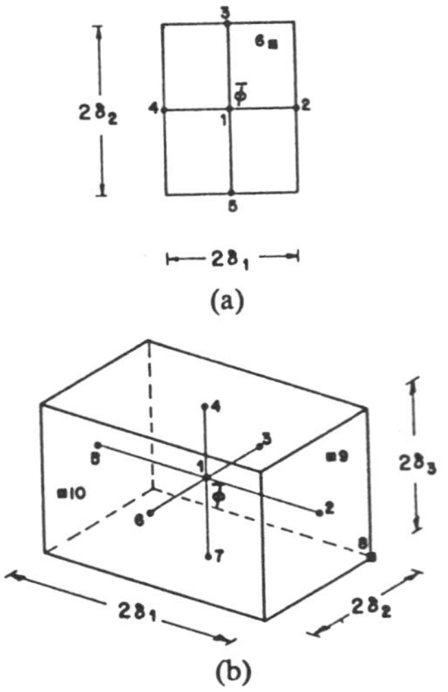

Fig. 3. Arrangement of the base points with respect to the centers of interpolation regions in (a) two dimensions (base point ϕ_b^6 is selected randomly) and (b) three dimensions (base points ϕ_b^8, ϕ_b^9, and ϕ_b^{10} are selected randomly).

is the Hessian matrix for the quadratic approximation.

Suppose ϕ^r and ϕ^s are adjacent vertices, i.e.,

$$\phi^r = \phi^s + 2\epsilon_i e_i \tag{29}$$

where e_i is the unit vector in the ith direction. In this case (25) and (26) reduce to

$$P(\phi^r) = P(\phi^s) + 2\epsilon_i \nabla_i P(\phi^s) + 2\epsilon_i^2 H_{ii} \tag{30}$$

$$\nabla P(\phi^r) = \nabla P(\phi^s) + 2\epsilon_i H_i \tag{31}$$

where ∇_i is the ith row of ∇, H_{ii} is the ith diagonal element of H, and H_i is the ith column of H.

Different approximations may be considered in different interpolation regions. To this end some relevant notation is introduced, as follows.

Let

$$I_\delta \triangleq \{i \,|\, \delta_i \geqslant \epsilon_i\} \tag{32}$$

$$I_\epsilon \triangleq \{i \,|\, \epsilon_i > \delta_i\} \tag{33}$$

and the number of elements of I_δ and I_ϵ be k_δ and k_ϵ, respectively. In an effort to describe the minimal number of interpolation regions N_{in} which collectively contain all the vertices we consider each element of I_ϵ in such a way that

$$N_{in} = 2^{k_\epsilon} \tag{34}$$

and (see, e.g., Fig. 4) that the centers of interpolation $\bar{\phi}^l$ are associated with $\phi^r \in R_v$ through

$$\bar{\phi}^l = \phi^0 + P(\phi^r - \phi^0) \tag{35}$$

where the projection matrix P is the diagonal matrix

$$P \triangleq \begin{bmatrix} p_1 & & & & \\ & p_2 & & & \\ & & \cdot & & \\ & & & \cdot & \\ & & & & \cdot \\ & & & & & p_k \end{bmatrix} \tag{36}$$

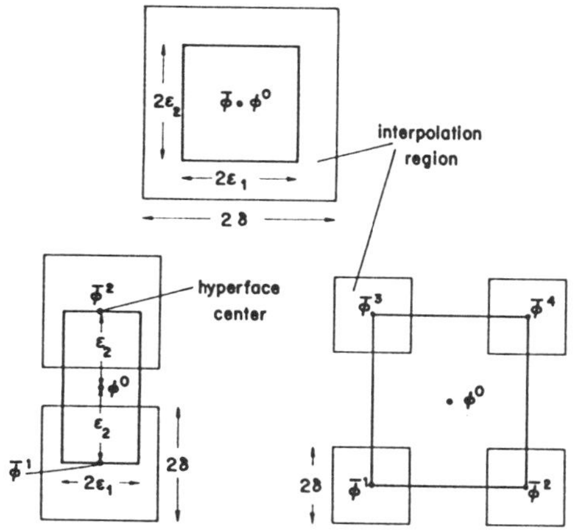

Fig. 4. Three situations created by certain step sizes $\delta=\delta_1=\delta_2$ and tolerances. The available interpolation regions and their centers are indicated as follows: (a) $I_\epsilon=\varnothing$, $I_v^1=\{1,2,3,4\}$. (b) $I_\epsilon=\{2\}$, $I_v^1=\{1,2\}$, $I_v^2=\{3,4\}$. (c) $I_\epsilon=\{1,2\}$, $I_v^1=\{1\}$, $I_v^2=\{2\}$, $I_v^3=\{3\}$, $I_v^4=\{4\}$.

and where

$$p_i=\begin{cases}0, & i\in I_\delta\\ 1, & i\in I_\epsilon.\end{cases} \tag{37}$$

A suitable numbering scheme for identifying vertices is [7]

$$r=1+\sum_{i=1}^{k}\left(\frac{\mu_i^r+1}{2}\right)2^{i-1}, \qquad \mu_i^r\in\{-1,1\} \tag{38}$$

so that adjacent vertices (i.e., vertices different in the ith parameter) satisfy

$$|r-s|=2^{i-1}. \tag{39}$$

An analogous numbering scheme for interpolation regions is given by

$$l=1+\sum_{i=1}^{k}p_i\left(\frac{\mu_i^r+1}{2}\right)2^{i_\epsilon-1}, \qquad \mu_i^r\in\{-1,1\} \tag{40}$$

where

$$i_\epsilon=\sum_{j=1}^{i}p_j. \tag{41}$$

Intuitively, i_ϵ is a renumbered index derived from i and the projection components p_j to include only the elements of I_ϵ in such a way that a doubling of the number of interpolation regions occurs for every such element. For example, if $p_i=1$, $i=1,2,\cdots,k_\epsilon$ we have

$$l=1+\sum_{i=1}^{k_\epsilon}\left(\frac{\mu_i^r+1}{2}\right)2^{i-1}, \qquad \mu_i^r\in\{-1,1\} \tag{42}$$

since $i_\epsilon=i$ follows from (41).

Since a given rth vertex belongs to a particular interpolation region l given by (40) we can, without ambiguity, let

$$P^r\triangleq P^l(\phi^r) \tag{43}$$

where $P^l(\phi)$ is the polynomial constructed for the lth

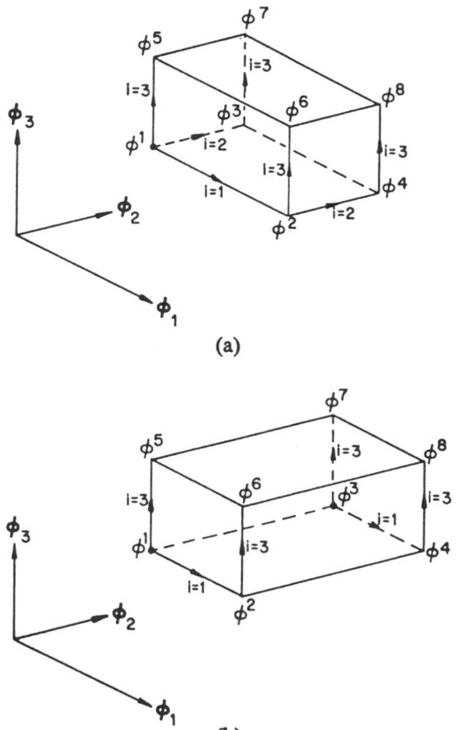

(b)

Fig. 5. Illustration of the efficient technique for evaluation of the approximations and their derivatives. (a) $k_\delta=3$, $N_{\text{in}}=1$, $I_v^1=I_v$ and, initially, $I=\{1\}$. (b) $k_\delta=2$, $N_{\text{in}}=2$, $I_v^1=\{1,2,5,6\}$, $I_v^2=\{3,4,7,8\}$ and, initially, $I=\{1,3\}$.

interpolation region. With this notation we rewrite (30) and (31) as

$$P^r=P^s+2\epsilon_i\nabla_iP^s+2\epsilon_i^2H_{ii}^l \tag{44}$$

$$\nabla P^r=\nabla P^s+2\epsilon_iH_i^l \tag{45}$$

where the superscript l identifies the relevant components of $\boldsymbol{H}$ (defined earlier) and where r and s are related by

$$r=s+2^{i-1}, \qquad i\in I_\delta \tag{46}$$

implying that ϕ^r and ϕ^s are adjacent (see (29) and belong to the same interpolation region, viz.,

$$r,s\in I_v^l\triangleq\{i\,|\,i\in I_v,\ \phi^i\in\bar{R}^l\} \tag{47}$$

where I_v is given by (4) and $\bar{R}^l$ is the lth interpolation region.

Algorithm for Polynomial Evaluation (APE)

This algorithm is illustrated in Fig. 5. The figure indicates two situations in three dimensions. Polynomial and gradient evaluations are made during each iteration at corresponding vertices in certain interpolation regions, starting with one vertex per region. New vertices are systematically considered in successive iterations, their number being doubled until the candidates have been exhausted.

This algorithm assumes that quadratic polynomial values P along with corresponding ∇P associated with a subset I_p of the N_{in} available interpolation regions are to be computed. The required subset will generally be de-

TABLE I

COMPUTATIONAL EFFORT FOR EVALUATION OF THE QUADRATIC POLYNOMIAL AND ITS DERIVATIVES

Description	Number of additions	Number of multiplications
At one vertex only	$\frac{1}{2}k(3k+5)$	$\frac{3}{2}k(k+1)$
At all the vertices using original formula	$2^{k-1}k(3k+5)$	$3 \times 2^{k-1}k(k+1)$
At all the vertices using the efficient scheme	$2^{k_\epsilon}[\frac{1}{2}k(3k+5)+(k+2)(2^{k_\delta}-1)]+k_\delta$	$2^{k_\epsilon}[\frac{3}{2}k(k+1)+k_\delta(k+1)+2^{k_\delta}-1]$
At all the vertices using the efficient scheme $k_\delta = k$	$\frac{1}{2}k(3k+7)+(k+2)(2^k-1)$	$\frac{5}{2}k(k+1)+2^k-1$

termined during worst case design in accordance with candidates for active vertices for the constraint under consideration.

Step 1: Evaluate P^s and ∇P^s for all $s \in I$, where

$$I = \left\{ i \mid i = \min_{j \in I_c^l} j,\ l \in I_p \right\}.$$

Comment: This is an initialization of the set of vertices, one vertex per interpolation region being considered, as required to start the computation of the polynomials and their gradients. Each vertex selected is the closest possible to the origin.

Step 2: $J \leftarrow I_\delta$.

Comment: J is a working set of indices, initialized here to correspond to all those designable parameters which can vary within each interpolation region.

Step 3: If $J = \varnothing$ stop.

Comment: This step tests whether there are any (remaining) candidates in J. If J is empty polynomials at all the vertices within the considered interpolation regions have been evaluated.

Step 4: $i \leftarrow \min_{j \in J} j$.

Comment: This ordering process selects the index i corresponding to the parameter to be varied in the following steps.

Step 5: $T \leftarrow \epsilon_i + \epsilon_i$.

Step 6: $G_i^l \leftarrow TH_i^l$ for all $l \in I_p$.

Step 7: For all $s \in I$

$$P^r \leftarrow P^s + T\nabla_i P^s + \epsilon_i G_{ii}^l$$

$$\nabla P^r \leftarrow \nabla P^s + G_i^l$$

where r and l are given by (46) and (40), respectively.

Comment: The values of the polynomials and the corresponding gradient vectors are calculated at all appropriate adjacent vertices. The number of vertices at which evaluations have been made are thus doubled in this step.

Step 8: $I \leftarrow I \cup \{ r \mid r = s + 2^{i-1},\ s \in I \}$.

Comment: The set of vertices already considered is updated.

Step 9: $J \leftarrow J \setminus \{i\}$.

Comment: The index i, already exploited, is removed from the working set J.

Step 10: Go to Step 3.

The computational effort required for considering all possible vertices, i.e., all N_{in} available interpolation regions, compared to that required for one vertex only is shown in Table I.

IV. WORST CASE DESIGN ALGORITHMS

The steps taken by these algorithms are shown in detail for the two-section transmission-line transformer example given in Section VII (refer to Fig. 10).

Phase 1: Single Interpolation Region

Step 1: Choose initial values for ϕ^0, ϵ, and $\delta \geqslant \epsilon$.

Step 2: $\bar{\phi} \leftarrow \phi^0$.

Step 3: Choose N base points to satisfy (21).

Step 4: Evaluate the constraint functions at these base points.

Step 5: Solve (17) to obtain the coefficients of the interpolating polynomials.

Step 6: Starting with the current ϕ^0 and ϵ solve the nonlinear program WCD for optimal values ϕ^{0*} and ϵ^*, employing the constraint approximations defined by Step 5.

Comment: Since values of design constraints as well as their sensitivities at vertices are required in solving WCD, the efficient technique for polynomial evaluation at vertices is used, namely, APE. Obviously, $I_p = \{1\}$ for all constraints, since there is only one interpolation region.

Step 7: $\phi^0 \leftarrow \phi^{0*}$ and $\epsilon \leftarrow \epsilon^*$.

Step 8: If $|\phi_i^0 - \bar{\phi}_i| \leqslant 1.5\ \delta_i$ for all $i = 1, 2, \cdots, k$, go to Step 10.

Comment: This tests whether the new nominal point ϕ^0 is close enough to $\bar{\phi}$ to ensure confidence in the accuracy of the approximations.

Step 9: Until $\delta_i > \epsilon_i$ for all $i = 1, 2, \cdots, k$, set $\delta_i \leftarrow 4\delta_i$. Go to Step 2.

Comment: Here, we ensure that all the vertices are contained in the interpolation region before repeating Phase 1.

Step 10: Stop if δ is sufficiently small.

Step 11: $\delta \leftarrow \delta/4$.

Step 12: If $\delta > \epsilon$ go to Step 2.

Comment: This check ensures that a single interpolation region is still applicable. If it is, Phase 1 is repeated.

Step 13: Go to Phase 2.

Phase 2: Multiple Interpolation Regions

This phase of the worst case design problem is executed if greater accuracy of the solution is required than is possible with the single interpolation region employed in Phase 1. The efficiency will be improved if suitable candidates for active constraints are determined so that not only would fewer interpolations be necessary but also fewer constraints would enter WCD. Step 1 of the present algorithm, therefore, calls for executing Phase 1, and collecting information about candidates for active vertices I_{av} and corresponding candidates for active constraints I_{ac}^s, $s \in I_{av}$.

Step 1: Choose δ_{ac} as a small positive number and execute Phase 1 to get

$$I_{av} \triangleq \{s | P_i^s \leqslant \delta_{ac}, i \in I_c, s \in I_v\}$$

$$I_{ac}^s \triangleq \{i | P_i^s \leqslant \delta_{ac}, i \in I_c, s \in I_v\}.$$

Comment: The set I_{av} is termed the set of candidates for active vertices. The set I_{ac}^s identifies the corresponding candidates for active constraints associated with the sth vertex.

Step 2: Use (35) to locate centers of interpolation $\bar{\boldsymbol{\phi}}^l$ for all $r \in I_{av}$.

Comment: Note that a subset of all possible interpolation regions is hereby identified because $I_{av} \subset I_v$.

Step 3: For each interpolation region $\bar{R}^l$ identified by $\bar{\boldsymbol{\phi}}^l$ and $\boldsymbol{\delta}$:

a) Choose N base points to satisfy (21).

b) $I_{ac}^l \leftarrow \cup_{s \in I_v^l} I_{ac}^s$.

c) Evaluate g_i for all $i \in I_{ac}^l$ at the N base points.

d) Solve (17) to obtain the coefficients of the corresponding polynomials for all $i \in I_{ac}^l$.

Comment: The set I_{ac}^l identifies all the constraints to be evaluated in $\bar{R}^l$.

Step 4: Starting with the current $\boldsymbol{\phi}^0$ and $\boldsymbol{\epsilon}$ solve the nonlinear program WCD for optimal $\boldsymbol{\phi}^{0*}$ and $\boldsymbol{\epsilon}^*$ employing the constraint approximations defined by Step 3. Algorithm APE is called for each constraint i to be evaluated by setting $I_p(i) \leftarrow \{l | i \in I_{ac}^l\}$.

Comment: Note that the set I_v^l replaces I_v and I_{ac}^l replaces I_c, thereby reducing the computational effort. Furthermore, $I_p(i)$ which becomes I_p on entry to APE concentrates evaluations in critical interpolation regions. (See the comment following Step 2.)

Step 5: $\boldsymbol{\phi}^0 \leftarrow \boldsymbol{\phi}^{0*}$ and $\boldsymbol{\epsilon} \leftarrow \boldsymbol{\epsilon}^*$.

Step 6: $I_{av} \leftarrow \{s | P_i^s \leqslant \delta_{ac}, i \in I_{ac}^l, s \in I_v^l\}$

$I_{ac}^s \leftarrow \{i | P_i^s \leqslant \delta_{ac}, i \in I_{ac}^l, s \in I_v^l\}$.

Comment: The set of candidates for active vertices and associated candidates for active constraints is updated by examining all the constraints used during Step 4. Refer to the comment following Step 4.

Step 7: If, for any $s \in I_{av}$, $|\phi_j^s - \bar{\phi}_j^l| > 2\delta_j$ for any j go to Step 2.

Step 8: Stop if $\boldsymbol{\delta}$ is sufficiently small.

Step 9: $\boldsymbol{\delta} \leftarrow \boldsymbol{\delta}/4$.

Step 10: Go to Step 2.

V. Yield Estimation and Yield Sensitivities

For a uniform distribution of outcomes inside the tolerance orthotope, computation of hypervolume plays the basic role in yield evaluation. A formula for the nonfeasible hypervolume (hypervolume outside the constraint region but inside the tolerance orthotope) is hereby derived. It is based upon linear cuts of the orthotope.

The Linear Cut and Evaluation of Hypervolume

Based upon either linearization or intersections (as elaborated on later in this section) of the hypersurface $g(\boldsymbol{\phi}) = 0$ with the tolerance orthotope R_ϵ, we construct the linear cut

$$\boldsymbol{q}^T\boldsymbol{\phi} - c \geqslant 0 \tag{48}$$

where $\boldsymbol{q}$ is a column vector of k components and c is a scalar. We will derive a general expression for the nonfeasible hypervolume defined by this linear cut and R_ϵ, denoted by $V(R)$, where

$$R = \{\boldsymbol{\phi} | g(\boldsymbol{\phi}) < 0\} \cap R_\epsilon. \tag{49}$$

Define a reference vertex

$$\boldsymbol{\phi}^r = \boldsymbol{\phi}^0 + \boldsymbol{E}\boldsymbol{\mu}^r \tag{50}$$

where

$$\mu_i^r = -\operatorname{sgn}(q_i), \qquad i = 1, 2, \cdots, k. \tag{51}$$

The general formula for the hypervolume can be written as

$$V = \left(\frac{1}{k!}\prod_{j=1}^{k}\alpha_j\right)\left(\sum_{s=1}^{2^k}(-1)^{\nu^s}(\delta^s)^k\right) \tag{52}$$

where

$$\delta^s = \max\left(0, 1 - \sum_{j=1}^{k}\frac{\epsilon_j}{\alpha_j}|\mu_j^s - \mu_j^r|\right) \tag{53}$$

$$\nu^s = \sum_{i=1}^{k}|\mu_i^s - \mu_i^r|/2 \tag{54}$$

and α_j is the distance between the intersections of the hyperplane $\boldsymbol{q}^T\boldsymbol{\phi} - c = 0$ and the reference vertex $\boldsymbol{\phi}^r$ along an edge of R_ϵ in the jth direction. It is to be noted that δ^s is positive if and only if the vertex $\boldsymbol{\phi}^s$ violates the linear cut (48).

Fig. 6 illustrates the evaluation of hypervolumes for two cases when $k = 3$.

Hypervolume Sensitivities

The hypervolume sensitivities can be expressed as

$$\frac{\partial V}{\partial \phi_i^0} = \frac{1}{k!}\left[\sum_{j=1}^{k}\frac{\partial \alpha_j}{\partial \phi_i^0}\prod_{\substack{p=1\\p\neq j}}^{k}\alpha_p\right]B + A\left(k\sum_{s=1}^{2^k}(-1)^{\nu^s}(\delta^s)^{k-1}\frac{\partial \delta^s}{\partial \phi_i^0}\right) \tag{55}$$

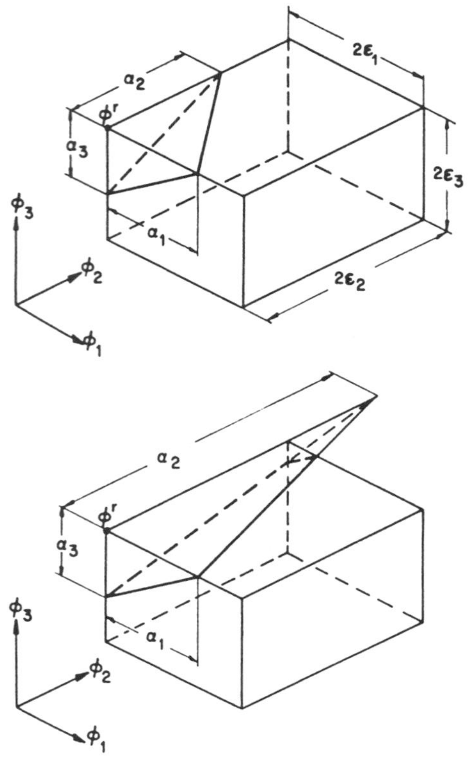

Fig. 6. The nonfeasible volume obtained by a linear cut (a) $V=(1/3!)\alpha_1\alpha_2\alpha_3$. (b) $V=((1/3!)\alpha_1\alpha_2\alpha_3)[1-(1-(2\epsilon_2/\alpha_2))^3]$.

and

$$\frac{\partial V}{\partial \epsilon_i}=\mu_i^r\frac{\partial V}{\partial \phi_i^0}-A\left(\frac{k}{\alpha_i}\sum_{s=1}^{2^k}(-1)^{\nu^s}|\mu_i^s-\mu_i^r|(\delta^s)^{k-1}\right) \tag{56}$$

where

$$A=\frac{1}{k!}\prod_{j=1}^{k}\alpha_j \tag{57}$$

$$B=\sum_{s=1}^{2^k}(-1)^{\nu^s}(\delta^s)^k \tag{58}$$

and

$$\frac{\partial \delta^s}{\partial \phi_i^0}=\begin{cases}0, & \text{if } \delta^s=0\\ \displaystyle\sum_{j=1}^{k}\frac{\epsilon_j}{(\alpha_j)^2}|\mu_j^s-\mu_j^r|\frac{\partial \alpha_j}{\partial \phi_i^0}, & \text{if } \delta^s>0.\end{cases} \tag{59}$$

It should be mentioned that the hypervolume and its sensitivities are defined when $\alpha_i \to \infty$ for any i, since the limit exists. But, the sensitivities are discontinuous whenever a vertex ϕ^s satisfies the equation

$$q^T\phi^s-c=0. \tag{60}$$

The Linear Constraints Case

Let the constraint region be defined by the m linear constraints

$$g_l(\phi)=\phi^Tq^l-c^l\geqslant 0, \qquad l=1,2,\cdots,m. \tag{61}$$

Assuming no overlapping of nonfeasible regions defined by different constraints inside the orthotope R_ϵ, i.e.,

$$R_i \bigcap_{i\neq j} R_j=\varnothing \tag{62}$$

where

$$R_l \triangleq \{\phi\in R_\epsilon | g_l(\phi)<0\} \tag{63}$$

the yield can be expressed as

$$Y=1-\sum_{l=1}^{m}V(R_l)/V(R_\epsilon). \tag{64}$$

Knowing that

$$V(R_\epsilon)=2^k\prod_{j=1}^{k}\epsilon_j \tag{65}$$

the yield sensitivities are given by

$$\frac{\partial Y}{\partial \phi_i^0}=-\sum_{l=1}^{m}\frac{\partial V^l}{\partial \phi_i^0}\Big/\left(2^k\prod_{j=1}^{k}\epsilon_j\right) \tag{66}$$

$$\frac{\partial Y}{\partial \epsilon_i}=\left(\frac{1}{\epsilon_i}\sum_{l=1}^{m}V^l-\sum_{l=1}^{m}\frac{\partial V^l}{\partial \epsilon_i}\right)\Big/\left(2^k\prod_{j=1}^{k}\epsilon_j\right) \tag{67}$$

where V^l denotes $V(R_l)$. The linear constraints can be used as linear cuts directly. Hence, the nonfeasible hypervolume V^l and its sensitivities can be obtained using (52), (55), and (56) for each constraint and where

$$\alpha_j^l=\mu_j^r g_l(\phi^r)/q_j^l$$
$$=\mu_j^r\left(\sum_{i=1}^{k}q_i^l(\phi_i^0+\mu_i^r\epsilon_i)-c^l\right)/q_j^l \tag{68}$$

$$\frac{\partial \alpha_j^l}{\partial \phi_i^0}=\mu_j^r q_i^l/q_j^l. \tag{69}$$

If $q_j^l=0$ we have $\alpha_j^l=\infty$, however, a limit exists as indicated after (59).

The Quadratic Constraints Case

Consider a vertex ϕ^r detected to be active with respect to a quadratic constraint $g_l(\phi)\geqslant 0$ after the worst case design process (see Section IV). If the tolerances are allowed to increase slightly beyond their worst case values, intersections between the orthotope edges passing through ϕ^r and the hypersurface $g_l(\phi)=0$ will arise. The number of these intersections is k, which is the number of edges passing through ϕ^r, if

$$\partial g_l(\phi^r)/\partial \phi_j\neq 0, \qquad \text{for all } j. \tag{70}$$

In order to find the intersection point along the jth edge, or its extension in the direction $-\mu_j^r e_j$, where e_j is a unit vector in the jth direction, we express $g_l(\phi)=0$ as

$$(\phi_j)^2+2\phi_j\xi_l(\phi_1^r,\phi_2^r,\cdots,\phi_{j-1}^r,\phi_{j+1}^r,\cdots,\phi_k^r)$$
$$+\eta_l(\phi_1^r,\phi_2^r,\cdots,\phi_{j-1}^r,\phi_{j+1}^r,\cdots,\phi_k^r)=0 \tag{71}$$

where ξ_l and η_l are constant functions and ϕ_j is the only variable. Hence, the point of intersection is

$(\phi_1^r, \phi_2^r, \cdots, \lambda_j^l, \cdots, \phi_k^r)$, where

$$\lambda_j^l = -\xi_l \pm \sqrt{\xi_l^2 - \eta_l}\,, \qquad \mu_j^r(\phi_j^r - \lambda_j^l) > 0 \tag{72}$$

is a real root of (71). The condition imposed on the root insures that it is in the direction $-\mu_j^r e_j$ with respect to ϕ^r. If both roots lie in this direction, the one closer to ϕ^r is chosen.

The equation in ϕ of the hyperplane, representing the linear cut, which passes through these k points of intersection is

$$\phi^T q^l - c^l = \det \begin{bmatrix} \phi_1 & \phi_2 & \cdots & \phi_k & 1 \\ \lambda_1^l & \phi_2^r & \cdots & \phi_k^r & 1 \\ \phi_1^r & \lambda_2^l & \cdots & \phi_k^r & 1 \\ \cdot & & & & \\ \cdot & & & & \\ \phi_1^r & \phi_2^r & \cdots & \lambda_k^l & 1 \end{bmatrix} = 0 \tag{73}$$

and ϕ^r is a reference vertex for this cut.

The yield sensitivities are calculated according to the gradients of the k intersections.

$$\frac{\partial \lambda_j^l}{\partial \phi_i} = -\frac{\partial \xi_l}{\partial \phi_i} \pm \frac{1}{2\sqrt{\xi_l^2 - n_l}}\left(2\xi_l \frac{\partial \xi_l}{\partial \phi_i} - \frac{\partial \eta_l}{\partial \phi_i}\right), \qquad i \neq j \tag{74}$$

$$\frac{\partial \lambda_i^l}{\partial \phi_i} = 0. \tag{75}$$

Thus if α_j^l is the distance from the vertex ϕ^r to the point of intersection with the lth constraint along the orthotope edge in the jth direction, then

$$\alpha_j^l = \mu_j^r(\phi_j^r - \lambda_j^l) \tag{76}$$

$$\frac{\partial \alpha_j^l}{\partial \phi_i^0} = -\mu_j^r \frac{\partial \lambda_j^l}{\partial \phi_i}, \qquad i \neq j \tag{77}$$

$$\frac{\partial \alpha_j^l}{\partial \phi_j^0} = \mu_j^r. \tag{78}$$

Equations (76), (77), and (78) are substituted directly into formulas (52), (55), and (56), whichever is relevant. Yield and its sensitivities are also obtained from (64), (66), and (67).

Overlapping Constraints

We discuss in this section an approach which is directed at solving some of the problems arising from constraints overlapping within the tolerance region. Since the analytical formulas for yield and yield sensitivities assume nonoverlapping linear cuts (see (62)), methods to avoid describing the boundary of the constraint region by overlapping cuts are required.

A single function of the least pth type [9] can be used to describe the boundary of the feasible region if the boundary is defined, as is usual, by more than one constraint. The least pth function is given by

$$G(g_i(\phi), I_c, p) = \begin{cases} 0, & M = 0 \\ M\left[\sum_{i \in J} (g_i(\phi)/M)^q\right]^{1/q}, & M \neq 0 \end{cases} \tag{79}$$

where

$$M = \min_{i \in I_c} g_i(\phi) \tag{80}$$

$$J = \begin{cases} \{i \mid g_i(\phi) < 0, i \in I_c\}, & \text{if } M < 0 \\ I_c, & \text{if } M > 0 \end{cases} \tag{81}$$

$$q = -p \operatorname{sgn} M$$

and p is given to be greater than 1.

The constraint $G \geqslant 0$ *exactly* describes the boundary of the constraint region R_c.

In order to define a linear cut based on G, we can either linearize G at an appropriate point or use intersections of the hypersurface $G = 0$ with the orthotope edges. A possible implementation is suggested in the appropriate steps of the following algorithms.

VI. Algorithm for Yield less than 100 Percent

It is assumed that Phase 1 and Phase 2 of the worst case design algorithm have been suitably executed. Information has, therefore, been gathered relating to active vertices I_{av}, associated active constraints I_{ac}^s at the sth vertex and also polynomial approximations $P_i(\phi)$ corresponding to the (generally) nonlinear $g_i(\phi)$. The least pth function $G_s(P_i(\phi), I_{ac}^s, p)$, $s \in I_{av}$, can be formulated according to the notation introduced by (79) and is associated with the sth vertex.

Note also that optimal values ϕ^{0*} and ϵ^* are known for worst case design. See Fig. 7.

Step 1: For $\kappa_i > 1$ set $\epsilon_i \leftarrow \kappa_i \epsilon_i^*$, $i = 1, 2, \cdots, k$.

Comment: This initializes the yield to be less than 100 percent. The κ_i are chosen such that all active constraints are violated, as indicated by Fig. 7(b).

Step 2: $\phi^0 \leftarrow \phi^{0*}$.

Step 3: Solve the nonlinear program YNP for optimal values ϕ^{0*} and ϵ^* employing algorithm YAN (which follows) for evaluating yield and yield sensitivities.

Algorithm for Dynamic Yield Analysis (YAN)

This algorithm is called for each evaluation of yield and its sensitivities as required during optimization.

Step 1: $S \leftarrow \{s \mid s \in I_{av}, G_s(P_i(\phi^s), I_{ac}^s, p) < 0\}$.

Comment: S is a working set of indices of reference vertices (1, 2, and 3 in Fig. 7(a)). We consider only those vertices which currently violate the design constraints for the nonfeasible hypervolume evaluation (1, 2, and 3 in Fig. 7(b)).

Step 2: $V \leftarrow 0$, $V_{\phi_i} \leftarrow 0$, and $V_{\epsilon_i} \leftarrow 0$, $i = 1, 2, \cdots, k$.

Comment: V, V_{ϕ_i}, and V_{ϵ_i} are to be updated to store the total nonfeasible hypervolume and its sensitivities with respect to ϕ^0 and ϵ, respectively.

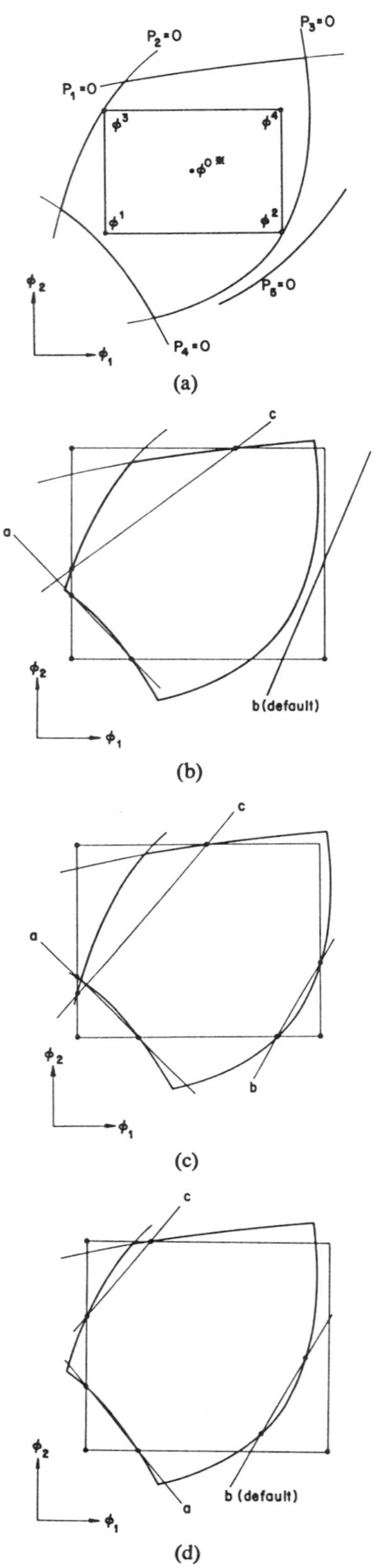

Fig. 7. Examples of yield determination. (a) Quadratic constraint approximations and a worst case design. $I_{av}=\{1,2,3\}$, $I_{ac}^1=\{4\}$, $I_{ac}^2=\{3,5\}$, $I_{ac}^3=\{1,2\}$. Notice that P_1 and P_2 are overlapping and P_5 is a redundant constraint. (b) Linear cuts arising after setting $\epsilon_i \leftarrow \kappa_i \epsilon_i^*$. Cuts a and c are based upon intersections, while cut b is based upon linearization. (c) A typical situation which may result during optimization, in which all cuts are based on intersections. (d) A situation which may arise during optimization. The linear cut b is not updated since less than k intersections exist, remaining as in (c).

Step 3: $r \leftarrow \min_{s \in S} s$.

Comment: This ordering process selects the index r corresponding to the reference vertex to be considered.

Step 4: For $j=1,2,\cdots,k$, execute Steps 5 and 6.

Comment: In this loop we consider the edges of the orthotope passing through $\boldsymbol{\phi}^r$ as indicated in Steps 5 and 6.

Step 5: Find λ_j^l, for all $l \in I_{ac}^r$, using (72).

Step 6: If λ_j^l is undefined for any $l \in I_{ac}^r$, go to Step 14.

Comment: The hypersurface $G_r=0$ has an intersection with an orthotope edge if P_l has an intersection with the edge for all $l \in I_{ac}^r$. We go to Step 14 if such intersections are not found for all k edges.

Step 7: If I_{ac}^r contains more than one element, go to Step 10.

Comment: In case I_{ac}^r contains one element only, l say, there is no need to consider G_r, since $G_r=P_l$.

Step 8: Find α_j^r and $\partial\alpha_j^r/\partial\phi_i^0$, $i,j=1,2,\cdots,k$, where $I_{ac}^r=\{l\}$, using (76), (77), and (78).

Comment: Notice that we will identify the cut by index of the reference vertex r rather than using l.

Step 9: Go to Step 12.

Step 10: $\alpha_j^r \leftarrow \max_{l \in I_{ac}^r} \alpha_j^l$, $j=1,2,\cdots,k$, where α_j^l is obtained by (76).

Comment: The furthest intersection, from $\boldsymbol{\phi}^r$, among the intersections of $P_l=0$, $l \in I_{ac}^r$, corresponds to the intersection of the hypersurface $G_r=0$.

Step 11: Find $\partial\alpha_j^r/\partial\phi_i^0$, $i,j=1,2,\cdots,k$, using (77) and (78).

Step 12: Set $\boldsymbol{q}^r$ and c^r for the rth linear cut according to (73).

Comment: In general, the explicit formulation of the linear cut is not necessary since information about α_j^r is the only requirement for hypervolume calculation. But this cut will be used later in the process as a default if less than k intersections are obtained (Fig. 7(d)).

Step 13: Go to Step 17.

Step 14: If this is not the first yield evaluation, go to Step 16.

Step 15: $\boldsymbol{q}^r \leftarrow \nabla G_r(P_l^r, I_{ac}^r, p)$

$$c^r \leftarrow (\boldsymbol{\phi}^r)^T \boldsymbol{q}^r - G_r(P_l^r, I_{ac}^r, p).$$

Comment: Initially if less than k intersections exist, linearization at the vertex $\boldsymbol{\phi}^r$ is used to provide a default cut. Cut b in Fig. 7(b) is an example.

Step 16: Find α_j^r and $\partial\alpha_j^r/\partial\phi_i^0$ using (68) and (69).

Comment: No updating of the rth cut is performed.

Step 17: $V \leftarrow V + V^r$

$$V_{\phi_i} \leftarrow V_{\phi_i} + V_{\phi_i}^r, \qquad i=1,2,\cdots,k$$
$$V_{\epsilon_i} \leftarrow V_{\epsilon_i} + V_{\epsilon_i}^r, \qquad i=1,2,\cdots,k$$

where V^r is given by (52), $V_{\phi_i}^r$ by (55) and $V_{\epsilon_i}^r$ by (56), respectively.

Step 18: $S \leftarrow S \backslash \{r\}$.

Comment: The index r, already exploited, is removed from the working set S.

Step 19: If $S \neq \varnothing$ go to Step 3.

Comment: This step checks if all reference vertices have been considered.

TABLE II
WORST CASE DESIGN OF THE TWO-SECTION 10:1 QUARTER-WAVE TRANSFORMER

Cost Function	Z_1^0	Z_2^0	ϵ_1/Z_1^0 (%)	ϵ_2/Z_2^0 (%)	δ	N.O.F.E.	CDC Time (sec)
C_1	2.5637	5.5048	14.678	9.007	0.4	18	7.2
	2.5234	5.4379	14.988	9.081	0.1	24	9.5
C_2	2.1515	4.7350	12.715	12.697	0.4	12	2.5
	2.1494	4.7305	12.687	12.700	0.1	18	3.0

Starting values $Z_1^0 = 2.2361$, $Z_2^0 = 4.4721$, $\epsilon_1 = 0.2$ and $\epsilon_2 = 0.4$

Frequency points used 0.5, 0.6, ..., 1.5 GHz

Objective cost functions $C_1 = \frac{1}{\epsilon_1} + \frac{1}{\epsilon_2}$, $C_2 = \frac{Z_1^0}{\epsilon_1} + \frac{Z_2^0}{\epsilon_2}$

Reflection coefficient specification $|\rho| \leq 0.55$

*N.O.F.E. denotes the number of function evaluations

Step 20: $Y \leftarrow 1 - V/V(R_\epsilon)$

$$\partial Y/\partial \phi_i^0 \leftarrow -V_{\phi_i}/V(R_\epsilon), \qquad i = 1,2,\cdots,k$$

$$\partial Y/\partial \epsilon_i \leftarrow [V/\epsilon_i - V_{\epsilon_i}]/V(R_\epsilon), \qquad i = 1,2,\cdots,k$$

where $V(R_\epsilon)$ is given by (65).

VII. EXAMPLES

Two-Section Transmission-Line Transformer

Consider the two-section 10:1 quarter-wave lossless transmission-line transformer used by Bandler *et al.* [1]. The specifications and results of the worst case tolerance optimization problem of the characteristic impedances Z_1 and Z_2 over 100-percent bandwidth are shown in Table II for two different objective functions. The constraint region and the resulting optimum solutions in two cases are shown in Fig. 8 and Fig. 9. An equal value of δ_1 and δ_2 was used. The figures show the interpolation regions and the resulting approximations for the constraint boundary. The results obtained are contrasted with the results obtained in [1]. Furthermore, the steps taken by the worst case design algorithm using the objective function C_1, shown in Table II, are detailed in Fig. 10.

Subsequently, the approximations obtained at the two active vertices for the worst case problem having the objective function C_1, shown in Table II and Fig. 8, were used for yield optimization. This problem is denoted $P0$. A rough estimate of δ used for stopping Phase 2 was obtained in the following way. For a yield constraint

$$Y \geqslant 90 \text{ percent}$$

the nonfeasible hypervolume (it is area in this example) is given approximately by

$$A \simeq (1-0.9)(2\epsilon_1)(2\epsilon_2).$$

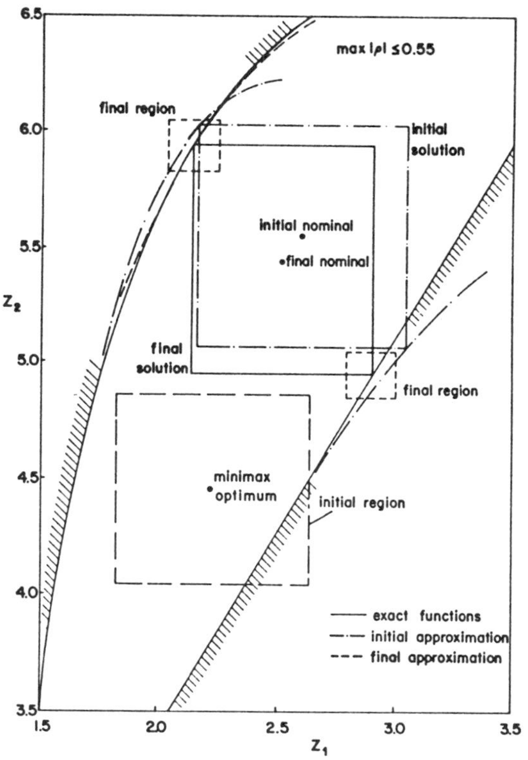

Fig. 8. Minimization of $1/\epsilon_1 + 1/\epsilon_2$ for the two-section transformer.

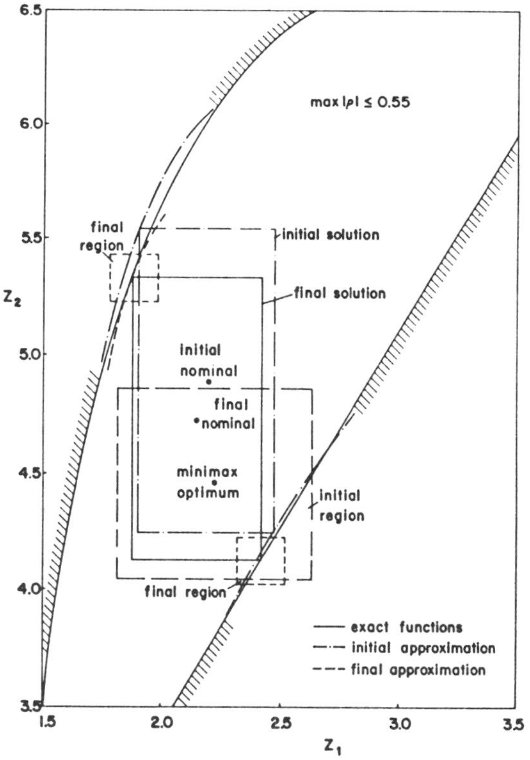

Fig. 9. Minimization of $Z_1^0/\epsilon_1 + Z_2^0/\epsilon_2$ for the two-section transformer.

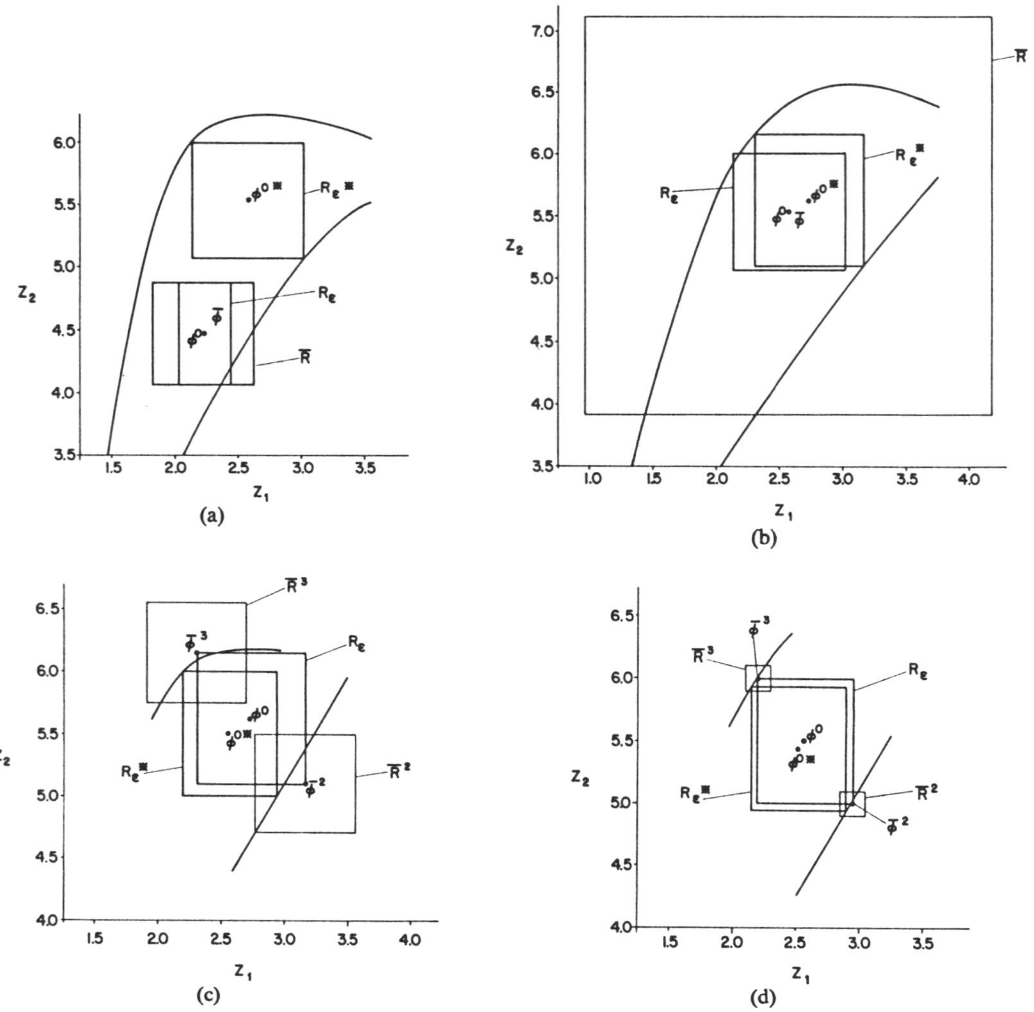

Fig. 10. Details for the problem of Fig. 8. (a) Initial interpolation region $\bar{R}$, tolerance region R_ϵ and the approximations to the boundary of the constraint region. R_ϵ^* is the optimum tolerance region using Phase 1 of the worst case design algorithm. (b) Enlarged interpolation region $\bar{R}$ and starting with the previous optimum R_ϵ we arrive at R_ϵ^*. (c) Reducing the interpolation region size and switching to Phase 2 of the algorithm. $I_{ac}^2 = \{$frequency point 1.0 GHz$\}$ and $I_{ac}^3 = \{$frequency points 0.5 and 1.5 GHz$\}$. (d) Further reduction in interpolation region size resulting in the final optimum solution R_ϵ^*.

The area cut off by each constraint is

$$A' \simeq \tfrac{1}{2}A.$$

But, assuming equal intersections $\alpha = \alpha_1 = \alpha_2$

$$A' = \tfrac{1}{2}\alpha^2.$$

Hence

$$\alpha \simeq \sqrt{0.1(2\epsilon_1)(2\epsilon_2)} = 0.27$$

where ϵ_1 and ϵ_2 are the worst case absolute tolerances. The approximation with $\delta = 0.1$ was used for solving problems

$$P1 \begin{cases} \text{minimize } 1/\epsilon_1 + 1/\epsilon_2 \\ \text{subject to} \\ \qquad Y \geqslant 90 \text{ percent} \end{cases}$$

$$P2 \text{ minimize } (1/\epsilon_1 + 1/\epsilon_2)/Y$$

assuming a uniform distribution of outcomes between tolerance extremes.

The optimum solutions for $P1$ and $P2$ are shown in Table III and contrasted with the worst case solution $P0$ in Fig. 11. The program FLNLP2 [15] was used for solving the resulting nonlinear programming problem. Since a convex constraint region appears in this problem, the values of yield obtained are lower bounds for the true yields.

Three-Component LC Low-pass Filter

A normalized three-component low-pass ladder network, terminated with equal load and source resistances of 1 Ω is shown in Fig. 12. The circuit was considered for worst case design by Bandler, Liu, and Chen [1]. Although this filter is symmetric a three-dimensional approximation was required in order to perform the yield technique described before.

TABLE III
YIELD DETERMINATION AND OPTIMIZATION OF THE TWO-SECTION 10:1 QUARTER-WAVE TRANSFORMER

Problem	Z_1^0	Z_2^0	ε_1/Z_1^0 (%)	ε_2/Z_2^0 (%)	Objective	Yield (%)	N.O.Y.E.***	CDC Time (sec)
P1*	2.5273	5.3998	21.09	13.51	3.2465	90.0	45	0.6
P2**	2.5290	5.1513	31.44	22.13	3.2597	65.5	15	0.3

* Minimize $1/\varepsilon_1 + 1/\varepsilon_2$ subject to yield $\leq$ 90%

** Minimize $(1/\varepsilon_1 + 1/\varepsilon_2)/Y$

*** N.O.Y.E. denotes the number of yield evaluations

Starting point for both P1 and P2 is $Z_1^0 = 2.5234$, $Z_2^0 = 5.4379$ (worst case nominal) and $\kappa_1 = \kappa_2 = 1.4$

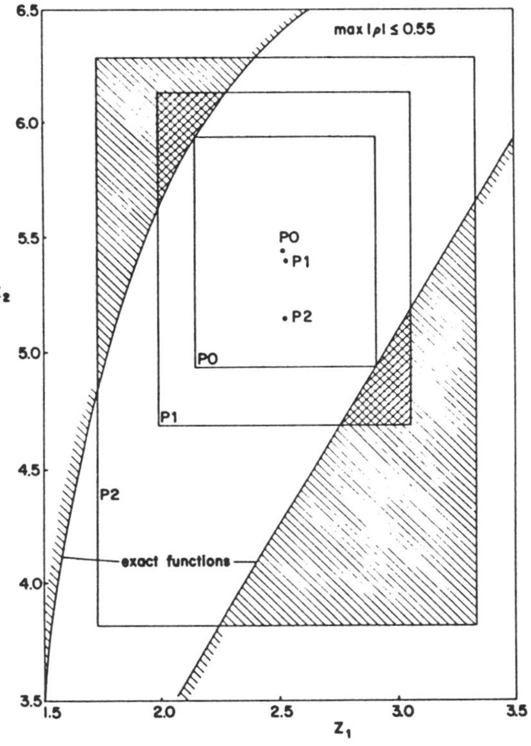

Fig. 11. The optimum tolerance regions and nominal values for the worst case, 90-percent yield and optimum yield designs.

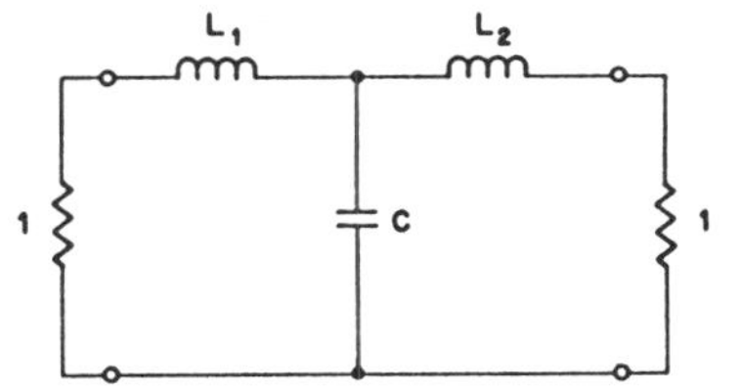

Fig. 12. The circuit for the *LC* filter.

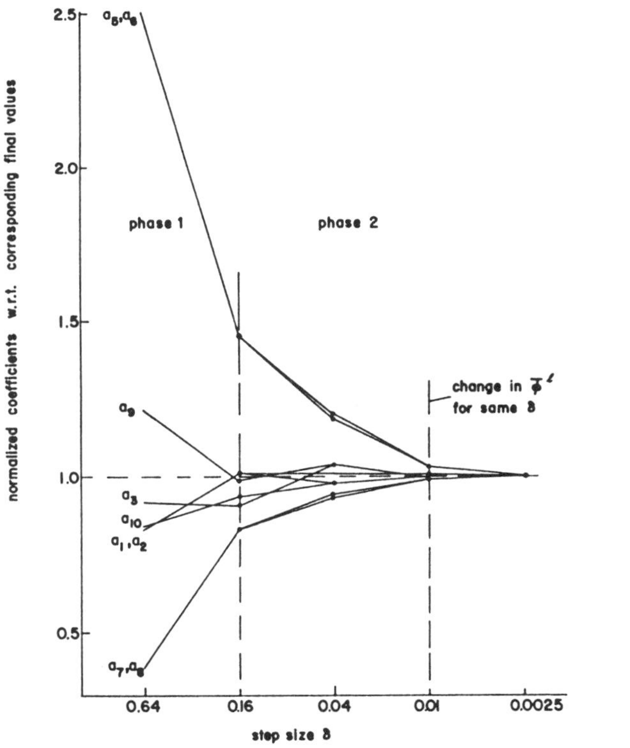

Fig. 13. Convergence for the *LC* filter of the quadratic approximation to the insertion loss constraint at 2.5 rad/s.

Using an equal step size δ for all components, a worst case design was first obtained with final $\delta = 0.01$. The base points used are given by (21) with

$$B = \begin{bmatrix} 0.5 & -0.5 & 1.0 \\ -0.5 & 0.5 & 1.0 \\ 0.8 & 0.8 & 1.0 \end{bmatrix}$$

consistent with the vector of components

$$\phi = \begin{bmatrix} L_1 \\ L_2 \\ C \end{bmatrix}.$$

The specifications and the objective function are given in Table IV. The convergence of the quadratic approximation coefficients as the step size δ is reduced is shown in Fig. 13 for the insertion loss constraint at the frequency point 2.5 rad/s. The coefficient a_4 is not shown in the figure. Its value is close to zero and hence the normalized

TABLE IV
WORST CASE AND CONSTRAINED YIELD RESULTS OF THE *LC* LOW-PASS FILTER

Yield (%)	L_1^0	L_2^0	C^0	ϵ_1/L_1^0 (%)	ϵ_2/L_2^0 (%)	ϵ_C/C^0 (%)	N.O.Y.E.*	CDC Time (sec)
100	1.999	1.998	0.9058	9.88	9.89	7.60	-	1.9
96	1.997	1.997	0.9033	11.23	11.23	12.46	38	1.0

* N.O.Y.E. denotes the number of yield evaluations

Frequency points used are 0.45, 0.5, 0.55, 1.0 in the passband and 2.5 in the stopband

Objective cost function is $L_1^0/\epsilon_1 + L_2^0/\epsilon_2 + C^0/\epsilon_C$

Insertion loss specification is $\leq$ 1.5 dB in the passband and $\geq$ 25 dB in the stopband

Starting point for the worst case design problem is $L_1^0 = L_2^0 = 2.0$, $C^0 = 1.0$ and $\epsilon_1/L_1^0 = \epsilon_2/L_2^0 = \epsilon_C/C^0 = 10\%$

Starting point for the 96% yield design problem is the optimal worst case nominal with $\kappa_1 = \kappa_2 = 1.06$ and $\kappa_3 = 1.45$

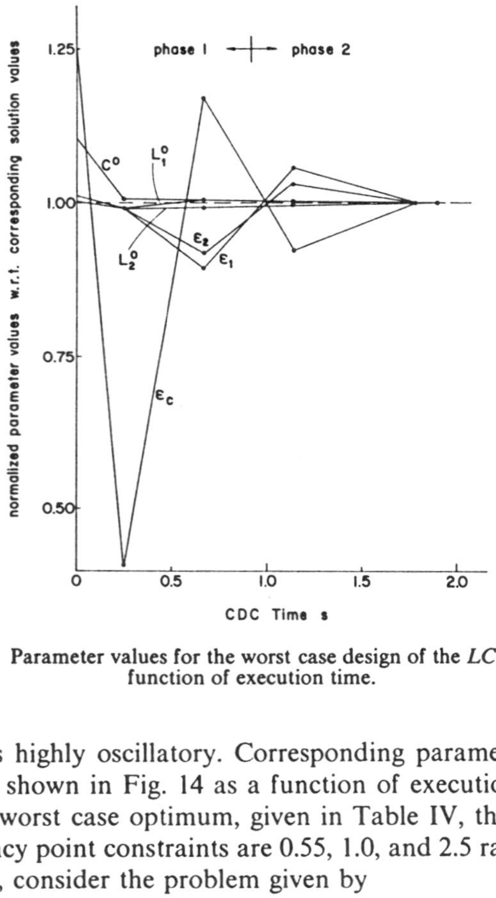

Fig. 14. Parameter values for the worst case design of the *LC* filter as a function of execution time.

value is highly oscillatory. Corresponding parameter values are shown in Fig. 14 as a function of execution time. At the worst case optimum, given in Table IV, the active frequency point constraints are 0.55, 1.0, and 2.5 rad/s.

Now, consider the problem given by

$$\text{minimize } L_1^0/\epsilon_1 + L_2^0/\epsilon_2 + C^0/\epsilon_C$$
$$\text{subject to}$$
$$Y \geqslant 96 \text{ percent.}$$

The quadratic approximation with $\delta = 0.04$, which was used in this problem, is shown in Table V after and before averaging symmetric coefficients. The diagonal elements of the Hessian matrix, as defined by the coefficients of the

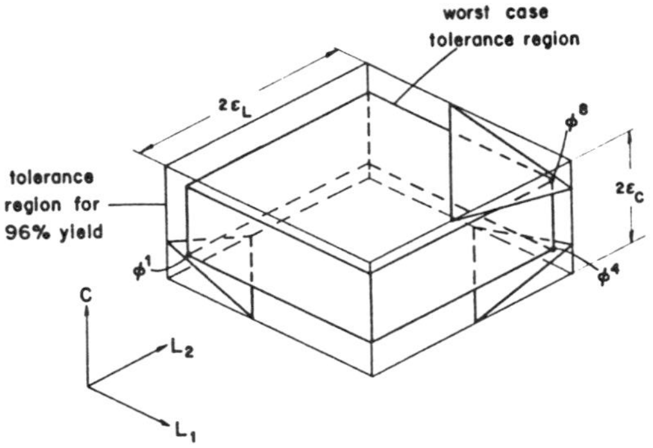

Fig. 15. The tolerance regions for the worst case design and the 96-percent yield for the *LC* filter. The linear cuts shown are based on the intersections of the active quadratic constraints approximations with edges of the tolerance region for 96-percent yield case. $I_{ac}^1 = $ {frequency point 2.5 rad/s}, $I_{ac}^4 = $ {frequency point 0.55 rad/s}, and $I_{ac}^8 = $ {frequency point 1.0 rad/s}.

approximating polynomial, suggest a one-dimensionally convex constraint region. Symmetry between L_1 and L_2 was used to reduce computation in finding the values and the gradients of the intersections between the orthotope edges and the quadratic constraints. The results are shown in Table IV and in Fig. 15. The tolerance for the capacitor ϵ_C was approximately doubled, with respect to its value for the worst case design, by allowing the yield to drop to 96 percent. (A Monte Carlo analysis at the solution indicated 96.6-percent yield by both the exact constraints and by the approximate ones.)

Two-Section Waveguide Transformer

The two-section waveguide transformer, investigated for a minimax (equal-ripple) response by Bandler [16] was selected to perform a tolerance assignment. The general configuration of such a structure is illustrated in Fig. 16. A design specification of a reflection coefficient of 0.05 over 500-MHz bandwidth centered at 6.175 GHz was chosen. Table VI shows the dimensions of the input and output waveguides and the widths of the two sections.

TABLE V

Coefficients of the Quadratic Approximation Around Active Vertices

Freq. point	State	L_1^2	L_2^2	C^2	L_1L_2	L_1C	L_2C	L_1	L_2	C	-
0.55	before	-0.06847	-0.06847	-0.57056	.33010	0.92247	0.93855	-1.67845	-1.69182	-0.46249	3.83750
	after	-0.06847	-0.06847	-0.57056	.33010	0.93051	0.93051	-1.68513	-1.68513	-0.46249	3.83750
1.00	before	-1.12188	-1.16702	-9.98122	.21439	-8.16357	-8.30295	10.21440	10.51832	44.18607	-33.86206
	after	-1.14445	-1.14445	-9.98122	.21439	-8.23326	-8.23326	10.36637	10.36637	44.18607	-33.86206
2.50	before	-1.38601	-1.42228	-9.90167	.39487	-0.92910	-0.94732	10.19142	10.32736	32.94001	-46.93184
	after	-1.40414	-1.40414	-9.90167	.39487	-0.93821	-0.93821	10.25939	10.25939	32.94001	-46.93184

Coefficients of the quadratic approximations obtained at active vertices with a step $\delta = 0.04$. The table shows the coefficients obtained by the algorithm and the coefficients used for yield determination after averaging symmetric coefficients.

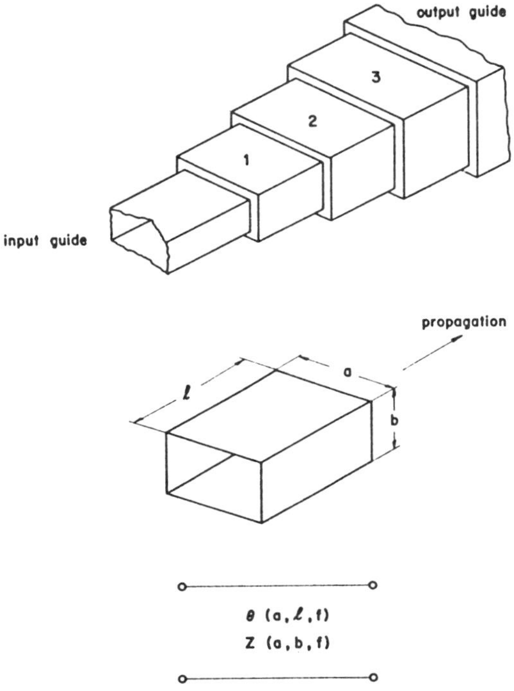

Fig. 16. Illustrations of an inhomogeneous waveguide transformer.

The program developed by Bandler and Macdonald [17] is used to obtain the reflection coefficient. No sensitivities are provided by this program. An equal absolute tolerance ϵ is assumed for the heights and the lengths of the two sections. The assumption seems reasonable if they are machined in the same manner.

The objective is to maximize the absolute tolerance ϵ. The optimum nominal point and associated tolerance, given in Table VII, were obtained by the worst case design algorithm presented in Section IV. The program FLOPT4 [18] was used for solving the nonlinear program:

$$\text{maximize } \epsilon$$
$$\text{subject to}$$
$$R_v \subset R_c.$$

TABLE VI

Fixed Parameters and Specifications for the Two-Section Waveguide Transformer

Description	Width (cm)	Height (cm)	Length (cm)
Input guide	3.48488	0.508	∞
First section	3.6	variable	variable
Second section	3.8	variable	variable
Output guide	4.0386	2.0193	∞

Frequency points used 5.925, 6.175, 6.425 GHz

Reflection coefficient specification $|\rho| \leq 0.05$

Minimax solution (no tolerances) $|\rho| = 0.00443$

TABLE VII

Results Contrasting the Toleranced Solution and the Minimax Solution with no Tolerances for the Two-Section Waveguide Transformer

Description	b_1 (cm)	b_2 (cm)	ℓ_1 (cm)	ℓ_2 (cm)
Toleranced optimum	0.72812	1.42432	1.55409	1.51153
Minimax optimum	0.71315	1.39661	1.56044	1.51621

Equal absolute value of tolerance = 0.02013 cm

Number of complete response evaluations = 45

CDC time (approximation and optimization) = 33 s

A tolerance of 0.02013 cm was obtained. The number of actual response evaluations to reach the optimum starting from the minimax optimum (no tolerances) is shown in Table VII. The execution time shown includes both approximation and optimization times.

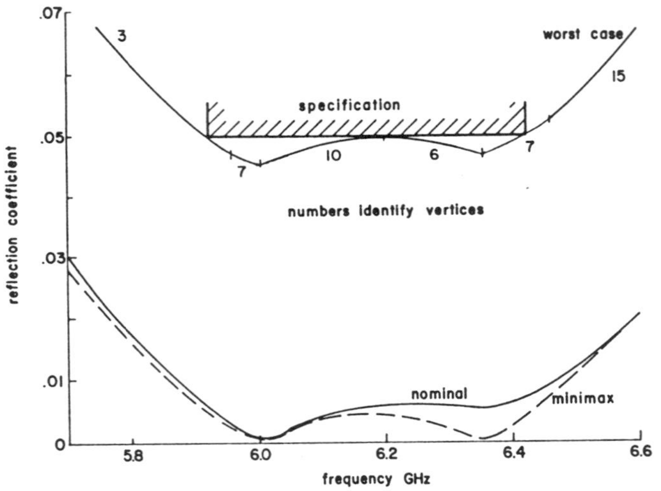

Fig. 17. Nominal, minimax and upper envelope of worst case responses for the two-section waveguide transformer.

The minimax, nominal and the upper envelope of worst case responses are shown in Fig. 17. The numbering scheme of the vertices is that given by (38) with the parameter vector

$$\phi = \begin{bmatrix} b_1 \\ b_2 \\ l_1 \\ l_2 \end{bmatrix}.$$

Vertices which fall within the worst case upper envelope are not indicated in Fig. 17. It was observed, however, that vertices 2, 6, 10, and 14 are either active or almost active with respect to the reflection coefficient constraint at band center. Furthermore, vertices 3, 7, 11, and 15 are either active or almost active near the band extremes. Hence, when b_1 is at its positive extreme while b_2 is at its negative extreme, the frequency point at the center of the band is more likely to be violated. The edges of the band are critical frequency points when b_1 is at its negative extreme while b_2 is at its positive extreme.

Retaining the approximations obtained by the worst case design procedure subsequently facilitates inexpensive Monte Carlo analyses. Hence, different statistical distributions of outcomes may be assumed and estimates of corresponding yields obtained. Assuming $\epsilon = 0.03$ cm, for example, while keeping the worst case nominal obtained, uniformly distributed Monte Carlo analyses were conducted with the approximation and with the actual functions: for 500 points, yields of 88 and 89 percent, respectively, are predicted. The approximation produces results 12 times faster.

Three-Section Waveguide Transformer

The three-section transformer with ideal junctions for which a minimax optimum was obtained by Bandler [16] is considered for tolerance assignment. Specifications and dimensions of input and output waveguides are given in Table VIII.

TABLE VIII
FIXED PARAMETERS AND SPECIFICATIONS FOR THE THREE-SECTION WAVEGUIDE TRANSFORMER

Description	Width (cm)	Height (cm)	Length (cm)
Input guide	3.48488	0.762	∞
First section	3.30581	variable	variable
Second section	3.12674	variable	variable
Third section	2.94767	variable	variable
Output guide	2.76860	1.60325	∞

Frequency points used 5.7, 6.1, 6.45, 6.8, 7.2 GHz
Reflection coefficient specification $|\rho| \leq 0.050$ (nonideal junctions)
Minimax solution (no tolerances) $|\rho| = 0.017$ (ideal junctions)

TABLE IX
RESULTS CONTRASTING THE TOLERANCED SOLUTION AND THE MINIMAX SOLUTION WITH NO TOLERANCES FOR THE THREE-SECTION WAVEGUIDE TRANSFORMER

Description	b_1 (cm)	b_2 (cm)	b_3 (cm)	ℓ_1 (cm)	ℓ_2 (cm)	ℓ_3 (cm)
Toleranced optimum	0.91034	1.36526	1.70189	1.45242	1.53875	1.63253
Minimax optimum	0.90318	1.37093	1.73609	1.54879	1.58375	1.64590

Equal absolute value of tolerance = 0.01383 cm
Number of complete response evaluations = 56
CDC time (approximation and optimization) = 167 s

Nonideal junctions were assumed and the widths of the three sections were fixed for convenience, so that the step changes are equal from one section to the next. An equal tolerance in the heights and lengths of the three sections was maximized for the reason already given.

Starting at the minimax optimum with equal steps of 0.02 for the interpolation region the results shown in Table IX were obtained. The program FLOPT4 [18] was used for solving the nonlinear programming problem formulated for the worst case design. Fig. 18 shows the upper envelope of worst case responses as well as the nominal design response. Although the envelope shows one vertex which is active at the lower frequency edge of the band, several other adjacent vertices, which restricted the increase in tolerance, are almost active. This appears to explain the fact that the envelope is substantially lower than the specification at other frequencies.

To verify the accuracy of the approximation an equal tolerance of 0.02 cm was assumed around the worst case nominal design and 500 uniformly distributed points were generated. The yield for the approximations was 96.4 percent and for the actual functions 96.0 percent. A twelve-fold improvement in execution time was again observed.

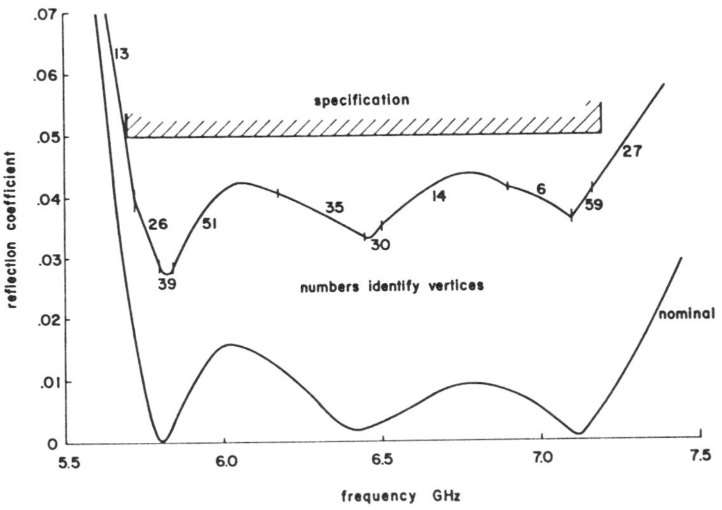

Fig. 18. Nominal and upper envelope of worst case responses for the three-section waveguide transformer.

VIII. Conclusions

A design centering technique based upon low-order multidimensional approximation and nonlinear programming is presented. The technique bridges the gap between available analysis programs, which may or may not be efficiently written and probably do not supply derivative information, and the advancing art of optimal centering, tolerancing and tuning. Efficient gradient methods, which are essential in such general design problems, can be usefully employed through the use of readily differentiable formulas and approximations.

In order to contrast various design centering techniques which rely on approximations, we point out that the method of Pinel and Roberts [3] and that of Karafin [4] are based upon truncated Taylor series expansions. Hence, not only sensitivities are required but also the validity of such an approximation for relatively large tolerances is uncertain. The simplicial approximation technique [6] does not require sensitivities, but the convexity assumption used is much more restrictive than the one-dimensional convexity assumption used in the present technique. Moreover, the approximation developed for the constraint region by Director and Hachtel [6] does not contain sensitivity information which allows the designer to check the effect of slightly relaxing some constraints. However, in the present technique, since there exists at least one quadratic approximation to each constraint it is possible to remove a constraint completely or slightly perturb its value (by changing the constant term in the quadratic approximation) to study such an effect on the design.

As expected, the design centering technique presented here facilitates subsequent inexpensive Monte Carlo analysis. For circuits which are expensive to analyze, such as switching circuits, this technique may be cheaper even for a single yield analysis using the Monte Carlo method in conjunction with the approximation. It is difficult to contrast our approach with the simplicial approximation approach from the point of view of Monte Carlo analysis. The fact that the simplicial approximation approach develops a relatively large number of linear constraints ($2^k + nk$, where k is the number of design parameters and n is the number of iterations required) while we develop quadratic constraints of the order of the number of actual constraints makes it hard to compare.

In addition, the quadratic approximations developed can be used for the new yield estimation and optimization technique developed. The yield estimation technique can also be used by itself if a reasonable worst case design is already known. The linear cuts may be obtained by linearizing active constraints at either associated active vertices or at the nominal point [19]. The technique can be extended to general nonlinear constraints. The efficient technique for calculation of the function and gradients at the different vertices (APE) may then be implemented with a suitable large-change sensitivity algorithm.

Yield estimation for other statistical distributions, different from the uniform distribution, can be done by regionalizing the space and associating a uniform distribution with each region [19].

Appendix
Preservation of One-Dimensional Convexity

As described in Section II, one-dimensional convexity is the property which makes the vertices candidates for the worst case. Hence, it is essential to preserve this property in the approximating polynomial $P(\phi)$ if it already exists in the exact function $g(\phi)$.

The following theorem indicates how to choose the base points in order to preserve one-dimensional convexity.

Theorem

If there exist three distinct base points $\boldsymbol{\phi}^1$, $\boldsymbol{\phi}^2$, and $\boldsymbol{\phi}^3$ in the ith direction, i.e.,

$$\boldsymbol{\phi}^j = \boldsymbol{\phi}^1 + c_j \boldsymbol{e}_i \tag{A1}$$

where c_j, $j = 2, 3$, are scalars and $\boldsymbol{e}_i$ is the unit vector in the ith direction, then the interpolating polynomial $P(\phi)$ is one-dimensionally convex/concave in the ith variable if the interpolated function $g(\phi)$ is so.

Proof: Assume that $P(\phi)$ is not one-dimensionally convex/concave, i.e.,

$$P(\lambda\phi^a + (1-\lambda)\phi^b) \gtrless \lambda P(\phi^a) + (1-\lambda)P(\phi^b), \qquad 0<\lambda<1 \tag{A2}$$

where

$$\phi^b = \phi^a + c\boldsymbol{e}_i \tag{A3}$$

and where c is a scalar. Hence

$$P(\phi^a + (1-\lambda)c\boldsymbol{e}_i) \gtrless \lambda P(\phi^a) + (1-\lambda)P(\phi^a + c\boldsymbol{e}_i). \tag{A4}$$

Expanding $P(\phi^a + (1-\lambda)c\boldsymbol{e}_i)$ and $P(\phi^a + c\boldsymbol{e}_i)$ as Taylor series and knowing that $P(\phi)$ is a quadratic polynomial, we have

$$P(\phi^a)+(1-\lambda)ce_i^T\nabla P(\phi^a)+\tfrac{1}{2}(1-\lambda)^2c^2e_i^T He_i$$
$$\gtrless P(\phi^a)+(1-\lambda)ce_i^T\nabla P(\phi^a)+\tfrac{1}{2}(1-\lambda)c^2e_i^T He_i. \tag{A5}$$

Thus

$$(1-\lambda)^2 e_i^T He_i \gtrless (1-\lambda)e_i^T He_i \tag{A6}$$

but since $0<(1-\lambda)<1$, hence

$$e_i^T He_i \lesseqgtr 0. \tag{A7}$$

Without any loss of generality we can number the three base points such that

$$\phi^3=\gamma\phi^1+(1-\gamma)\phi^2, \qquad 0<\gamma<1. \tag{A8}$$

and

$$\phi^2=\phi^1+\beta e_i \tag{A9}$$

where β is a scalar. Then

$$\begin{aligned}
P(\phi^3) &= P(\gamma\phi^1+(1-\gamma)\phi^2)\\
&= P(\phi^1+(1-\gamma)\beta e_i)\\
&= P(\phi^1)+(1-\gamma)\beta e_i^T\nabla P(\phi^1)+\tfrac{1}{2}(1-\gamma)^2\beta^2 e_i^T He_i\\
&= \gamma P(\phi^1)+(1-\gamma)\left[P(\phi^1)+\beta e_i^T\nabla P(\phi^1)+\tfrac{1}{2}\beta^2 e_i^T He_i\right]\\
&\quad -\tfrac{1}{2}(1-\gamma)\beta^2 e_i^T He_i+\tfrac{1}{2}(1-\gamma)^2\beta^2 e_i^T He_i\\
&= \gamma P(\phi^1)+(1-\gamma)P(\phi^2)-\tfrac{1}{2}\gamma(1-\gamma)\beta^2 e_i^T He_i.
\end{aligned}$$

But, using (A7)

$$P(\phi^3)\gtrless \gamma P(\phi^1)+(1-\gamma)P(\phi^2) \tag{A10}$$

and since $g=P$ at the base points, then

$$g(\phi^3)\gtrless \gamma g(\phi^1)+(1-\gamma)g(\phi^2) \tag{A11}$$

which contradicts the fact that $g(\phi)$ is one-dimensionally convex/concave in the ith variable. Hence, the assumption (A2) is never true. Q.E.D.

Corollary

A quadratic polynomial is one-dimensionally convex/concave if and only if all of the diagonal elements of its Hessian matrix are nonnegative/nonpositive.

The proof follows since inequality (A7) is never true.

The previous corollary allows an easy check on one-dimensional convexity of any quadratic function. In addition, the choice of base points as given in (21) satisfies the requirement of locating three base points in each direction.

ACKNOWLEDGMENT

The authors thank R. Biernacki, K. Madsen, M. R. M. Rizk, and H. Tromp for suggesting some improvements in our presentation.

REFERENCES

[1] J. W. Bandler, P. C. Liu, and J. H. K. Chen, "Worst case network tolerance optimization," *IEEE Trans. Microwave Theory Tech.*, vol. MTT-23, pp. 630–641, Aug. 1975.

[2] J. W. Bandler, P. C. Liu, and H. Tromp, "A nonlinear programming approach to optimal design centering, tolerancing and tuning," *IEEE Trans. Circuits Syst.*, vol. CAS-23, pp. 155–165, Mar. 1976.

[3] J. F. Pinel and K. A. Roberts, "Tolerance assignment in linear networks using nonlinear programming," *IEEE Trans. Circuit Theory*, vol. CT-19, pp. 475–479, Sept. 1972.

[4] B. J. Karafin, "The optimum assignment of component tolerances for electrical networks," *Bell Syst. Tech. J.*, vol. 50, pp. 1225–1242, Apr. 1971.

[5] N. Elias, "New statistical methods for assigning device tolerances," in *Proc. 1975 IEEE Symp. Circuits Syst.*, Newton, MA, pp. 329–332, Apr. 1975.

[6] S. W. Director and G. D. Hachtel, "The simplicial approximation approach to design centering and tolerance assignment," in *Proc. 1976 IEEE Symp. Circuits Syst.*, Munich, Germany, pp. 706–709, Apr. 1976.

[7] J. W. Bandler, "Optimization of design tolerances using nonlinear programming," *J. Optimization Theory and Appl.*, vol. 14, pp. 99–114, July 1974.

[8] J. W. Bandler and P. C. Liu, "Some implications of biquadratic functions in the tolerance problem," *IEEE Trans. Circuits Syst.*, vol. CAS-22, pp. 385–390, May 1975.

[9] J. W. Bandler and C. Charalambous, "Practical least *p*th optimization of networks," *IEEE Trans. Microwave Theory Tech.*, vol. MTT-20, pp. 834–840, Dec. 1972.

[10] H. S. M. Coxeter, *Regular Polytopes*, 2nd ed., New York: MacMillan, 1963, chap. 7.

[11] R. K. Brayton, A. J. Hoffman, and T. R. Scott, "A theorem of inverses of convex sets of real matrices with application to the worst case dc problem," *IEEE Trans. Circuits Syst.*, vol. CAS-23, pp. 409–415, Aug. 1977.

[12] S. L. Sobolev, "On the interpolation of functions of *n* variables" (transl.), *Sov. Math. Dokl.*, vol. 2, pp. 343–346, 1961.

[13] H. C. Thacher, Jr., and W. E. Milne, "Interpolation in several variables," *SIAM J.*, vol. 8, pp. 33–42, 1960.

[14] H. C. Thacher, Jr., "Generalization of concepts related to linear dependence," *SIAM J.*, vol. 6, pp. 288–299, 1959.

[15] W. Y. Chu, "Extrapolation in least *p*th approximation and nonlinear programming," Faculty of Engineering, McMaster University, Hamilton, Canada, Rep. SOC-71, Dec. 1974.

[16] J. W. Bandler, "Computer optimization of inhomogeneous waveguide transformers," *IEEE Trans. Microwave Theory Tech.*, vol. MTT-17, pp. 563–571, Aug. 1969.

[17] J. W. Bandler and P. A. Macdonald, "Response program for an inhomogeneous cascade of rectangular waveguides," *IEEE Trans. Microwave Theory Tech.*, vol. MTT-17, pp. 646–649, Aug. 1969.

[18] J. W. Bandler and D. Sinha, "FLOPT4—A program for least *p*th optimization with extrapolation to minimax solutions," Faculty of Engineering, McMaster University, Hamilton, Canada, Rep. SOC-151, Jan. 1977.

[19] H. L. Abdel-Malek and J. W. Bandler, "Yield estimation for efficient design centering assuming arbitrary statistical distributions," in *Conf. Computer-Aided Design of Electronic and Microwave Circuits and Systems*, Hull, England, pp. 66–71, July 1977; also to be published in *Int. J. Circuit Theory and Applications*.

Yield Maximization and Worst-Case Design with Arbitrary Statistical Distributions

ROBERT K. BRAYTON, SENIOR MEMBER, IEEE, STEPHEN W. DIRECTOR, FELLOW, IEEE, AND GARY D. HACHTEL, FELLOW, IEEE

***Abstract*—We describe a method by which a variety of statistical design problems can be solved by a linear program. We describe three key aspects of this approach. 1) The correspondence between the level contours of a given probability density function and a particular *norm*, which we shall call a *pdf-norm*. 2) The expression of distance in this norm from a given set of hyperplanes in terms of the *dual* of the pdf-norm. 3) The use of a linear program to inscribe a maximal pdf-norm-body into a simplicial approximation to the feasible region of a given statistical design problem. This work thus extends the applicability of a previously published algorithm, to the case of arbitrary pdf-norms and consequently to a wide variety of statistical design problems including the common mixed worst-case-yield maximization problem.**

I. Introduction

Of PRIME concern in the design of any electronic circuit, especially integrated circuits, is that the yield be high. Yield is defined as the ratio of the number of manufactured circuits which meet all design specifications to the total number of manufactured circuits. Because of the statistical fluctuations inherent in the manufacturing process, the yield is usually less than 100 percent. Design Centering is the process of choosing a nominal design point so as to tend to maximize yield.[1] Recently, two somewhat different approaches have been developed for solving the Design Centering and other related statistical design problems.

One of these approaches (see, for example, [1–3]) is based upon the use of quadratic approximations to the constraint surfaces in conjunction with nonlinear programming. The second approach [4] is based upon the generation of polyhedral approximations to the feasible region and the use of linear programming. While both methods have their strong points and weaknesses, meaningful comparison between them is still somewhat premature because of the continually evolving nature of the two approaches. For example, the procedure described in [4], called simplicial approximation, applied only to design centering for yield maximization problems with joint Gaussian distributions. More specifically, the design centering problem solved was the problem of inscribing a maximal n-ellipsoid into the (assumed convex) n-dimensional body of the feasible region. Results were then obtained by a linear program stated in terms of the Euclidean norm.

In this paper we generalize the simplicial approximation approach so that it may directly solve a variety of statistical design problems. Now included are worst-case tolerance assignment problems (maximum norm) with arbitrary (e.g. asymmetrical and/or mixed independent and joint) distributions, and mixed worst-case and yield maximization problems (hybrid maximum and arbitrary norms). The latter problem arises when some of the design parameters are statistical in nature while other parameters, such as power supply voltages, are worst case. In other words, the circuit is required to operate satisfactorily for all values of power supply voltages within some interval.

We assume at the outset that all statistical parameter distributions have equal variance or spread. No loss of generality is incurred by this assumption, and we show how differences in variance may be handled by appropriate scaling of the norm. We assume also that the feasible region of the statistical design problem is convex. Thus any point which lies inside the simplicial approximation (cf. (7), below) is a feasible point. The plan of the paper is as follows. In Section II we establish notation and briefly review the simplicial approximation process in a suitable framework for the work of this paper. In Section III we show how norms can be associated with probability density functions in a natural way. Thus we arrive at the concept of a *norm body*, which enables us, in Section IV, to generalize the "yield maximization via linear programming" approach of [4] to the case of arbitrary statistical distributions. In Section V we show that by defining norms in Cartesian product spaces, we are able to further generalize this approach to mixed worst-case and yield maximization problems. In Section VI we give some related results on the tolerance assignment problem and in Section VII we treat a practical example of the statistical design of a current source driver circuit (this type of circuit often appears as the "current source" of a current switch emitter follower circuit).

Manuscript received February 2, 1979; revised November 1, 1979. This work was supported in part by NSF Grant 7720895.

R. K. Brayton and G. D. Hachtel are with the Mathematical Sciences Department, IBM Thomas J. Watson Research Center, Yorktown Heights, NY 10598.

S. W. Director is with the Department of Electrical Engineering, Carnegie-Mellon University, Pittsburgh, PA 15213.

[1]Some confusion exists between the terms *design centering* and *yield maximization*. We choose to use the term design centering as a geometrical one referring to the inscription of a convex n-body into a region of acceptability. Design centering is only an approximation to the yield maximization problem. The distinction between these two is clarified in recent research by Lightner [11].

Reprinted from *IEEE Trans. Circuit Syst.*, vol. CAS-27, no. 9, pp. 756–764, September 1980.

II. The Simplicial Approximation

In this section we will briefly review some important concepts, including that of a simplicial approximation. Let the network under consideration obey the set of differentialalgebraic equations

$$f(\xi, \dot{\xi}, x, t) = 0, \qquad 0 \leq t \leq t_f \tag{1}$$

where ξ is a vector of node voltages, branch voltages, and branch currents and x is an n-vector of statistically varying parameters with a joint probability density function (pdf)

$$g(x, x_0, \sigma).$$

Here x_0 is the nominal value of x and σ is the standard deviation, or spread, about the nominal. Further, let X denote the n-dimensional parameter space so that $x \in X$. A network with parameter values x is considered to be acceptable if the solution $\xi(t)$ of (1) obeys a given vector of n_c constraints, whose components are defined by

$$\Phi_i(x) = \int_0^{t_f} \phi_i(\xi, \dot{\xi}, x, t)\, dt \leq 0, \quad i = 1, 2, \cdots, n_c. \tag{2a}$$

In addition the parameter vector x is assumed to lie between some lower and upper limits, or *box constraints*

$$x^l \leq x \leq x^u. \tag{2b}$$

Note that the constraints Φ_i depend on x implicitly through the solutions $\xi_j(t)$ of (1) as well as through their role as explicit arguments of the ϕ_i.

The set of constraints (2) defines the *region of acceptability* R:

$$R = \{x \mid \Phi_i(x) \leq 0, \text{ for all } i \in \{1, 2, \cdots, n_c\} \text{ and } x^l \leq x \leq x^u\}. \tag{3}$$

Note that R is closed, and is bounded if the limits in (2b) are finite. Of particular interest will be the boundary of R, denoted by ∂R and defined by

$$\partial R = \{x \mid \Phi(x) \leq 0, x^l \leq x \leq x^u, \text{ and } \Phi_i = 0, \text{ or } x_i = x_i^l, \text{ or } x_i = x_i^u, \text{ for some } i\}. \tag{4}$$

The simplicial approximation to R, denoted by $\hat{R}$, is a polyhedron whose faces are $(n-1)$-dimensional simplices. (An n-simplex is an n-dimensional polyhedron with exactly $n+1$ vertices which do not all lie in an $(n=1)$ dimensional hyperplane.) Such an approximation to R can be obtained by using the procedure described in [4], in which the convex hull of a set of $m \geq n+1$ points on ∂R is formed. The points on ∂R are found by searching along rays emanating from feasible points in R. This convex hull is characterized by the set of n_f inequalities

$$\eta_k^T x \leq b_k, \qquad k = 1, 2, \cdots, n_f \tag{5}$$

where η_k is an outward pointing vector normal to the kth hyperplane defined by

$$\eta_k^T x = b_k, \qquad k = 1, 2, \cdots, n_f \tag{6}$$

and b_k is related to the distance of the kth hyperplane

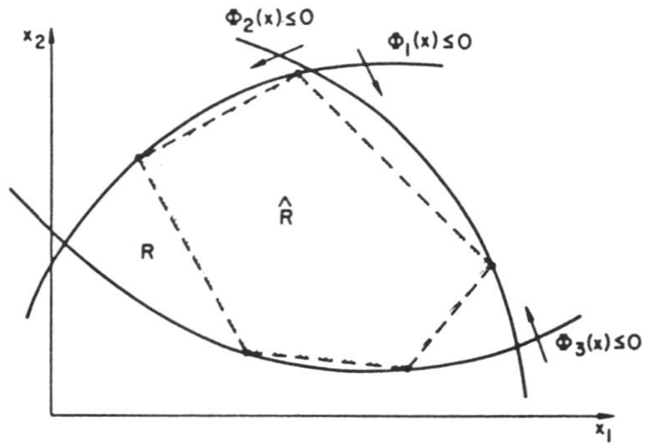

Fig. 1. Illustration of the feasible region R and simplicial approximation $\hat{R}$ for a 2-dimensional problem with three constraints.

from the origin. If R is convex this convex hull is an *interior* approximation to R. If we define the matrix of normals N

$$N \equiv (\eta_1, \eta_2, \cdots, \eta_{n_f})$$

and the vector of right-hand sides as

$$b^T = (b_1, b_2, \cdots, b_{n_f})$$

then the convex hull can be concisely defined by

$$\hat{R} \equiv \{x \mid N^T x \leq b\}. \tag{7}$$

The boundary of the hull is given by

$$\partial \hat{R} \equiv \{x \mid N^T x \leq b \text{ and } \eta_k^T X = b_k, \text{ for at least one } k \in \{1, 2, \cdots, n_f\}\} \tag{8}$$

which is an approximation to ∂R. Without loss of generality we can assume that the origin is interior to the feasible region. This can always be accomplished by translating the initial nominal (hence feasible) point to the origin. In this case each $b_k > 0$.

A simplicial approximation to a hypothetical feasible region is illustrated for two dimensions in Fig. 1.

III. Norm Bodies and pdf Norms

In general, for arbitrary joint statistical distributions in n dimensions, a level contour on the surface of the pdf defines a closed body. In the sequel, we consider only distributions where this closed body is well approximated by a convex body. To any such convex body we can associate a norm $n(x)$ with the properties[2]

i) $n(x) \geq 0$ for all x
ii) $n(x+y) \leq n(x) + n(y)$
iii) $n(\alpha x) = \alpha n(x)$ for $\alpha > 0$.

In particular, the 2-norm

$$n_2(x) = \|x\|_2 = \left(\sum_{i=1}^{n} x_i^2 \right)^{1/2}$$

whose level contours are n-spheres, can be used to char-

[2]Generally a norm, $n(x)$, has the property that $n(\alpha x) = |\alpha| n(x)$, sometimes termed an absolute or equilibrated norm. In this paper the term *norm* denotes a larger class of functions in which $n(x) \neq n(-x)$ necessarily [5]. This means that asymmetrical distributions can be allowed in what follows.

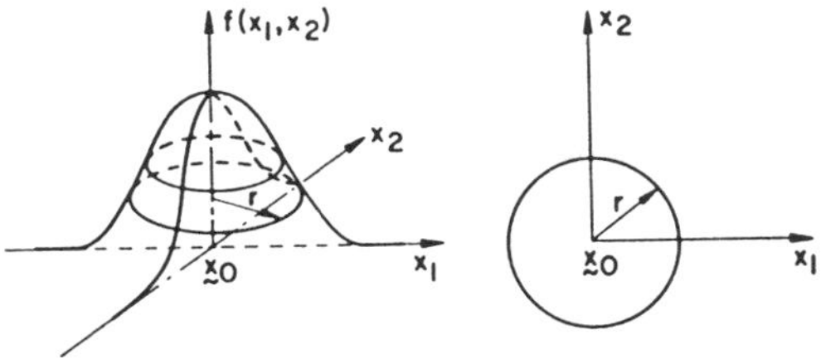

Fig. 2. (a) Illustration of the level contour for a 2-dimensional equivariant, joint normal distribution. (b) Projection of a level contour on the parameter space.

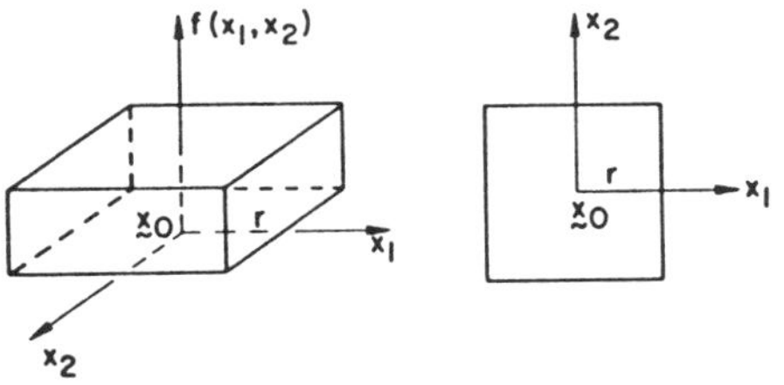

Fig. 3. (a) Illustration of the level contours for a 2-dimensional, equal width, joint uniform distribution. (b) Projection of the level contour on the parameter space.

acterize the level contours of an n-dimensional, equivariant, joint *normal* distribution with mean x_0, by defining the surface

$$n_2(x-x_0)=r.$$

The associated closed convex body is defined by

$$\{x|n_2(x-x_0)\leqslant r\}$$

and is illustrated for two dimensions in Fig. 2. Similarly, the max (or infinity) norm

$$n_\infty(x)=\|x\|_\infty=\max_i\{|x_i|\}$$

which corresponds to an n-cube, can be used to characterize the level contours of an n-dimensional, equal width, joint *uniform* distribution with a mean of x_0, by defining the surface

$$n_\infty(x-x_0)=r.$$

The associated closed convex body is defined by

$$\{x|n_\infty(x-x_0)\leqslant r\}$$

and is illustrated for two dimensions in Fig. 3. A norm which is used to characterize the level contour of a pdf is referred to as a *pdf-norm* and the convex body associated with this norm will be referred to as a *norm body* [5].

In addition to the concept of norm bodies as defined above, we will need the concept of a dual norm. Associated with each norm $n(x)$ there is a corresponding *dual norm* $n^*(x)$, [5], defined by

$$n^*(x)=\max_y\{y^Tx|n(y)\leqslant 1\}. \tag{9}$$

While in general (9) is an optimization problem with a linear objective function and nonlinear convex constraint, it is well known that if

$$n_p(x)=\|x\|_p\equiv\left(\sum_{i=1}^n|x_i|^p\right)^{1/p}$$

the dual norm is

$$n_p^*(x)=\|x\|_q$$

where

$$\frac{1}{p}+\frac{1}{q}=1.$$

In other words, the dual norm of the 2-norm is the 2-norm, while the dual norm of the max- (or infinity) norm is the 1-norm

$$\|x\|_i=\sum_{i=1}^n|x_i|$$

and vice versa. Furthermore, for the case of an arbitary norm body, we can approximate the norm body itself by a simplicial approximation, in which case (9) is a linear program.

Of key importance, in the development of the algorithm for yield maximization described in [4], and generalized below, is the ability to compute the distance from a point to a hyperplane in any arbitrary norm. The distance in norm $n(x)$ from a point x to a hyperplane π, is defined [6] by

$$d_n(x,\pi)\equiv\min_y\{n(y-x)|y\in\pi\}. \tag{10}$$

The following theorem establishes that $d_n(x,\pi)$ is linear in x thus making it possible to easily extend the yield maximization procedure of [4] to handle arbitrary norm bodies.

Theorem I (Normed Distance to a Hyperplane)

Let $n(x)$ be a norm of x and $n^*(x)$ be the corresponding dual norm of x defined by (9). Then the distance in n-norm from point x_0 to hyperplane $\pi\equiv\{x|\eta^Tx=b\}$ is

$$d_0\equiv d_n(x_0,\pi)=\frac{|b-\eta^Tx_0|}{n^*(\gamma\eta)}$$

where

$$\gamma=(b-\eta^Tx_0)/|b-\eta^Tx_0|=\operatorname{sgn}(b-\eta^Tx_0). \tag{11}$$

Proof: We assume first that $b-\eta^Tx_0\geqslant 0$. Define

$$\beta=\max_y\{\eta^Ty|n(y-x_0)\leqslant d_0\}. \tag{12}$$

Then

$$\beta=\eta^Tx_0+\max_{y-x_0}\{\eta^T(y-x_0)|n((y-x_0)/d_0)\leqslant 1\}$$

$$=\eta^Tx_0+d_0n^*(\eta)$$

or

$$d_0=\frac{(\beta-\eta^Tx_0)}{n^*(\eta)}.$$

It is, therefore, sufficient to show that $\beta=b$.

Assume that $\beta>b$. Let y^* denote the maximizer of (12). Choose $0<\lambda<1$ such that $\eta^T(\lambda x_0+(1-\lambda)y^*)=b$. Then

$$n(\lambda x_0+(1-\lambda)y^*-x_0)=(1-\lambda)n(y^*-x_0)<d_0$$

since $\lambda x_0+(1-\lambda)y^*\in\pi$, this contradicts the fact that

$d_0=\min\{n(y-x_0), y\in\pi\}$. Conversely, assume that $\beta<b$. Let y^* be such that $n(y^*-x_0)=d_0$, $y^*\in\pi$. Then $\eta^T y^*=b>\beta$ and $n(y^*-x_0)\leq d_0$, contradicting the fact that β is the maximum such value. thus $\beta=b$.

Next assume that $b-\eta^T x_0<0$. Replace b by γb and η by $\gamma\eta$, and note that π is unchanged by this. Using the first half of this proof we obtain that

$$d_0=\frac{(\gamma b-\gamma\eta^T x_0)}{n^*(\gamma\nu)}=\frac{|b-\eta^T x_0|}{n^*(\gamma\eta)}. \quad \blacksquare$$

IV. Design Centering

Let us define the norm body

$$B(x_0,r)\equiv\{x\,|\,n(x-x_0)\leq r\}.$$

By equating this norm body with the set contained by the level contour of a joint probability density function for the statistically varying parameters, we can approximate the yield maximization problem by that of inserting the maximal norm body inside the feasible region:[3]

$$\underset{x_0,r}{\text{maximize}}\ r$$
$$\text{subject to the constraints}$$
$$B(x_0,r)\subset R.$$

This problem is transformed to a computationally tractable problem by employing the simplicial approximation to R:

$$\hat{R}=\{x\,|\,\eta_i^T x\leq b_i, i=1,2,\cdots,n_f\}$$

which allows us to transform the above problem to a linear programming problem of the form

$$\underset{x_0,r}{\text{maximize}}\ r$$
$$\text{subject to the constraints}$$
$$d_n(x_0,\pi_j)\geq r, \qquad j=1,2,\cdots,n_f,\ x_0\in\hat{R}$$

where π_j is the jth hyperplane of the simplicial approximation defined by

$$\eta_j^T x=b_j.$$

Clearly, if $d_n(x,\pi_j)\geq r$, and $x_0\in\hat{R}$, then $B(x_0,r)\subseteq\hat{R}$. Using Theorem 1, this can be written as a linear program

$$\underset{x_0,r}{\text{maximize}}\ r$$
$$\text{subject to the constraints} \tag{13}$$
$$\eta_j^T x_0+rn^*(\eta_j)\leq b_j, \qquad j=1,2,\cdots,n_f.$$

Observe that if $n(x)=\|x\|_2$, so that $n^*(x)=\|x\|_2$, we are inscribing an n-sphere in $\hat{R}$ and (13) reduces to the linear program described in [4]. Similiarly, if $n(x)=\|x\|_\infty$, so that $n^*(x)=\|x\|_1$, we are inscribing n-cubes in $\hat{R}$.

Upon solution of the linear program (13) we can further refine the approximating polytope $\hat{R}$ as was done in [4]. Specifically, begin by determining which of the n_f faces of $\hat{R}$ are touching the inscribed norm body. This information is obtained by examining the slack variables associated with the linear programming solution procedure. Those constraints for which the slack variables are zero correspond to the touching faces. Of the faces which touch we choose the largest (in volume as is done in [4]) and search in an outward normal direction from this face to find a new point on ∂R. The approximating polytope $\hat{R}$ is then inflated to include this new point as a vertex. The above procedure can then be repeated to determine a new maximum yield point.

We have thus far assumed that the statistical parameters have equal variances or spreads. In the event that the variances or limits of the statistical parameter distributions vary in a given problem, an additional step must be taken in the aforementioned procedures. This step amounts to a scaling of the n-dimensional parameter space with an $n\times n$ nonsingular scaling matrix T. Equivalently, one can scale instead the norm itself, as may be seen from the following:

Corollary

Let x be interior to the convex hull (7) and let $n(x)$ and $n_T(x)$ be norms satisfying $n_T(x)=n(T^{-1}x)$. Then

$$n_T^*(x)=n^*(T^T x) \tag{14}$$

and the distance in n_T norm from point x to hyperplane $\pi=\{x\,|\,\eta^T x=b\}$ is

$$d_{n_T}(x,\pi)=\frac{|b-\eta^T x|}{n^*(\gamma T^T\eta)}.$$

where $\gamma\equiv\text{sgn}(b-\eta^T x)$.

The proof of this corollary is similar to the proof of Theorem I and is omitted.

Thus a yield maximization problem with differing limits or variances of the statistical parameters may be brought into the framework developed above by defining a scaling matrix T and using the corresponding scaled norm. More specifically we define the norm body:

$$B_T(x_0,r)\equiv\{x\,|\,n_T(x-x_0)\leq r\}$$

so that with scaling the yield maximization problem is

$$\underset{x_0,r}{\text{maximize}}\ r$$
$$\text{subject to the constraints}$$
$$B_T(x_0,r)\subseteq R.$$

The corresponding linear program which results from using the simplicial approximation $\hat{R}$ for R is

$$\underset{x_0,r}{\text{maximize}}\ r$$
$$\text{subject to the constraints}$$
$$d_{n_T}(x_0,\pi_j)\geq r, \qquad j=1,2,\cdots,n_f,\ x_0\in\hat{R}$$

[3] It may be that as r changes, we would want to change the norm used to approximate the level contours of the pdf. This may have to be done if the level contours are radically different for different level values. However, there are many pdf's (e.g., normal or uniform) where different level contours are similar and associated with the same norm. Since design centering is only an approximation to the yield maximization problem, we need only require that the largest norm body imbedded approximates a level contour of the pdf within some allowed error.

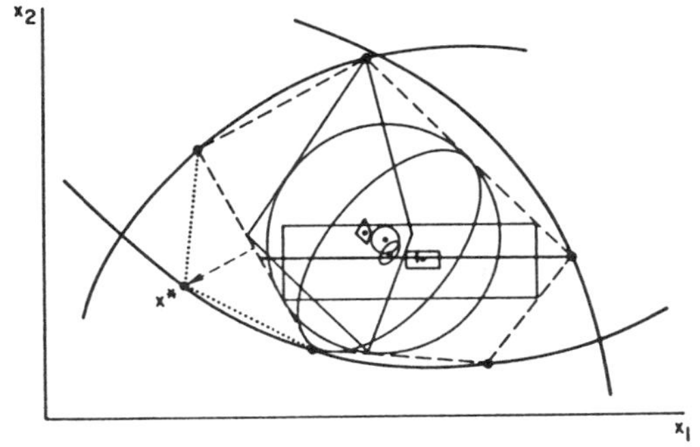

Fig. 4. Examples of maximum norm body inscription in a polygon. The dotted line shows the effect of a typical inflation step, which expands the polygon to include the point x^* as a vertex.

or, from the above corollary

$$\underset{x_0, r}{\text{maximize}}\, r$$

subject to the constraints

$$\eta_j^T x_0 + r n^*(T^T \eta_j) \leq b_j, \qquad j=1,2,\cdots,n_f.$$

Fig. 4 illustrates the result of inscribing several different types of norm bodies in the simplicial approximation of Fig. 1. Observe that if an inflation step were to be taken the approximating polygon would expand as shown by the dotted lines. Note that asymmetrical, polygonal, and/or covariant (i.e., rotated by a nondiagonal scaling matrix T) norm bodies are easily handled by the present theory, as are degenerate (not of full dimensionality) norm bodies such as the horizontal line in Fig. 4.

V. Mixed Worst-Case Yield Maximization

As mentioned in the introduction, an important situation which arises in many types of designs is one in which some of the parameters are statistical in nature with given joint pdf's and others are worst-case parameters, i.e., the circuit must work for all values of the worst-case parameters which lie within some interval. A typical problem of this type is a circuit which is required to operate for all values of some power supply voltage within a specified range and over some range of temperatures. Thus power supply voltages and temperature are worst case parameters. In this section we describe a procedure for handling this situation.

Assume that of the n design parameters, n_w are worst case and n_s are statistical. It is convenient to partition the parameter vector x as follows:

$$x = \begin{bmatrix} w \\ s \end{bmatrix}$$

where w is the n_w-vector of worst-case parameters and s is the n_s-vector of statistical parameters. We can thus view the n-dimensional parameter space X as the Cartesian product of the n_w-dimensional worst case parameter space W and the n_s-dimensional statistical parameter space S. The parameters w are subject to worst-case interval restrictions which, in general, can be expressed as

$$w \in W_I \subset W. \tag{15}$$

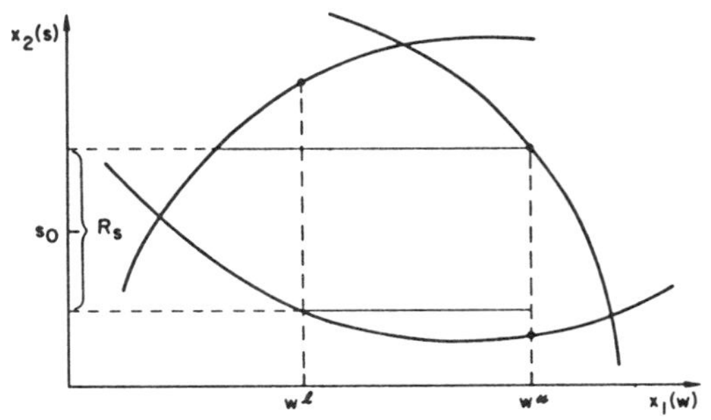

Fig. 5. A mixed worst-case yield maximization problem with solution s_0, the midpoint of the interval R_S. R_S is the feasible region projected onto the subspace of statistical parameters.

Typically, W_I is an orthotope (i.e., a tolerance box), in which case

$$W_I = W_B \equiv \{ w \mid w^l \leq w \leq w^u \}. \tag{16}$$

A given circuit is acceptable if the constraints on circuit performance (2a) are satisfied, i.e.,

$$\Phi_i(w,s) \leq 0, \qquad i=1,2,\cdots,n_c \tag{17}$$

for all $w \in W_I$. Thus the feasible region is

$$R = \{ w, s \mid \Phi_i(w,s) \leq 0, i=1,2,\cdots,n_c; \text{ for all } w \in W_I \}. \tag{18}$$

An alternate statement of acceptable performance is obtained by defining the functions

$$\tilde{\Phi}_i(s) \equiv \max_{w \in W_I} \Phi_i(w,s), \qquad i=1,2,\cdots,n_c. \tag{19}$$

Thus acceptable performance is given by all s for which

$$\tilde{\Phi}_i(s) \leq 0, \qquad i=1,2,\cdots,n_c \tag{20}$$

and the feasible region can be described in the space S by

$$R_S \equiv \{ s \mid \tilde{\Phi}_i(s) \leq 0, i=1,2,\cdots,n_c \}. \tag{21}$$

Fig. 5 illustrates R_s for the hypothetical 2-dimensional case of Figs. 1 and 4. Given this definition of the feasible region we could employ the procedure of Section II to find a simplicial approximation $\hat{R}_S$ to R_S and use the procedure of Section IV to obtain a maximum yield point. However, this approach is burdened by the fact that each step in the procedure requires solution of the worst-case problem (19). Thus an alternate approach is needed. Towards this end we state the following.

Theorem II

Assume that the region of acceptability R as given by (18) is convex and that the region

$$\hat{R} \equiv \{ (w,s) \mid \eta_{jw}^T w + \eta_{js}^T s \leq b_j, j=1,2,\cdots,n_f \}$$

is an interior simplicial approximation to R, i.e.,

$$\hat{R} \subseteq R.$$

(Note that the outward pointing normals η_j have been partitioned as $\eta_j^T = (\eta_{jw}^T, \eta_{js}^T)$.) Let $\tilde{\Phi}_i(s)$ be defined by (19). Then

$$\hat{R}_S \equiv \{ s \mid \eta_{js}^T s \leq b_j - \eta_{jw}^T w_0 - n_W^*(\eta_{jw}) \} \tag{22}$$

is an interior simplicial approximation to R_S as defined by (21) where w_0 is any point *interior* to W and

$$n_W(x) \equiv \min\left\{\alpha \mid \alpha > 0, \frac{x}{\alpha} + w_0 \in W\right\}. \tag{23}$$

Proof: We need to show that $\hat{R}_S \subset R_S$. Suppose that $s^* \in \hat{R}_S$ so that

$$\eta_{js}^T s^* \leqslant b_j - \eta_{jw}^T w_0 - n_W^*(\eta_{jw}), \qquad j = 1, 2, \cdots, n_f$$

or

$$\frac{b_j - \eta_{js}^T s^* - \eta_{jw}^T w_0}{n_W^*(\eta_{jw})} \geqslant 1, \qquad j = 1, 2, \cdots, n_f. \tag{24}$$

The left-hand side of (24) can be interpreted as the distance of the point w_0 from a hyperplane $\pi_j(s^*)$ measured in n_W norm where

$$\pi_j(s^*) = \left\{w \mid \eta_{jw}^T w = b_j - \eta_{js}^T s^*\right\}$$

which is the jth hyperplane

$$\eta_j^T x = b_j$$

restricted to the subspace $s = s^*$. Therefore, (24) is equivalent to

$$d_W(w_0, \pi_j(s^*)) \geqslant 1. \tag{25}$$

Since (25) holds for all j, the convex body W_I is contained in $\hat{R} \cap \{x \mid s = s^*\}$. (See definition (23).) Thus for any $s^* \in \hat{R}_s$ and $w \in W$, we have $\Phi_j(w, s^*) \leqslant 0$, $j = 1, 2, \cdots, n_c$, which implies that

$$\max_{w \in W} \Phi(w, s^*) \leqslant 0$$

or that $s^* \in R_S$. ■

Theorem II enables us to solve the mixed worst-case yield maximization problem by first determining a simplicial approximation to the feasible region in the full parameter space by the usual method, and using it to determine the simplicial approximation $\hat{R}_S$ of expression (22). Given $\hat{R}_S$, the procedure of Section IV could be used to maximize the yield. More specifically, we have the following linear program.

$$\underset{s_0, r_S}{\text{maximize}}\ r_S$$

subject to the constraints

$$\eta_{js}^T s_0 + r_S n_S^*(\eta_{js}) \leqslant b_j - \eta_{jw}^T w_0 - n_W^*(\eta_{jw}), \qquad j = 1, 2, \cdots, n_f. \tag{26}$$

Note that (22–26) apply independently of the choice made for the S-space norm n_S or the tolerance region W_I. However, if $W_I = W_B$ (see (16)), and we choose

$$w_0 = \frac{1}{2}(w^l + w^u)$$

$$T_B = \frac{1}{2}(\text{diag}(w^u - w^l))$$

then we get the special case

$$n_W(w) \equiv n_\infty(T_B^{-1} w) = \max_i \left\{\frac{2|w_i|}{w_i^u - w_i^l}\right\} \tag{27}$$

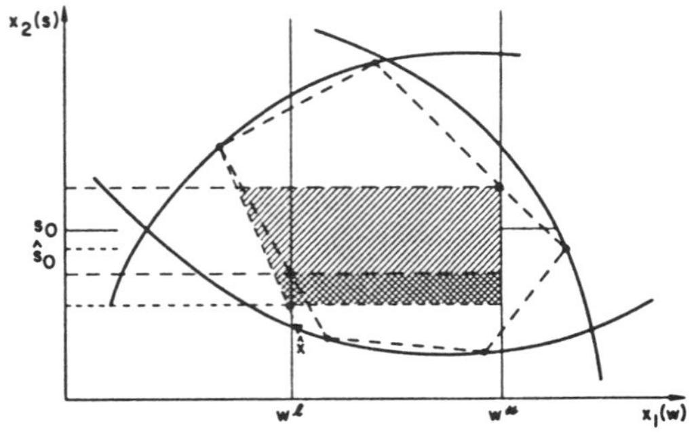

Fig. 6. Illustration of the dependence of the solution s_0 of the mixed worst-case yield maximization problem on the simplicial approximation $\hat{R}$. As $\hat{R}$ is improved as a simplicial approximation to R by the addition of the point $\hat{x}$, s_0 decreases to level $\hat{s}_0$.

and from (14)

$$n_W^*(\eta_{jw}) \equiv n_1(T_B \eta_{jw}) = \frac{1}{2} \sum_{i=1}^{n_w} (w_i^u - w_i^l) |(\eta_{jw})_i|. \tag{28}$$

Therefore, the linear program (26) becomes

$$\underset{s_0, r_S}{\text{maximize}}\ r_S$$

subject to the constraints

$$\eta_{js}^T s_0 + r_S n_S^*(\eta_{js}) \leqslant b_j - \sum_{i=1}^{n_w} \max\left\{(\eta_{jw})_i w_i^l, (\eta_{jw})_i w_i^u\right\}. \tag{29}$$

Several comments are in order. First, the solution that results from (26) or (29) is very much dependent upon the simplicial approximation $\hat{R}$, as illustrated in Fig. 6. The vertical line $w = w^l$ of Fig. 6 establishes the lower (circled corners) boundary of the projected feasible region $\hat{R}_S$. Thus when the simplicial approximation is inflated by the addition of the point $\hat{x}$, $\hat{R}_S$ expands, and its center s_0 descends to a new level $\hat{s}_0$. Thus in practice the approximation R should be updated and the process repeated as indicated earlier. Second, it is possible that the right-hand side of the inequality constraints in (29) can be negative, in which case r_S will be negative which indicates that (29) does not have a feasible solution. Such a situation can result from two causes: 1) either the worst case constraints are too severe to allow for a solution, or 2) the simplicial approximation $\hat{R}$ to R is too crude and needs further refinement. These cases can be illustrated with Fig. 4, which was described previously. In Fig. 4, let us now regard the vertical dimension as the S space and the horizontal dimension as the W space. For the inscribed rectangle the radius r_S is positive, but for the inscribed straight line norm body r_S is zero. If the required worst-case interval is longer than the inscribed line in Fig. 4, the radius r_S would be negative. However, in this case the refinement of the simplicial approximation illustrated in Fig. 4 would lead to a positive radius on the next iteration.

Finally, consider the special case in which $n_s = 0$, $W_I = W_B$, which is the classical worst case analysis problem of determining if, for a fixed set of tolerances, whether or not the corresponding n_w-cube fits inside the feasible region.

In this case the above formulation provides a significant savings in computational effort. Whereas the usual procedure would require testing for feasibility of each of the 2^{n_w} vertices of the n_w-cube, all we need to do is determine whether or not the right hand side of the inequality constraints of (29) are positive.

VI. Tolerance Assignment

The basic idea behind the yield maximization procedure described above was to incribe a maximal norm body inside the feasible region. The norm body had a fixed aspect ratio, defined by the scaling matrix T (see Section IV). We can view the tolerance assignment problem as a generalization of the yield maximization problem in which the aspect ratio of the inscribed norm body is allowed to vary so as to minimize an auxiliary cost function $C(t)$ [7], [8]. To see this, let $t^T=(t_1,t_2,\cdots,t_n)$ be the vector of component tolerances, $T=\text{diag}\,(t)$ a diagonal matrix, and assume that $C(t)$ is a decreasing function of t so that $C(\alpha t)<C(t)$ for $\alpha>1$. Then the solution of the problem

$$\begin{gathered}\underset{x_0,t}{\text{minimize}}\,C(t)\\ \text{subject to the constraints}\\ \left(B_T(x_0,1)\equiv\left\{x|n\left(T^{-1}(x-x_0)\right)\leqslant 1\right\}\right)\subset\hat{R}\end{gathered} \tag{30}$$

approaches the solution of the tolerance assignment problem as $\hat{R}$ approaches R. Note that as t_i increases, the allowable deviation of x_i about its nominal increases and the cost of manufacturing component i decreases. Furthermore, observe that if all the t_i are to be identical, say, equal to r, then minimizing $C(t)$ is the same as maximizing r and this problem is the same as the problem considered in Section IV.

In general (30) is a nonlinear constrained optimization problem, and could be solved by use of appropriate algorithms. However, in this section we only consider the special case of inscribing a orthotope, i.e., a tolerance box, and show how, using a piecewise linear approximation to $C(t)$, (30) can be solved using linear programming.

It is a natural extension of the ideas of simplicial approximation to approximate the cost function $C(t)$ by a piecewise linear function. For this we assume $C(t)$ is convex and consider the convex region in the $(n+1)$-dimensional (y,t)-space, $\{(y,t)|y\geqslant C(t)\}$. We then approximate this region using the ideas of simplicial approximation by

$$\left\{(y,t)|A^T\begin{pmatrix}y\\t\end{pmatrix}\leqslant d\right\}$$

i.e., we determine a set of points $\{(y^j,t^j)\}$ such that $y^j=C(t^j)$ and then using a convex hull algorithm to find the hyperplanes which bound the convex hull of these points (the matrix A contains the hyperplane normals).

The piecewise linear approximation to $C(t)$, say $\hat{C}(t)$, is then given by

$$\hat{C}(t)=\min\left\{y|A^T\begin{pmatrix}y\\t\end{pmatrix}\leqslant d\right\}. \tag{31}$$

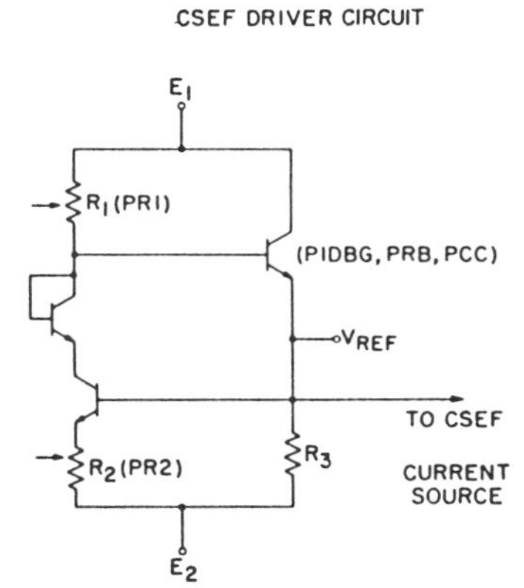

Fig. 7. A driver circuit for a CSEF logic gate.

Thus (30) is approximated by

$$\begin{gathered}\underset{x_0,t}{\text{minimize}}\,\hat{C}(t)\\ \text{subject to the constraints}\\ B_T(x_0,1)\subset\hat{R}.\end{gathered} \tag{32}$$

Using (31), we can rewrite (32) as

$$\begin{gathered}\underset{x_0,y,t}{\text{minimize}}\,Y\\ \text{subject to the constraints}\\ A^T\begin{pmatrix}y\\t\end{pmatrix}\leqslant d\\ d_{n_T}(x_0,\pi_j)\geqslant 1,\qquad x_0\in\hat{R}.\end{gathered}$$

Now we use the Corollary and Theorem I to rewrite the constraints as

$$\begin{gathered}A^T\begin{pmatrix}y\\t\end{pmatrix}\leqslant d\\ b_j-\eta_j^Tx_0\geqslant n_T^*(\eta_j),\qquad j=1,\cdots,n_f\end{gathered}$$

or

$$\begin{gathered}A^T\begin{pmatrix}y\\t\end{pmatrix}\leqslant d\\ \eta_j^Tx_0+n^*(T^T\eta_j)\leqslant b_j,\qquad j=1,\cdots,n_f.\end{gathered}$$

Since $n_T(x)=\max_i\{|t_i^{-1}x_i|\}$ and $t_i>0$, then (32) is equivalent to (using the duality of the 1-norm and the ∞-norm)

$$\begin{gathered}\underset{x_0,y,t}{\text{minimize}}\,y\\ \text{subject to the constraints}\\ A^T\begin{pmatrix}y\\t\end{pmatrix}\leqslant d\\ \eta_j^Tx_0+\sum t_i|\eta_{ji}|\leqslant b_j,\qquad j=1,\cdots,n_f\end{gathered}$$

which we recognize as a *linear* program.

VII. An Example

As a practical example we demonstrate the mixed worst-case design procedure by applying it to the CSEF "current source driver" circuit of Fig. 7. The design objective is to make the reference voltage V_{REF} minimally sensitive to statistical fluctuations of the designable parameters. Here the "statistical" parameters are the resistor parameters PR1, PR2, and R3, and the transistor process parameters PIDBG, PRB, PCC. The worst-case parameters are the temperature (TC), and the power supply voltages E1 and E2. The design specifications are

$$1.30 = V_{DN} \leq V_{REF} \leq V_{UP} = 1.35 \tag{33}$$

for *any* E1, E2, TC satisfying

$$1.175 \leq E1 \leq 1.250$$
$$2.910 \leq E2 \leq 3.090$$
$$50 \leq TC \leq 100. \tag{34}$$

That is, we specify that 100-percent yield is required, and seek the nominal values of R3, PR1, PR2, PIDBG, PRB, PCC which meet this requirement while maximizing the allowable tolerances on these parameters. In this maximization, we assume as before that all tolerances grow proportionately, that is, the aspect ratio of the tolerance cube is constant. We assume that all the design parameters have independent, truncated distributions, but make no assumption regarding the shape of the distributions.

The procedure is initiated by performing line searches in the positive and negative coordinate directions, yielding $2n$ points on the boundary, ∂R, of the feasible region. These points are used to obtain an initial simplicial approximation which is an n-diamond truncated by the worst case specification planes (34). The dual norms of the face normals of this approximation are then computed and the linear program (29) solved, with the scaling matrix T set to the appropriate identity matrix. However, a negative radius is thus obtained.

The procedure now undergoes a worst-case space expansion phase of the type suggested by the straight-line inscription of Fig. 4. However, after 20 expansions, the radius r_S was still negative. We investigated this situation graphically, by plotting (using the multivariate piecewise linear interpolation results of [10]) various 2-dimensional cross sections of the approximating polygon. This investigation led to the suspicion that the worst-case temperature specification was the difficulty. Fig. 8 shows the intersection of the polygon of the simplicial approximation with a plane parallel to the PR1 and TC axes which passes through the nominal design. Inside this intersecting plane a rectangle is shown, whose height is determined by a plus and minus 100-percent tolerance interval on statistical parameter PR1. The intersection of this rectangle with the simplicial approximation shows clearly the impossibility of obtaining 100-percent tolerance on PR1. A closer look at the rightmost and leftmost points of the approximation also shows the unlikelihood of any nonzero tolerance (note the fair degree of refinement at the edges of the simplicial approximation).

This analysis suggested relaxation of the worst case temperature specification. We found that a relaxation of 5 degrees on either side of the worst case interval was sufficient to yield an acceptable solution of the problem (cf. the ±20 degree lines of Fig. 8).

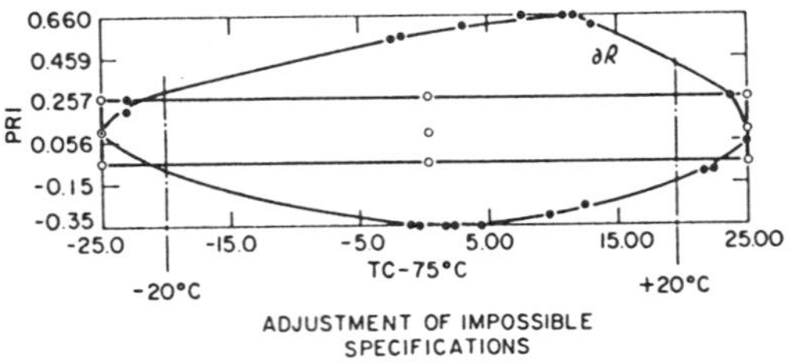

Fig. 8. PR1–TC cross section of the simplicial approximation (after 20 negative-radius expansion steps).

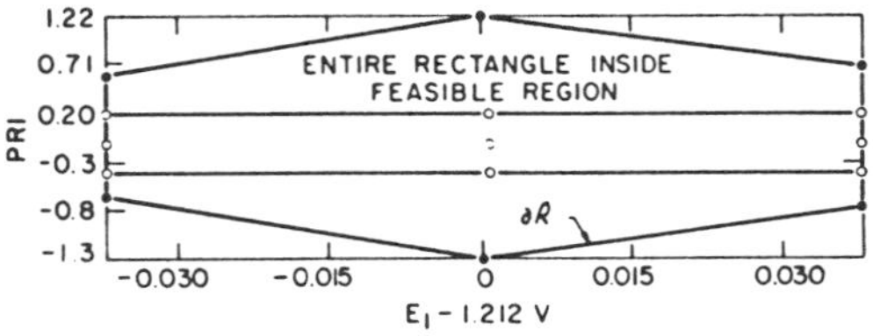

Fig. 9. PR1–E1 cross section of the simplicial approximation (after initialization).

To illustrate that temperature was the main difficulty in this design problem, we show in Fig. 9 the intersection of the PR1–E1 plane with the initial (i.e., $2n$-point) simplicial approximation. Note that if these were the only two dimensions of the design problem, an initial value of about 0.6 might be expected for r_S.

VIII. Conclusions

By introducing the concepts of norm-bodies, pdf-norms and dual norms, we have proven a theorem which extends a previously published, [4], statistical design technique, based on simplicial approximation, into a much more powerful tool. The previously reported approximate yield maximization procedure now applies for truncated and/or asymmetrical and/or polygonal pdf's. Further, the new procedure also handles worst-case problems and problems with mixed yield maximization and worst-case specifications. An example of a mixed worst-case design problem has been discussed which illustrates the physical significance of the negative-radius case and a remedy. This practical problem was successfully treated and illustrates that *a priori* convexity guarantees are not necessary.

The tolerance assignment problem was discussed for the case of a convex tolerance cost function and a rectangular tolerance region. The dual norm associated with this problem was linear in the tolerances. By replacing the cost function with linear inequalities, we demonstrated how this problem becomes a linear program. Further work, especially on practical applications, is needed in this area.

Acknowledgment

The authors are grateful to M. S. Martin for helpful consultations in the early stages of this work, and to A. Jiminez, T. Scott, and T. Walker for valuable discussion and suggestions.

References

[1] J. W. Bandler *et al.*, "A nonlinear programming approach to optimal design centering, tolerancing, and tuning," *IEEE Trans. Circuits Syst.*, vol. CAS-23, pp. 155–165, Mar. 1976.

[2] J. W. Bandler and H. L. Abdel-Malek, "Optimal centering, tolerancing and tuning via updated approximations and cuts," *IEEE Trans. Circuits Syst.*, vol. CAS-25, 1978.

[3] J. W. Bandler and P. C. Liu, "Automated network design with optimal tolerances," *IEEE Trans. Circuits Syst.*, vol. CAS-21, pp. 219–222, 1974.

[4] S. W. Director and G. D. Hachtel, "The Simplicial approximation approach to design centering and tolerance assignment," *IEEE Trans. Circuits Syst.*, vol. CAS-24, pp. 363–371, 1977.

[5] J. Stoer and C. Witzgall, *Convexity and Optimization in Finite Dimensions*. New York: Springer, 1970.

[6] D. Luenberger, *Optimization by Vector Space Methods*. New York: Wiley, 1969.

[7] B. J. Karafin, "The general component tolerance assignment problem in electrical networks," Ph.D. dissertation, Univ. of Pennsylvania, Philadelphia, PA, 1974.

[8] J. F. Pinel and K. A. Roberts, "Tolerance assignment in linear networks using nonlinear programming," *IEEE Trans. Circuit Theory*, CT-19, pp. 475–479, Sept., 1972.

[9] R. K. Brayton, "Tolerance assignment and design centering in the convex case," IBM Research report 5964, IBM Thomas J. Watson Research Center, P. O. Box 218, Yorktown Heights, NY, 10598, Apr. 1976.

[10] G. D. Hachtel and S. W. Director, "A point basis for simplicial approximation," in *Proc. of the Hull Conference on Circuits and Systems*, (Hull, England), 1977.

[11] M. R. Lightner, "Multiple criterion optimization and statistical design for electronic circuits," Ph.D. dissertation, Carnegie-Mellon Univ., Pittsburgh, PA., Feb. 1979.

A Design Centering Algorithm for Nonconvex Regions of Acceptability

LUIS M. VIDIGAL, MEMBER, IEEE, AND STEPHEN W. DIRECTOR, FELLOW, IEEE

Abstract—Random variations inherent in any fabrication process may result in very low production yield. This is especially true in the fabrication of integrated circuits. Several methods have been proposed to help the circuit designer minimize the influence of these random variations. Most of these methods are deterministic and try to maximize yield by centering the nominal value of the designable parameters in the so-called region of acceptability. However, these design centering techniques require an assumption of convexity which is not valid in many real design situations. To overcome this problem a new convergent method is proposed which is based on the sequential solution of subproblems for which the convexity assumption is valid. A practical implementation of the algorithm is shown by examples to be computationally efficient.

I. Introduction

DUE TO INHERENT fluctuations is the integrated circuit manufacturing the actual element values, and hence, circuit performance will deviate from some specified nominal value. Because of these fluctuations some of the manufactured circuits will fail to meet design specifications and yield will be less than 100 percent. Since the cost associated with the mass production of any circuit is directly related to yield, the area of yield maximization has been particularly active. One approach to yield maximization is to employ a design centering procedure. In order to formally describe such a procedure, we must introduce the concept of a *region of acceptability* R_a, which is defined as the set of values of the n designable parameters p for which a circuit performs according to specifications. Formally if $f(p)$ is an m-vector[1] and f^L and f^U are vectors the jth component of which represents the maximum or the minimum values that the performance function can assume, then R_a is defined as

$$R_a = \{p \mid f^L \leqslant f(p) \leqslant f^U\}. \quad (1)$$

We assume that R_a is a simply connected set in the n-dimensional space defined by the designable parameters. It will be also assumed henceforth that the performance functions f_i are twice differentiable.

Manuscript received March 30, 1981; revised August 3, 1981. This work was supported in part by the National Science Foundation under Grant ECS 79-23191.

L. M. Vidigal was with the Department of Electrical Engineering, Carnegie-Mellon University, Pittsburgh, PA 15213. He is now with Instituto Superior Tecnico, Lisbon, Portugal.

S. W. Director is with the Department of Electrical Engineering, Carnegie-Mellon University, Pittsburgh, PA 15213.

[1] Vectors will be represented by lower case letters with (when needed) a superscript, e.g., f, f^1, f^2. Subscripts indicate the component, thus f_j^k represents the jth component of vector f^k.

It is now convenient to define the *level set*, $L_c(\alpha)$ associated with the known unimodal joint distribution function of the designable parameters, $\Phi(p, c)$

$$L_c(\alpha) = \{p \mid \Phi(p, c) \geqslant \alpha\}$$

where c denotes the mean value of p. If the level sets corresponding to different values of α, are homothetic figures, i.e., are related through a scaling property (see [5], [32]), then it is always possible to associate a Minkowski norm n [22] with them and define the corresponding *norm body* as

$$\begin{aligned} B_c(r) &= \{p \mid p \in L_c(\alpha)\} \\ &= \{p \mid n(p - c) \leqslant r\} \\ &= \{p \mid p = c + r\omega, n(\omega) \leqslant 1\}. \end{aligned} \quad (2)$$

Then the *Design Centering* (or *DC*) procedure can be expressed as the mathematical programming problem

$$\text{DC:} \quad \max_{c,\, r} r \quad \text{s.t. } B_c(r) \subset R_a. \quad (3)$$

Unfortunately the DC problem cannot be solved because R_a is only known implicitly. In order to make this problem tractable the Simplicial Approximation method was proposed [5], [14] under the assumption that the region of acceptability is convex. However in many situations R_a is nonconvex [2], [6], [19], [26], and the solution obtained using the Simplicial Approximation algorithm can be deceptive. To see this consider the case in which a designer must guarantee that the performance of a network falls within certain tolerances. This requirement can be expressed by imposing an upper and a lower boundary to the constraints as in (1). Unless the $f_j(p)$ are linear functions, the region of acceptability will not be convex. Note that even if the $f_j(p)$ are convex functions, so that the sets

$$C_j = \{p \mid f_j(p) \leqslant f_j^U\} \quad (4)$$

are convex [34, p. 31], the sets

$$D_j = \{p \mid f_j^L \leqslant f_j(p)\} \quad (5)$$

are nonconvex, as they are the complements of open convex sets. Therefore, the region of acceptability,

$$R_a = \left(\bigcap_{j=1}^{n} C_j\right) \cap \left(\bigcap_{j=1}^{n} D_j\right) \quad (6)$$

will not in general be convex.

Reprinted from *IEEE Trans. Computer-Aided Design of Integrated Circuits System,* vol. CAD-1, no. 1, pp. 13–24, January 1982.

While the above convexity assumption is often a poor one, we have concluded that in many situations it is reasonable to assume that R_a results from the intersection of sets which are either *convex* or *complementary convex* (a complementary convex set is one whose complement is convex, see [1] where a similar development is made within a different context). In this paper we develop a DC procedure which can be employed in this more general situation. Specifically, this new technique is based upon generating approximations to the constraints rather than to R_a.

Before leaving this section we show that the most general class of performance functions which result in feasible regions which are intersections of sets which are convex and complementary convex are those that are *quasi-convex* or *quasi-concave*.

Definition 1: A function $f: R^n \to R^1$ is *quasi-convex* if given $p^1, p^2 \in R^n$, then for any value of the scalar θ, $0 \leqslant \theta \leqslant 1$

$$f(\theta p^1 + (1 - \theta) p^2) \leqslant \max [f(p^1), f(p^2)]. \tag{7}$$

A function f is *quasi-concave* if its negative, $-f$, is quasi-convex. □

If the function f is differentiable it can be shown [23] that this definition implies that if $f(p^2) \leqslant f(p^1)$ then the directional derivative at p^1 is nonincreasing, i.e., that $\nabla f(p^1)^T \cdot (p^2 - p^1) \leqslant 0$.

Our interest in quasi-convex functions derives from the following

Theorem 2: Let f be a real-valued function defined on a convex set. Then f is quasi-convex if and only if the set

$$\Lambda_\alpha = \{p | f(p) \leqslant \alpha\} \tag{8}$$

is convex for each $\alpha \in R$. □

See [23] for a proof of Theorem 2.

Observe that for the class of quasi-convex and quasi-concave functions the constraints (1) imply that the region of acceptability will be the intersection of the convex sets C_j and complementary convex sets D_j where

$$\begin{aligned} C_j &= \{p | f_j(p) \leqslant f_j^U\}, \quad \text{if } f_j \text{ is quasi-convex} \\ &= \{p | f_j(p) \geqslant f_j^L\}, \quad \text{if } f_k \text{ is quasi-concave} \end{aligned} \tag{9}$$

and

$$\begin{aligned} D_j &= \{p | f_j(p) \geqslant f_j^U\}, \quad \text{if } f_j \text{ is quasi-convex} \\ &= \{p | f_j(p) \leqslant f_j^L\}, \quad \text{if } f_j \text{ is quasi-concave.} \end{aligned} \tag{10}$$

To ease the notational burden, and simplify the description of the design centering algorithm which is presented in Section III we define two new functions $g_j(p)$ and $h_j(p)$ such that

$$\begin{aligned} C_j &= \{p | g_j(p) \leqslant g_j^U\}, \quad j \in J_g = \{1, 2, \cdots, n_c\} \\ D_j &= \{p | h_j(p) \leqslant h_j^U\}, \quad j \in J_h = \{1, 2, \cdots, n_c\} \end{aligned} \tag{11}$$

where the g_j are the quasi-convex functions and the $h_j(p)$ are the quasi-concave functions. We will refer to g_j and h_j as *convex constraints* and *complementary convex* constraints, respectively. Observe that g_j and h_j are both in absolute value equal to f_j. Note that in a real design problem the functions f_j are only known implicitly through the solution of a system of nonlinear differential equations and, therefore, determining if a given f_j is quasi-convex or quasi-concave is a nontrivial problem.

In order to have the boundaries of R_a well defined, we make one final assumption about the performance functions, specifically, that the gradients of f_j do not vanish at the boundary of R_a.

II. Optimal Conditions for Yield Maximization

In the following discussion we derive necessary conditions for a locally optimal solution to the DC problem (3). These conditions can also be shown to be sufficient if the region of acceptability is convex. However, as this situation is unlikely, we will not exploit these sufficient conditions for optimality. From the outset it should be noted that because the region R_a is not in general convex, local maxima can occur and convergence to the globally optimal solution cannot be guaranteed.

We begin by using (6), (11), and (2) to rewrite the DC problem (3) as

$$\begin{aligned} \text{DC:} \quad & \max_{c,r} r \\ & \text{s.t. } g(c + r\omega) \leqslant g^U, \quad n(\omega) \leqslant 1 \\ & \qquad h(c + r\omega) \leqslant h^U. \end{aligned} \tag{12}$$

While (12) is an accurate statement of the DC problem it cannot be solved directly due to the fact that the constraints which must be satisfied for all values of ω such that $n(\omega) \leqslant 1$, implies that there exists an infinite number of constraints.

The following two lemmas will permit us to modify the statement of the DC problem by reducing the set of points ω that need to be considered.[2]

Lemma 3: A global maximum of a quasi-convex functions over the set $B_c(r)$ occurs on the boundary of $B_c(r)$. □

Lemma 4: If the constraint $h_j(p) \leqslant h_j^U$ is not trivial, i.e. the set $\{p | h_j(p) > h_j^U\}$ is not empty, then for any feasible solution (r^*, c^*) of (12), the maximum of h_j over the convex set $B_{c^*}(r^*)$ is attained at a point on the boundary of $B_{c^*}(r^*)$. □

Note that while Lemmas 3 and 4 do not prevent the existence of a local maximum in the interior of $B_c(r)$, they do establish in this case the existence of a boundary point with the same function value. As a consequence of Lemmas 3 and 4 we need only impose the constraints in (12) over values of ω which are on the boundary of the norm body. Unfortunately this set of ω is still infinite. To overcome this situation we define the *constraint maximization* (CM) subproblems

$$\begin{aligned} \text{CM:} \quad & g_i(c + r\bar{\omega}^i) = \max_{\{\omega | n(\omega) = 1\}} g_i(c + r\omega), \quad i \in J_g \\ & h_j(c + r\bar{\omega}^{n_c + j}) = \max_{\{\omega | n(\omega) = 1\}} h_j(c + r\omega), \quad j \in J_h. \end{aligned} \tag{13}$$

Assume for the moment that the solutions to (13) are unique. Then the DC problem involves $2n_c$ constraints and

[2] The proof of all lemmas and theorems can be found in Appendix.

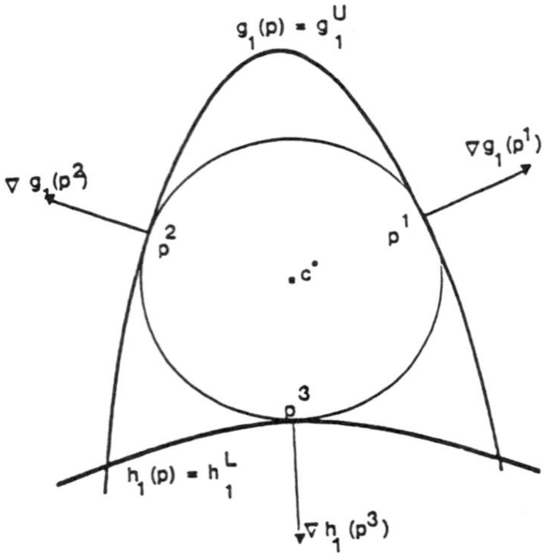

Fig. 1. Points p^1 and p^2 are solutions to problem CM.

can be written as

$$\text{DC: } \max_{c,r} r$$
$$\text{s.t. } g_i(c + r\bar{\omega}^i) \leq g_i^U, \quad i \in J_g$$
$$h_j(c + r\bar{\omega}^{n_c+j}) \leq h_j^U, \quad j \in J_h. \tag{14}$$

The Kuhn-Tucker necessary conditions for a stationary point of (14) are

$$\sum_{i \in J_g} \lambda_i \nabla g_i(p^{i*}) + \sum_{j \in J_h} \lambda_{n_c+j} \nabla h_j(p^{j*}) = 0$$
$$1 + \sum_{i \in J_g} \lambda_i \bar{\omega}^{i*T} \nabla g_i(p^{*i}) + \sum_{j \in J_h} \lambda_{n_c+j} \bar{\omega}^{n_c+j*T} \nabla h_j(p^{j*}) = 0 \tag{15}$$

where

$$\lambda_i \geq 0, \quad i = 1, \cdots, 2n_c$$
$$\lambda_i(g_i(p^{j*}) - g_i^U) = 0, \quad i \in J_g$$
$$\lambda_{n_c+j}(h_j(p^{j*}) - h_j^U) = 0, \quad j \in J_h$$

where $p^{i*} = c^* + r^*\bar{\omega}^{j*}$ represents the point where the g_j (or h_j) reach their unique extreme value and c^*, r^* are the solution to (14).

Observe that (15) implies that at the solution the convex hull of the gradients of the active convex constraints, evaluated at the solutions to (13), and the gradients to the active complementary convex constraints must contain the origin. Thus we can conclude that a point $(c^*, r^*) \in R^{n+1}$ is a stationary solution to the problem (14) if the nearest points to c^* (measured with norm $n(\cdot)$) on the boundary of R_a are at distance r^*, and the convex hull of the gradients (∇g or ∇h) of the active constraints evaluated at these near points contains the origin. Hence, a possible algorithm to solve the DC problem would identify the points on the boundary of R_a closest to c, as they are the limiting factors to the expansion of the inscribed norm body. We pursue this idea in the next section.

We now show that the above necessary conditions are still valid even if the solution to CM is not unique. Consider the simple geometrical example illustrated in Fig. 1. If the subproblem (13) has more than one solution (at points p^1 and p^2 in Fig. 1) additional constraints to (12) are created. (Observe that it is possible for the number of constraints to be infinite and the Kuhn-Tucker conditions could not be applied.) In order to establish the necessary conditions for optimallity in this situation we must show that the functions defined by (13) are Lipschitz continuous [33]. This condition is established by the following:

Theorem 5: Consider the functions γ: $R^{n+1} \to R, \eta$: $R^{n+1} \to R$ defined by

$$\gamma(c, r) = \max_{\omega \in B_0(1)} g(c + r\omega)$$
$$\eta(c, r) = \max_{\omega \in B_0(1)} h(c + r\omega). \tag{16}$$

Then γ and η are Lipschitz locally continuous. In other words, for any $x, y \in X \subseteq R^{n+1}$, where X is bounded, there is a $K > 0$ such that

$$|\gamma(x^2) - \gamma(x^1)| \leq K \|x^2 - x^1\|$$

where x^1, x^2 are the $n + 1$ vectors $(c^1, r^1)^T$ and (c^2, r^2) (see [27, appendix 0]). □

Using expression (16) the DC problem (14) can now be stated as

$$\text{DC: } \max_{c,r} r$$
$$\text{s.t. } \gamma_i(c, r) \leq g_i^U, \quad i \in J_g$$
$$\eta_j(c, r) \leq h_j^U, \quad j \in J_h. \tag{17}$$

In order to overcome the fact that γ and η are not differentiable everywhere, we employ a theorem due to Clarke [7], presented here in a simplified form

Theorem 6: If $\bar{c}$, $\bar{r}$ is a local solution to (17) then there exists a set of numbers λ_0, λ_i, $i \in \{1, 2, \cdots, n_c\}$, not all zero, such that

a) $\lambda_0 \geq 0, \lambda_i \geq 0$

b) $\lambda_i(\gamma_i - g_i^U) = 0, \quad i \in J_g$

$\lambda_i(\eta_i - h_i^U) = 0, \quad i \in J_n$

c) $0 \in \lambda_0 e_{n+1} + \sum_{i \in J_g} \lambda_i \partial\gamma_i(\bar{c}, \bar{r}) + \sum_{i \in J_n} \lambda_i \partial\eta_i(\bar{c}, \bar{r})$. □

(18)

This theorem can be viewed as a generalization of the Fritz-John necessary conditions for *optimality* of a mathematical programming problem or under suitable constraint qualifications [7] of the Kuhn-Tucker conditions (i.e., if $\lambda_0 > 0$). Note that (18) does not require γ and η to be differentiable, but solely locally Lipschitz. Because the gradient of Lipschitz functions is not defined everywhere, Clarke introduces the concept of a *generalized gradient.*

Definition 7: ([7, proposition 5]). For a locally Lipschitz function the generalized gradient $\partial f(x)$ is the convex hull of all points y of the form

$$y = \lim_{i \to \infty} \nabla f(x^i)$$

where $\{x^i\}$ is a sequence converging to x such that f is differentiable at each x_i. □

From the above definition it can be seen that the generalized gradient reduces to a single point (the gradient) whenever the function is differentiable.

We conclude from Theorem 6 that if we can identify the elements of the generalized gradient, and if the convex hull of the elements contains the origin, then the solution is locally optimal. Unfortunately for our problem it is difficult, or impossible, to identify all of the elements of the generalized gradient. Most practical algorithms which can be used to solve (17) will return only one element of the solution set. We postpone considerations on the way convex constraints will be handled. However, for complementary convex constraints, this difficulty will not arise because due to the following Theorem, the solution to (13) is unique.

Theorem 8: If the complementary convex constraint h_j is active at the solution $(c^*, r^*)^T$ to problem DC, then the generalized gradient of $\eta(c^*, r^*)$ contains a unique element.

III. Main Algorithm

We now show that the solution of the DC problem for a nonconvex region of acceptability can be carried out in terms of a convergent sequence of subproblems each of which has a convex region of acceptability. Recall that convergent algorithms exist for the solution of these subproblems, e.g., [14], [33]. Towards this end observe that if the complementary convex constraints are replaced by supporting hyperplanes to the sets D_j defined in (11), the resulting approximation R_a' to R_a is convex. This situation results because R_a' is the intersection of the convex sets C_j with half spaces, and approximates R_a interiorly. Furthermore, for a given nominal point c, the size of the largest norm body which can be inscribed in R_a is limited by the points on the boundary of R_a which are closest to c. (We assume that the measure used for the distance is the norm associated with the distribution $\Phi(p, c)$.)

These considerations suggest the following algorithm:

Step 0. Let c^0 be an acceptable point, i.e., $c^0 \in R_a$.

Let $k = 0, J = \{1, 2, \cdots, n_c\}$.

Step 1. For all $j \in J$ solve the following near point (NP) problems:

$$\begin{aligned} \mathrm{NP}_{k,j}:\ & \min n(p - c^k) \\ & \text{s.t. } g_i(p) \leqslant g_i^U, \quad i = 1, 2, \cdots, n_c \\ & h_j(p) = h_j^U \\ & h_k(p) \leqslant h_k^U, \quad k = 1, \cdots, j-1, \ j+1, \cdots, n_c. \end{aligned} \tag{19}$$

Step 2. a) If $\mathrm{NP}_{k,j}$ has no solution, drop constraint j from further consideration, i.e., $J = J - j$

b) Else, if $p^{k,j}$ is a solution to $\mathrm{NP}_{k,j}$, evaluate the approximations

$$\tilde{h}_{k,j}(p) = \nabla h_j(p^{k,j})^T (p - p^{k,j}) + h_j^U.$$

Step 3. Solve the yield maximization problem for c^{k+1}, while the convex region of acceptability

$$\begin{aligned} \mathrm{YMC}_k:\ & \max_{c,r} r \\ & \text{s.t. } g_i(c + r\omega) \leqslant g_i^U, \quad i \in J_g \\ & \tilde{h}_{k,j}(c + r\omega) \leqslant h_j^U, \quad i \in J. \end{aligned} \tag{20}$$

Step 4. If the solution to YMC_k is locally optimal to DC(12), then stop; else set $k = k + 1$ and go to Step 1.

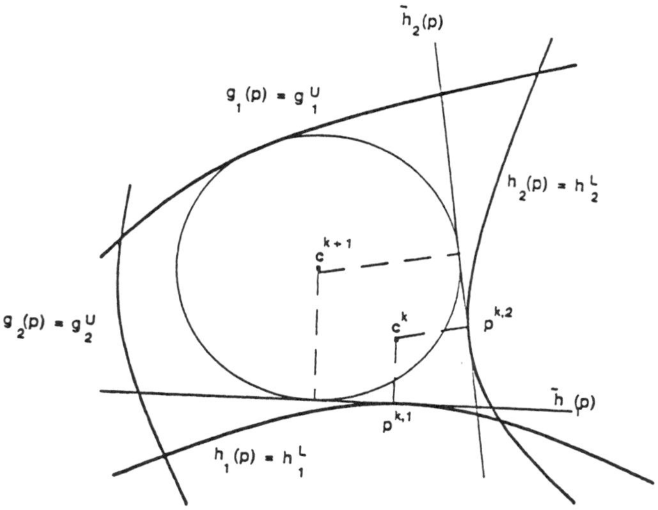

Fig. 2. Illustration of a step of Main Algorithm.

One pass through of the above algorithm is illustrated in Fig. 2 where the norm body is assumed to be a circle. Given the center of the last inscribed circle, c^k, subproblems $\mathrm{NP}_{k,1}$ and $\mathrm{NP}_{k,2}$ are solved to find the points $p^{k,1}$ and $p^{k,2}$ on the surfaces $h_1(p) = h_1^U$ and $h_2(p) = h_2^U$. The convex approximation to R_a is then constructed by introducing the two half spaces $\tilde{h}_1(p) \leqslant h_1^U$ and $\tilde{h}_2(p) \leqslant h_2^U$. The new center c^{k+1} is then determined by inscribing the largest circle in the convex region.

Remarks

1) The above algorithm begins with an initial feasible design c^0, which exists in the interior of R_a.

2) If no solution can be found to $\mathrm{NP}_{k,j}$, then the constraint h_j is *superfluous* in the sense that it does not constrain the design, and can be dropped, giving useful information to the designer.

3) Step 3 requires the use of a convergent method such as the simplicial approximation method for the case where R_a is convex.

4) A sufficient condition to check that the solution is locally optimal in Step 4 is to verify if the points where the norm body touches the hyperplane $\tilde{h}$ is on the boundary of R_a, thus satisfying the conditions of Theorem 6.

5) This algorithm is not immediately implementable because we do not know which constraints are convex and which are complementary convex.

In the Appendix we establish that under the assumptions made earlier the algorithm converges to a local solution of the DC problem.

IV. An Implementable Algorithm

There are two difficulties with the algorithm introduced above. The first concerns the solution of subproblem YMC_k, in Step 3, as most methods available for solving this problem require a large number of expensive circuit simulations. Furthermore, if this subproblem has to be solved repeatedly the computational requirements, even for small circuits, will make the method unattractive. The second difficulty concerns the identification of the quasi-convex or quasi-concave characteristics of the performance functions. These functions are known only implicitly through the solution of a system of equations that describes the network's behavior. Hence, using a criterion like (7) would require the verification of an infinite number of inequalities. Methods to finitely test quadratic functions are known, however, and can be employed.

Significant computational savings can be obtained by approximating the constraints by, for example, the first three terms of a Taylor expansion. Global approximations to the constraints are not in our opinion as attractive as local ones. From the main algorithm introduced in the previous section it can be infered that ideally we would like to approximate the constraints at those points which minimize the distance to the initial design center c^0. If the initial design is a reasonable one, the final solution is not very distant from it, although significant increases in yield can be obtained by the process of centering. If this is not the case the designer could always re-evaluate the approximation. A modified version of the above algorithm which employs such an approximation can be summarized by the following steps:

Step 0. Let $c^0 \in R^a$, $k = 0$, $J = \{1, 2, \cdots, n_c\}$.

Step 1A. Solve the following near point problems for $i = 1, \cdots, n_c$

$$\begin{aligned} NP_{U,i}\!: \; & \min n(p - c^0) \\ & \text{s.t. } f_i(p) = f_i^U \\ & f_i(p) \geqslant f_i^L \\ & f_j^L \leqslant f_j(p) \leqslant f_j^U, \quad j \neq i \end{aligned} \tag{21}$$

$$\begin{aligned} NP_{L,i}\!: \; & \min n(p - c^0) \\ & \text{s.t. } f_i(p) \leqslant f_i^U \\ & f_i(p) = f_i^L \\ & f_j^L \leqslant f_j(p) \leqslant f_j^U, \quad j \neq i. \end{aligned} \tag{22}$$

Step 1B.

1) If there is no solution to one of the near point problems drop the constraint

2) Else, evaluate the gradient $\nabla f_i(\cdot)$ and the Hessian $\nabla^2 f_i(\cdot)$ at the solutions $p^{U,i}$ and $p^{L,i}$ of each problem.

Step 1C. From the Hessian and the gradient infer the convexity characteristics of the constraints (see comments below).

Step 1D. Determine local quadratic approximations to the constraints, e.g.,

$$\begin{aligned} q_i^U = f_i^U &+ \nabla f_i(p^{U,i})^T (p - p^{U,i}) \\ &+ \tfrac{1}{2}(p - p^{U,i})^T \nabla^2 f_i(p^{U,i}) (p - p^{U,i}) \end{aligned} \tag{23}$$

and with the information from Step 1C, classify it as g_i or h_i Go to Step 2.

Step 1E. Solve the near point problem as in Step 1 of the main algorithm.

Step 2. Linearize the complementary convex constraints as in Step 2 of the main algorithm.

Step 3. Solve the yield maximization problem YMC_k.

Step 4. If the solution is locally optimal for the quadratic approximations, then stop; else go to Step 1E.

Observe that Steps 1E–4 correspond to the Steps 1 to 4 in the original algorithm proposed in Section III, except for the substitution of the constraints by their quadratic approximations. Steps 1A–1D are necessary to evaluate the approximations and to discriminate between the quasi-convex and quasi-concave constraints, as discussed below. Furthermore, note that evaluation of the Hessians can be done efficiently, if the optimization algorithm used for Step 1A employs gradient information. Note that while convergence has been established for the main algorithm, we cannot establish convergence for this implementation. In fact it is possible to construct situations in which this technique will not converge as was pointed by one of the reviewers. However this difficulty does not appear to arise in the examples we have tried. Devising a better implementation is an area for further research.

In our implementation a variation of the Han-Powell algorithm [29]–[31] for constrained optimization is used. This algorithm is the most efficient of the algorithms we experimented with and requires the gradient of every constraint at each iteration. Knowledge of these gradients can clearly be used to build an approximation to the Hessian by using an update formula, similar to the ones used by quasi-Newton methods. However, two important differences must be accounted for. The first difference concerns the direction of the search step, which is not defined by the update but solely by the optimization algorithm. The update should therefore be capable of handling arbitrary directions. It is well known that the symmetric rank one update

$$Q^* = Q + \frac{(y - Qs)(y - Qs)^T}{s^T(y - Qs)}, \qquad s = p^* - p, \quad y = \nabla f(p^*) - \nabla f(p) \tag{24}$$

only requires that the directions of search be linearly independent [11]. If the new search direction is not independent from the previous ones, and the function is quadratic, then the denominator of (24) will vanish [28] making the independence test particularly easy to implement.

To keep the update well conditioned we do not update if

$$\left|s^T(y - Qs)\right| \leqslant \epsilon. \tag{25}$$

The second difference results from the fact that for most optimization procedures the approximation to the Hessian has to be kept positive definite. In our case we want the update to approximate as closely as possible the Hessian even if it contains negative eigenvalues. The following theorem establishes a suitable property of the update (24). This theorem is basically a generalization of theorem 41 in [17], where positive definiteness is not required and we update Q and not Q^{-1}.

Theorem 9: If the performance function $f(p)$ is quadratic, i.e.,

$$f(p) = \tfrac{1}{2} p^T H p + g^T p + a$$

then after at most n updates $Q^n = H$. □

Determining if a function is a quasi-convex function, as is required in Step 1C of the algorithm is an unsolved problem [10]. Therefore, we replace the problem of establishing quasi-convexity of the performance function by one of establishing quasi-convexity of its quadratic approximation q (see (23)). In other words (see Definition 1) we must determine if $q(x^2) \leq q(x^1)$ implies

$$\nabla q(x^1)^T (x^2 - x^1) \leq 0$$

or

$$(Qx^1 + g)^T (x^2 - x^1) \leq 0. \tag{26}$$

Notice that (26) does not allow the property of quasi-convexity to be finitely tested. We now summarize the work described in [8], [16], [24], [25] which made possible the test of quasi-convexity.[3] Martos [24], [25] was able to relate quasi-convexity of a quadratic form and of a quadratic function[4] to a class of real, symmetric square matrices which he designated as *positive subdefinite:*

Definition 10: A real symmetric matrix Q is positive subdefinite if for any vector x the following implication holds:

$$x^T Q x < 0 \Rightarrow Qx \geq 0 \quad \text{or} \quad Qx \leq 0. \quad \square \tag{27}$$

Notice that a positive semidefinite matrix is also subdefinite as the inequality on the left of relation (27) cannot occur. Martos showed that a quadratic form

$$f(x) = \tfrac{1}{2} x^T Q x$$

is quasi-convex if and only if Q is positive subdefinite, and suggested rules to verify this property. His work was extended in [8] and from this paper we extract the following two theorems, which make the verification of quasi-convexity practical:

Theorem 11: The quadratic form

$$f(x) = x^T D x$$

is quasi-convex, but not convex in the nonnegative orthant if and only if

1) $D \leq 0$, i.e., all elements are nonpositive.
2) The spectrum of D contains exactly one negative element. □

It is a well-known property of convex quadratic forms that the function obtained by adding a linear term to a quadratic form will also be convex. This property does not hold for quasi-convex functions in the nonnegative orthant. The following theorem gives a criterion for this case:

Theorem 12: If a quadratic function

$$f(x) = \tfrac{1}{2} x^T Q x$$

is not convex in E^n, then $f(x)$ is quasi-convex in the nonnegative orthant if and only if the quadratic form

$$\psi(x) = \tfrac{1}{2}(x^T, \alpha) \begin{bmatrix} Q & g \\ g^T & 0 \end{bmatrix} \begin{bmatrix} x \\ \alpha \end{bmatrix} \tag{28}$$

is quasi-convex in the nonnegative orthant of E^{n+1}. □

Theorems 11 and 12 can be used straightforwardly, to recognize the convex properties of a quadratic function. Our procedure is based on the algorithm represented in the following steps:

Algorithm to check if a quadratic function is quasi-convex or quasi-concave

Step Q0. Given the real symmetric $n \times n$ matrix Q and n-vector g, a vector x^n and a constant a:

$$f(x) = \tfrac{1}{2} x^T Q x + g^T x + a.$$

Step Q1. (Check if convex or concave) Evaluate the eigenvalues of Q: $\lambda_1, \cdots, \lambda_n$

1) If $\lambda_j \geq 0, j = 1, \cdots, n,$ then $f(x)$ is: convex.
2) If $\lambda_j \leq 0, j = 1, \cdots, n,$ then $f(x)$ is: concave.
3) If $\lambda_j > 0$ for some i, $\lambda_j < 0$ for some j, then go to Step Q2.

Step Q2. (Check if quasi-convex or quasi-concave) If all elements of Q are nonnegative or nonpositive then evaluate the eigenvalues of

$$\mathbb{Q} = \begin{bmatrix} Q & Qx^n + g \\ (Qx^n + g) & 0 \end{bmatrix}. \tag{29}$$

1) If all elements of Q and g are nonpositive and only one eigenvalue of $\mathbb{Q}$ is negative then f is quasi-convex. (This is a consequence of successive applications of Theorems 12 and 11.)

2) If all elements of Q and g are nonnegative and only one eigenvalue is positive $\mathbb{Q}$ is quasi-concave.

3) If neither of the above holds, $f(x)$ does not satisfy the quasi-convex quasi-concave hypothesis. Requirement of Theorem 12 does not hold and, therefore, f is neither quasi-convex nor quasi-concave.

V. Numerical Examples

The algorithm described above has been applied to numerous problems. Due to space limitations this section summarizes the results which were obtained when using the algorithm to solve the first three and the fifth examples described in [4].

[3]We need not limit ourselves to considering functions which are quasi-convex or quasi-concave in the whole space. It is enough that they exhibit this property in the nonnegative orthant. This is particularly important as otherwise we would be limiting ourselves to the set of convex functions, because quasi-convex quadratic functions in R^n are also convex [25]. In our applications the designable parameters are either always positive or always negative, therefore, with a proper change of sign we can consider that all parameters are in the nonnegative orthant. As a result the performance functions do not have to be quasi-convex or quasi-concave in the whole space, it is enough to require them to be quasi-convex or quasi-concave in the nonnegative orthant.

[4]A quadratic form is a relation of the type

$$f(x) = \tfrac{1}{2} x^T Q x.$$

A quadratic function is the sum of a quadratic form with a constant and a term linear in x:

$$f(x) = \tfrac{1}{2} x^T Q x + g^T x + a.$$

TABLE I
DATA FOR EXAMPLE 1

pstart	0.5,0.6
pfinal	0.0,0.0
numb. func. eval.	6
numb. grad. eval.	6
box constraints	-2 < p1 < 2 -2 < p2 < 2
near point (n. p.)	0.33763,0.77089
gradient at n. p.	-0.42216,0.44434
hessian at n. p.	1.01000 -0.99000 -0.99000 1.0100

TABLE II
DATA FOR EXAMPLE 2

pstart	0.7 0.9
pfinal	0.850992 1.01212
radius	0.258805
numb. func. eval	15
num. grad. eval.	15 ×3
box constraints	0 ≤ p1 ≤ 2 0 ≤ p2 ≤ 2
near points (n. p.)	0.60075 0.875234 0.801093 0.697814 0.970725 1.24808
grad. at n. p.	-1.50002 -0.374294 1.50000 -3.000000 1.94145 2.49615
hessian at n. p.	1.51520E0 1.76832E-1 1.76832E-1 3.68777E0 1.51208E0 -3.02278E0 -3.02278E0 6.04294E0 2.00000E0 4.06550E-17 4.06550E-17 2.00000E0
eigenvalues of hessian	

TABLE III
DATA FOR EXAMPLE 3

pstart		1.3 0.3
pfinal		1.33159 0.142014
radius		0.624182
numb. func. eval		24
num. grad. eval.		4×5
box constraints		0 ≤ p1 ≤ 4 -2 ≤ p2 ≤ 2
near points (n. p.)	(2) (3) (4) (6)	1.00000 -0.5000000 1.17879 0.357354 1.95501 -0.534514 3.99999 0.250000
grad. at n. p.	(2) (3) (4) (6)	-0.749999 -1.00000 -0.954344 0.451737 1.534490 -1.954875 0.937499 -2.00000
hessian at n. p.	(2)	-1.58966E-1 -1.04112 -1.04112 1.98946
	(3)	2.62051E-2 3.77580E-1 3.77580E-1 2.55022
	(4)	5.54470E-18 -1.00000 -1.00000 4.58735E-17
	(6)	7.90974E-5 -5.04749E-1 -5.04749E-1 -7.71033
eigenvalues of hessian	(2) (3) (4) (6)	-0.580708 2.41120 -0.029068 2.60550 -1.00000 1.00000 0.032981 -7.74324

For Example 3 the assumption of quasi-convexity (quasi-concavity) does not hold, but it is interesting to note that even in this case the solution obtained is close to optimal and represents a significant improvement when compared to the starting nominal point.

Example 1

$$f_1(p) = 0.505(p_1^2 + p_2^2) - 0.99 p_1 p_2. \tag{30}$$

The constraint is $f_1(p) \leqslant 0.1$. Table I summarizes the results for this problem. Notice the close agreement between the Hessian to $f_1(p)$ and the approximation at the near point, due to the property of the rank one update established in Theorem 9.

Example 2

$$f_1 = e^{-p_1+1}((p_2 - 1)^2 + 1)$$

$$f_2 = e^{p_1 - 2p_2 + 1}$$

and

$$f_3 = p_1^2 + p_2^2 - 1. \tag{31}$$

The constraint for this problem is $\max_i f_i \leqslant 1.5$. In Table II, observe the approximations to the Hessians have positive eigenvalues so that all quadratic approximations are convex.

Example 3

$$\begin{aligned} f_1 &= 1.5 - p_1(1 - p_2) & f_4 &= -f_1 \\ f_2 &= 2.25 - p_1(1 - p_2^2) & f_5 &= -f_2 \\ f_3 &= 2.625 - p_1(1 - p_2^3) & f_6 &= -f_3. \end{aligned} \tag{32}$$

In this example the algorithm determined that constraints f_1 and f_6 were superfluous, in the sense defined above, after six function evaluations and four gradient evaluations. From the eigenvalues it can be seen that none of the constraints is either convex or concave (see Table III).

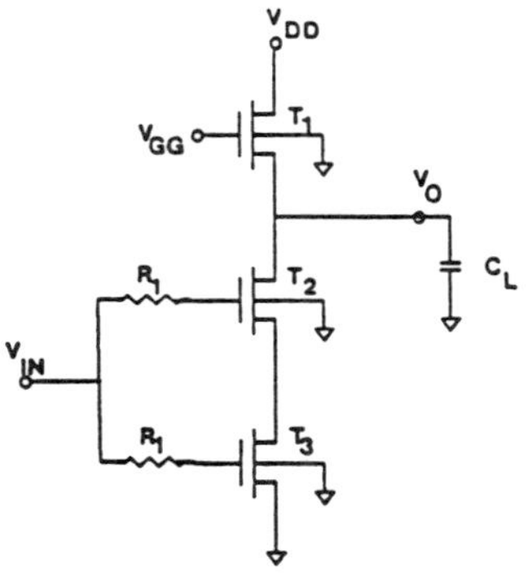

Fig. 3. Two input MOSFET NAND gate.

TABLE IV
DATA FOR EXAMPLE 4

pstart	0.8 0.9
pfinal	0.999843 1.01583
radius	0.084170
numb. func. eval	11
num. grad. eval.	11× 2
box constraints	0 ≤ p1 ≤ 2 0 ≤ p2 ≤ 2
near points (n. p.)	0.731672 0.934164 0.816026 0.875960
grad. at n. p.	-0.536657 0.268328 0.367942 -0.551923
hessian at n. p.	2.00000 -1.21895E-16 -1.21895E-16 2.00000 2.00000 7.15607E-16 7.15607E-16 2.00000
eigenvalues of hessian	2 2 -2 -2

Example 4

$$f_1 = 0.96 + (p_1 - 1)^2 + (p_2 - 0.8)^2$$

$$f_2 = 1.16 - (p_1 - 1)^2 - (p_2 - 0.6)^2. \quad (33)$$

In this example there are two performance functions, one convex and the other concave. The level contours are circles for both functions (see Table IV).

VI. CIRCUIT EXAMPLE

In this section we describe the results of applying the algorithm to the statistical design of a MOSFET NAND gate which has been extensively studied [3], [14], [18], [21]. The circuit representation of the two input NAND gate is shown in Fig. 3. The three transistors are enhancement, p-channel devices. T_1 acts as a load resistor and has a longer channel length than the two driver devices T_2 and T_3, to originate a higher ON resistance. The two drivers have identical channels. Since these devices are to be implemented in an integrated circuit a four terminal model is required for simulation purposes. As in [21] we use the MOSFET model shown in Fig. 4, which is essentially the model derived in [9].

The designable parameters are the channel width of the two drivers: W_{23} and the flatband voltage V_{FB}. We impose the following limits on the parameter values:

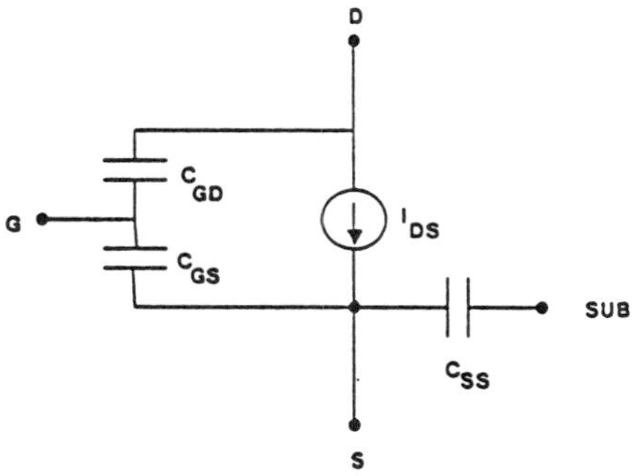

Fig. 4. Model used for MOS transistor in NAND example.

$$5\ \mu m \leqslant W_1 \leqslant 50\ \mu m$$

$$50\ \mu m \leqslant W_{23} \leqslant 250\ \mu m$$

$$-2\ V \leqslant V_{FB} \leqslant -1\ V. \quad (34)$$

The initial values for these parameters are $W_1 = 12\ \mu m$, $W_{23} = 230\ \mu m$, and $V_{FB} = -1.2$ V and the two remaining geometries of the channel are assumed constant: $L_1 = 12.7\ \mu m$ and $L_{23} = 5.08\ \mu m$.

The design constraints are expressed in terms of the channel area f_A (roughly proportional to the total area of the NAND gate), the output voltage when both drivers are conducting (ON state) f_V and the propagation delay f_p when the devices change from the ON to the OFF state, which is for this type of gates much larger than the propagation delay corresponding to the inverse transition. Specifically a circuit is considered acceptable if the following constraints are met:

$$f_A \leqslant 2500\ \text{mil}^2$$

$$f_V \leqslant 0.7\ V$$

$$f_p \leqslant 110\ \text{ns}. \quad (35)$$

The time delay can be approximated by [21]

$$f_p = A_F = \frac{L_1}{W_1} \tau\alpha \quad (36)$$

where A_F is a constant employed to match the delay found using an accurate transient simulation and τ and α are given in [9]. The two other performance functions are given by

$$f_A = W_1 L_1 + 2L_{23} W_{23}$$

$$f_V = -V_0(\text{ON}). \quad (37)$$

With this formulation to evaluate the performance functions, only a dc analysis with $V_{IN} = -6$ V is needed. To evaluate the node voltages we simply write the nodal equations and apply the Newton-Raphson method to the resulting system of nonlinear equations. The gradients were found by solving the adjoint equations [13], with the formulation proposed in [20]. Lightner [21] pointed out that the region of acceptability for this problem is not convex.

In order to allow comparison with the results published in [14] the scaling matrix $S = \text{diag}\ (4.57, 50.5, 1)$ was used and, therefore, we should expect the algorithm to converge to the solution obtained by the Simplicial Algorithm in [14]. Starting from $p_{start} = (12, 230, -1.2)$ the following near points were found on the constraints:

TABLE V
DATA FOR NAND EXAMPLE

		Hessian	Eigen Values	Eigen Vectors
f_p	Pert	57.40 0.0468 7.102		0.9916 0.0011 0.1295
		0.0468 0.1300 .1400	58.238, 0.1238, 3.0353	0.0048 0.9990 -0.0454
		7.102 0.1400 3.957		-.1294 -.0454 0.9905
	Update	58.20 -.0325 7.222		0.9920 -.0005 0.1259
		-.0325 0.0787 -.0005	59.116, 0.0786, 1.2954	0.0005 0.9999 0.00007
		7.222 -.0005 2.212		-1.1259 0.0000 .9920
f_V	Pert	-.0067 -.0437 -.0439		0.9064 0.3379 0.2534
		-.0437 0.0600 0.0283	-.0372, 0.0416, 0.1277	-.1263 .7894 -6007
		-0.439 0.0283 0.0809		-.4030 0.5125 0.7582
	Update	-.0289 -.0487 -.0287		0.9232 0.3710 0.1002
		-.0487 0.0593 0.0379	-.0516, 0.0311, 0.1235	-.1830 0.6538 -.7341
		-.0287 0.0379 0.0727		-.3379 0.6594 0.6716
f_A	Pert	0.0 0.0 0.0		
		0.0 0.0 0.0		
		0.0 0.0 0.0		
	Update	0.0 0.0 0.0		
		0.0 0.0 0.0		
		0.0 0.0 0.0		

For

$$f_p\text{:}\quad x^p = (8.856, 229.975, -2.000)$$

$$f_A\text{:}\quad x^A = (12.106, 230.932, -1200)$$

$$f_V\text{:}\quad x^v = (13.403, 229.110, -1.978).$$

The Hessians, one for each constraint, as computed by the update method described earlier, are shown in Table V. For comparison purposes we include the Hessians of the constraints as evaluated by perturbation and the eigenvalues and eigenvectors of all the matrices. In Fig. 5, we illustrate in two dimensions, the region of acceptability, by representing three cuts corresponding to $V_{FB} = -1$ V, -1.5 V, -2.0 V. The small triangles indicate the closest points to the initial nominal p_{start}.

We will analyze these results in more detail below, but it is clear that there is good agreement between the matrices, especially when comparing their eigenvectors and eigenvalues. Note that f_V is not convex because the first eigenvalue is negative.

The Hessian information was used to generate the quadratic approximation to the three constraints and $p_{final} = (10.6, 209.0, -1.47)$ is the center of the largest ball inscribed in the approximation to the region of acceptability. Table VI summarizes the main results obtained here and the equivalent ones obtained in [14]. A Monte Carlo analysis with 500 samples was made for both design centers, showing that they are essentially equivalent. The number of function evaluations and gradient evaluations is however, significantly smaller for the quadratic approximation approach.

As can be seen from Table VI, the Simplicial Approximation yielded a similar solution in spite of the fact that one of the constraints f_V was nonconvex. From Table V it can be seen that the first eigenvalue is negative while the other two are positive (in the three-dimensional space we are considering, this surface is an hyperboloid of one sheet [15]). The curvature associated with this surface is, however, very small as may be judged from the value of the Hessian in Table V, and, therefore, the nonconvexity has a negligible influence on the region of acceptability.

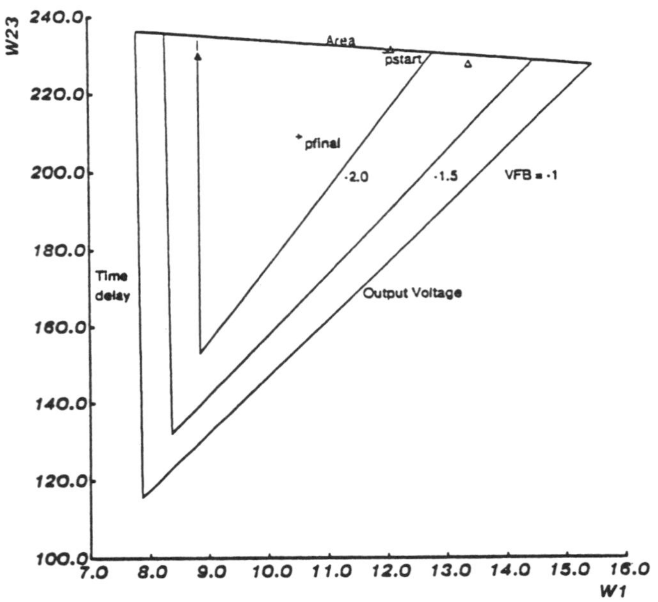

Fig. 5. Region of acceptability for NAND gate.

TABLE VI
COMPARISON OF THE RESULTS OBTAINED WITH THE PROPOSED METHOD AND THOSE PUBLISHED IN [14]

Method	p^{final}	Number of functions	Number of gradients	Yield (500 samples) s.d. 2.2, 24, 0.47
Simp. Appr	10.6, 209.0, -1.47	>300		0.370
Impl. Alg.	10.55, 208.50, -1.48	26	3x17	0.372

Of interest are the conclusions a designer may draw from the information contained in the Hessians. Specifically, since $\nabla^2 f_A$ is zero, and the Hessian $\nabla^2 f_V$ is small, the designer can conclude that f_A is either independent or linear in the parameters and f_V is "almost" linear in p. In fact we can for this simple example check these conclusions. First,

$$f_A = (2L_{23}W_{23}) + W_1 L_1$$

which is clearly linear in the designable parameters W_1 and W_{23} and independent of V_{FB}. The output voltage in the ON state is computed by solving the DC network equations, for a quick analysis, the output voltage can be evaluated as follows [32]:

$$V_0 \simeq = k' \frac{W_1}{L_1} \times \frac{L_{23}}{W_{23}} \frac{1}{(V_{IN} - V_T)}.$$

Thus the output voltage depends as all of the parameters, but this dependence can be seen to be linear in W_1 and inversely proportional to W_{23} and $(V_{IN} - V_T)$, which explains the small values in the Hessian.

Finally the Hessian of the propagation delay has the second row and second columns very small, indicating that the propagation delay is either independent of W_{23} or dependent on W_{23} linearly. Again inspection of (36) confirms this result. It should be kept in mind that for this simple problem one can easily check the expressions to confirm the results obtained with the Hessian. However, in most cases, simple closed-form expressions for the performance functions cannot be found.

VII. Conclusions

In this paper we have presented a new method for solving the DC problem, which is not limited by assumptions of convexity. Rather than approximating the region of acceptability, as is done by the simplicial approximation method, this algorithm approximates each one of the relevant constraints. It is shown that in the case where the performance functions are either quasi-convex or quasi-concave, two types of constraints have to be considered: convex constraints and complementary convex constraints. In general complementary convex constraints will cause the region of acceptability to be nonconvex.

In the proposed algorithm, the complementary convex constraints are approximated by a set of supporting hyperplanes. If these hyperplanes are properly chosen we show that the resulting algorithm will converge to a locally optimal solution of the DC problem. It was also shown how approximation of the constraints with quadratic surfaces can result in a dramatic reduction in the computational complexity of the procedure. Furthermore, these approximations contain valuable information which a designer may be able to use to gain a better understanding of the way the circuit behaves.

Appendix

Proof of Lemma 3

Proof: The proof derives immediately from the definition of quasi-convex function: given any two points p^1, p^2 on the boundary of $L_c(r)$, then

$$g(\theta p^1 + (1 - \theta)p^2) \leqslant \max\,[g(p^1), g(p^2)] \tag{38}$$

i.e., the values of g for interior points is at most equal to one of the extreme values. □

Proof of Lemma 4

Proof: By contradiction. Assume the maximum of h_j is reached at a point $\bar{p} \in \text{int}\,(B_{c^*}(r^*))$ and, for any point p^* on the boundary $h(p^*) < h(\bar{p})$ then $\bar{p}$ is the unconstrained maximum of h_j. However, because $\bar{p}$ must satisfy the constraints to (12) we have $h_j(p^*) \leqslant h_j^U$. This contradicts our assumption of h_j being nontrivial. Therefore, a maximum must exist on the boundary of $L_{c^*}(r^*)$. □

Proof of Theorem 8

Proof: It was established by Lemma 4 that the solution to $\max_{\omega \in L_0(1)} h(c + r\omega)$ is on the boundary of the set $B_{c^*}(r^*)$. Furthermore, because h_j, by assumption, is active, at a solution $\bar{\omega}$, $h_j(c^* + r^*\bar{\omega}) = h_j^U$ then $\bar{p} = c^* + r^*\bar{\omega}$ exists on the boundary of the convex set

$$G_j = cl(-D_j) \tag{39}$$

i.e., the closure of the complement of D_j.

Because both G_j and $B_{c^*}(r^*)$ are convex sets, and they do not contain common interior points, we can use one of the forms of the Separation Theorem for convex sets ([34, section 5.12]) to establish the existence of a hyperplane $\langle s, x\rangle = b$ such that

$$\langle s, x\rangle \leqslant b, \qquad \text{for } x \in B_{c^*}(r^*)$$

$$\langle s, x\rangle \geqslant b, \qquad \text{for } x \in G_j$$

which contains a point on the boundary of G_j. This plane is therefore a supporting hyperplane to the set G_j. Because h_j is differentiable, and the gradient does not vanish at the boundary of R_a, the gradient must be colinear with the supporting plane normal. □

Convergence Proof for the Main Algorithm

We establish here that the main algorithm converges to a point which is locally optimal for problem DC. Abstractly, any algorithm can be viewed as a point to set mapping A: $R^n \to R^n$, which at each iteration evaluates a new point

$$z^{k+1} \in A(z^k).$$

Zangwill [34] shows that the following conditions are sufficient to establish the convergence of an algorithm:

Theorem 13: General Convergence Theorem: Let a point-to-set mapping A: $R^n \to R^n$ determine an algorithm that given a point $z^1 \in R^n$, generates the sequence $\{z^k\}$. Also let a solution set $\Omega \subset R^n$ be given. Suppose that:

1) All points z^k are in a compact set $X \subset R^n$.

2) There is a continuous function Z: $R^n \to R$ such that:

a) if z^k is not a solution, then for any $y \in A(z^k)$

$$Z(y) > Z(z^k)$$

b) if z^k is a solution then either the algorithm terminates or for any $y \in A(z^k)$

$$Z(y) \geqslant Z(z).$$

3) The map A is closed at z^k if z^k is not a solution.

Then either the algorithm stops at a solution, or the limit of any convergent subsequence is a solution. □

We now show that our algorithm can be interpreted as a map composition and satisfies the requirements of the General Convergence Theorem. Without loss of generality it will be assumed that no constraint is superfluous, i.e., there is a solution to every problem $\text{NP}_{k,j}$. For each center c^k finding the solution to $\text{NP}_{k,j}$ can be viewed as a mapping $N_j(c^k)$ from c^k to the set of points on the constraint surface which minimize the distance to c^k. The linearization described in Step 2(b) can be viewed as a mapping L_j from the point $p^{k,j}$ to the coefficient set $\{\nabla h_j(p^{k,j})\}$. Since the performance functions are twice differentiable, L_j is continuous and, therefore, closed (see [34]). Finally the subproblem YMC_k described in Step 3 of the algorithm can be viewed as a mapping C from these coefficient sets to the set of optimal solutions of YMC_k. Thus the algorithm can be considered as a mapping composition $A = \text{CNL}$: $R^{n+1} \to R^{n+1}$. It remains to prove that the conditions of the convergence theorem are met. Towards this end we prove the following theorem.

Theorem 14: The set of points $\{c^k\}$ exist in R_a, is compact, i.e., it is closed and bounded.

Proof: By assumption c^0 is an interior point of R^a. The region of acceptability R_k defined by the constraints to problem YMC_k is a subset of R_a, i.e.,

$$R_k = \{p|g(p) \leqslant g^U, \tilde{h}(p) \leqslant h^U\} \subset R_a$$

as for any point p such that $\tilde{h}(p) \geqslant h^U, h(p) \geqslant h^U$. This results from the assumption that h is quasi convex and, therefore, the set $h(p) \leqslant h^L$ is convex.

By construction every point $c^{k+1} \in R_k \subset R_a$. □

We take $Z = r$ where Z was defined in Theorem 13 and show that this function satisfies condition (2) by proving that at each iteration of the algorithm the radius (size) of the norm body increases unless the center c^K is R_a solution of the problem.

Theorem 15: Let $(c^{k+1}, r^{k+1}) \in A(c^k, r^k)$. Then either $(c^k, r^k)^T$ is a solution to the problem DC or $r^{k+1} > r^k$.

Proof: If c^k does not exist in the solution set to problem DC, then at least one linearization $\tilde{h}_j(p)$ is active and the point $y^{k,j}$ where the norm body $L_{c^k}(r^k)$ touches $\tilde{h}_j(p)$ does not exist on the surface $h_j(p) = h_j^L$.

We want to establish that the new linearization say $\bar{h}_3$ does not touch $L_{c^k}(r^k)$ and, therefore, r^{k+1} will be strictly greater than r^k. This results from the observation that $\bar{h}_3$ is a supporting hyperplane to the set G_3 and the distance from c^k to $\bar{h}_3$ is

$$d(c^k, \bar{h}_3) = \min_{y \in \bar{h}_3} n(y - c^k) = n(p^{k,3} - c^k) > r^k. \quad \square$$

It remains to establish condition (3), i.e., that the mapping A = CLN is closed.

Theorem 16: The mapping N_j is closed.

Proof: Assume the sequence $c^k \to \bar{c}$, for k in some subsequence K, $p^{k,j} \in N_j(c^k)$, $k \in K$ and $p^{k,j} \to \bar{p}^j$. Then N is closed if $\bar{p}^j \in N_j(\bar{c})$.

Exploiting the continuity of the performance functions and because for every $p^{k,j}$ we have $h_j(p^{k,j}) = h_j^L, h_i(p^{k,j}) \geqslant h_i^L, i \neq j, g_i(p^{k,j}) \leqslant g_i^U$, taking the limit

$$\lim_{k \in K} h(p^{k,j}) = h(\bar{p}^j) = h_j^L$$

$$\lim_{k \in K} h_i(p^{k,j}) = h_i(\bar{p}^j) \geqslant h_i^L$$

$$\lim_{k \in K} g(p^{k,j}) = g(\bar{p}^j) \leqslant g^U$$

i.e., $\bar{p}^j$ satisfies the constraints to problem NP, and it can be shown by contradiction that it minimizes the objective function: assume the existence of a solution $p^* \neq \bar{p}$ such that

$$n(p^* - \bar{c}) < n(\bar{p} - \bar{c}) = \lim_{k \in K} n(p^k - c^k).$$

Because the norm n(•) is a continuous function, and $c^k \to \bar{c}$ and $p^k \to \bar{p}$, we must have a neighborhood of $\bar{c}$ and of $\bar{p}$ such that for $k > M$

$$n(p^* - c^k) < n(p^k - c^k)$$

contradicting the assumption that $p^k \in N(c^k)$. □

The closeness of mapping C can be established by the following theorem, which synthesizes theorems II.1.4 and I.2.2 of Dantzig [12]. The notation was altered to suit our exposition.

Theorem 17: Assume that the set $\{p|g(p) \leqslant g^U, \tilde{h}(p) \geqslant h^L\}$ has a nonempty interior and that none of the linear constraints $\tilde{h}$ is identically satisfied. Then the point-to-set mapping C is closed. □

By construction, and because c^k is an interior point to R_a, the distance from the linear constraints to c^k is nonzero, therefore, c^k is an interior point to the set mentioned in the theorem. By assumption the gradient $h(p)$ does not vanish at the boundary of R_a, therefore, the constraints $\nabla h_j(p - p^{k,j}) \geqslant 0$ are not identically satisfied.

It was assumed earlier then the region of acceptability is a closed set, hence the mapping A = NLC is closed (see [34]). This establishes condition 3 of the General Convergence Theorem and completes the proof of convergence for the main algorithm.

Proof of Theorem 9

Proof: Let Q^1, Q^2, Q^n be the updates, $s^1, \cdots, s^n$ be n linearly independent vectors and notice that for a quadratic function

$$y^i = Hs^i.$$

By induction we establish that a similar relation holds for Q^n:

$$Q^1 s^1 = \left[Q^0 + \frac{(y^1 - Q^0 s^1)(y^1 - Q^0 s^1)^T}{s^{1T}(y^1 - Q^0 s^1)}\right] s^1$$

$$= Q^0 s^1 + \frac{(y^1 - Q^0 s^1)(y^{1^T} s^1 - s^{1^T} Q^0 s^1)}{s^{1^T} y^1 - s^{1^T} Q^0 s^1}$$

$$= Q^0 s^1 + y^1 - Q^0 s^1 = y^1.$$

Assume now that $Q^i s^j = y^i$ for $0 \leqslant j \leqslant 1$, then by substitution into (24) we can establish, similarly

$$Q^{i+1} s^{i+1} = y^{i+1}$$

and for $j < i + 1$

$$Q^{i+1} s^j = Q^i s^j + \frac{(y^{i+1} - Q^i s^{i+1})(y^{i+1^T} s^j - s^{i+1^T} Q^i s^j)}{s^{i+1^T}(y^{i+1} - Q^i s^{i+1})}$$

$$= y^j + \frac{(y^{i+1} - Q^i s^{i+1})(y^{i+1^T} s^j - s^{i+1^T} y^j)}{s^{i+1^T}(y^{i+1} - Q^i s^{i+1})}$$

$$= y^j + \frac{(y^{i+1} - Q^i s^{i+1})(s^{j+1^T} H s^j - s^{i+1^T} H s^j)}{s^{i+1T}(y^{i+1} - Q^i s^{i+1})}$$

$$= y^j.$$

Defining two square matrices S and Y whose columns are s^j and y^i, respectively, we have

$$Y = QS$$

$$\Rightarrow QS = HS$$

$$Y = HS$$

but because the columns of s are linearly independent S^{-1} exists and $Q = H$. □

Acknowledgment

We are indebted to R. Brayton, G. Hachtel, and A. Sangiovanni-Vincentelli for helpful discussions during the course of this work. We are also especially grateful to one of the reviewers for his detailed comments and his efforts to verify our results.

References

[1] Mordecai Avriel, *Nonlinear Programming, Analysis and Methods.* Englewood Cliffs, NJ: Prentice-Hall, 1976.

[2] John W. Bandler, Peter C. Liu, and James H. K. Chen, "Worst case network tolerance optimization," *IEEE Trans. Microwave Theory Tech.*, vol. MTT-23, pp. 630–640, Aug. 1975.

[3] Robert K. Brayton and Stephen W. Director, "Computation of delay time sensitivities for use in time domain optimization," *IEEE Trans. Circuits Syst.*, vol. CAS-22, pp. 910–920, Dec. 1975.

[4] Robert K. Brayton, S. W. Director, Gary D. Hachtel, and L. M. Vidigal, "A new algorithm for statistical circuit design based on quasi-Newton methods and function splitting," *IEEE Trans. Circuits Syst.*, vol. CAS-26, pp. 784–794, Sept. 1979.

[5] Robert K. Brayton, Stephen W. Director, and Gary D. Hachtel, "Yield maximization and worst case design with arbitrary statistical distributions," *IEEE Trans. Circuits Syst.*, vol. CAS-27, pp. 756–764, Sept. 1980.

[6] Edward M. Butler, "Realistic design using large-change sensitivities and performance contours," *IEEE Trans. Circuit Theory*, vol. CT-18, pp. 58–65, Jan. 1971.

[7] Frank H. Clarke, "A new approach to Lagrange multipliers," *Math. Oper. Res.*, vol. 1, no. 2, pp. 165–174, May 1976.

[8] Richard W. Cottle and Jacques A. Ferland, "Matrix–Theoretic criteria for quasi-convexity and pseudo-convexity of quadratic functions," *Linear Algebra and its Applications*, vol. 5, pp. 123–136, 1972.

[9] Robert H. Crawford, "MOSFET in circuit design," Texas Instruments Inc., 1967.

[10] Jean-Pierre Crouzeix, "On second order conditions for quasi-convexity," *Math. Programming*, vol. 18, pp. 349–352, 1980.

[11] Jane Cullum and R. K. Brayton, "Some remarks on the symmetric rank one update," Tech. Rep. RC 6157, IBM, Aug. 1976.

[12] George Dantzig, Jon Folkman, and Norman Shapiro, "On the continuity of the minimum set of a continuous function," *J. Opt. Theory Appl.*, vol. 17, pp. 519–548, 1967.

[13] Stephen W. Director and Ronald A. Rohrer, "Automated network design–The frequency domain case," *IEEE Trans. Circuit Theory*, vol. CT-16, pp. 330–337, Aug. 1969.

[14] Stephen W. Director and Gary D. Hachtel, "The simplicial approximation approach to design centering," *IEEE Trans. Circuits Syst.*, vol. CAS-24, pp. 363–372, July 1977.

[15] Arnold Dresden, *Solid Analytical Geometry and Determinants.* New York: Wiley, 1948.

[16] Jacques A. Ferland, "Maximal domains of quasi-convex and pseudo-convexity for quadratic functions," *Math. Programming*, vol. 3, pp. 178–192, 1972.

[17] Anthony V. Fiacco and Garth P. McCormick, "Computational aspects of unconstrained minimization algorithms," in *Nonlinear Programming, Sequential Uncontrained Minimization Techniques.* New York: Wiley, 1968, pp. 156–197, ch. 8.

[18] D. L. Fraser Jr., "Modeling and optimization of MOSFET LSI circuits," Ph.D. dissertation, Univ. Florida, 1977.

[19] Manfred Glesner and Manfred Newmuller, "Investigation and approximation of feasible regions for statistical design of electrical networks," *Proc. Int. Symp. Circuits and Systems*, 1980.

[20] Gary Hachtel, Robert K. Brayton, and Fred G. Gustavson, "The sparse tableau approach to network analysis and design," *IEEE Trans. Circuit Theory*, vol. CT-18, pp. 101–113, Jan. 1971.

[21] Michael R. Lightner, "Multiple criterion optimization and statistical design for electronic circuits," Ph.D. dissertation, Carnegie-Mellon University, 1979.

[22] D. G. Luenberger, *Optimization by Vector Space Methods.* New York: Wiley, 1969.

[23] O. L. Mangasarian, *Nonlinear Programming.* New York: McGraw-Hill, 1969.

[24] Bela Martos, "Subdefinite matrices and quadratic forms," *SIAM J. Appl. Math.*, vol. 17, no. 6, pp. 1215–1223, Nov. 1969.

[25] ——, "Quadratic programming with a quasiconvex objective function," *Operations Res.*, vol. 19, pp. 87–97, 1971.

[26] J. Ogrodzki, L. Opalski, and M. Styblinski, "Acceptability regions for a class of linear networks," in *Proc. Int. Symp. Circuits and Systems*, pp. 187–190, 1980.

[27] Elijah Polak and Alberto Sangiovanni-Vincentelli, "Theoretical and computational aspects of the optimal design centering, tolerancing, and tuning problem," *IEEE Trans. Circuits Syst.*, vol. CAS-26, pp. 795–813, Sept. 1979.

[28] M. J. D. Powell, "Rank one method for unconstrained optimization," in J. Abach (editor), *Nonlinear and Integer Programming.* North-Holland, Amsterdam, 1970.

[29] ——, "A fast algorithm for nonlinearity constrained optimization calculations," in A. Dold and B. Eckmann (editor), *Lecture Notes in Math.*, Springer-Verlag, 1977, pp. 144–157.

[30] ——, "Variable metric methods for constrained optimization," in *Third Int. Symp. Computing Models in Applied Sciences and Engineering*, 1977.

[31] ——, "Algorithms for nonlinear optimization that use Lagrangian functions," *Math. Programming*, vol. 14, pp. 224–248, 1978.

[32] L. M. Vidigal and S. W. Director, "Design centering: The quasi-convex, quasi-concave performance function case," in *IEEE Int. Symp. Circuits and Systems*, 1980.

[33] A. L. Sangiovanni-Vincentelli, "Circuit simulation," NATO Course (Italy), 1980.

[34] Willard I. Zangwill, "Convergence conditions for nonlinear programming algorithms," *Manag. Sci.*, vol. 16, no. 1, pp. 1–13, Sept. 1969.

A Radial Exploration Approach to Manufacturing Yield Estimation and Design Centering

KIRPAL S. TAHIM, MEMBER, IEEE, AND ROBERT SPENCE, FELLOW, IEEE

Abstract—**An integrated approach to manufacturing yield estimation and design centering is presented in which linear searches along "radial" directions within the multiparameter component space are used to locate points on the boundary of the feasible region for a given nominal design and a set of component tolerances. This set of boundary points enables the volume of the feasible region, and hence the manufacturing yield, to be estimated by using a newly developed computational technique. For linear frequency-domain circuit behavior, application of the tracking-sensitivity algorithm further enhances the searching efficiency. The boundary points are also used within a practical design centering algorithm which minimizes the asymmetry of the feasible region around the nominal design. Iterative application of this process leads to a better centered design. Examples of the application of this algorithm to circuit design are given. The results indicate that the radial exploration approach to yield estimation and design centering is effective in practice, computationally cheap, and applicable to many situations in practical circuit design.**

I. Introduction

PRACTICAL circuits have to be designed in the face of uncertainties. These may be the statistical variations in component values due to manufacturing tolerances, or environmental factors such as temperature or humidity. The effect of the ensuing variations in component values is that the circuit's response will also exhibit variation from one sample to another. In this case the manufacturing yield—which is the proportion of manufactured circuits which meet the performance specifications—is of considerable interest to the designer. Monte Carlo methods may be used to simulate component variation in order to esimate the yield, but can be rather expensive in terms of computing time. In any case, such simulations only enable the designer to estimate the yield of a given design. The general problem of obtaining a better design (e.g., with a higher yield) still remains essentially unanswered.

Although a considerable amount of research is now in progress [1]–[12], the solutions presented so far are by no means general or demonstrably the most appropriate to practical circuit design. Most of the effort has been concentrated on the first half of the problem, by developing more efficient alternatives to Monte Carlo methods for yield estimation [10]–[17]. For the second half of the problem—that of *increasing* the yield—the solution can involve a change in nominal component values with the tolerances held fixed (design centering), or vice versa (tolerance assignment), or a combination of the two approaches. Design centering is important in that it involves no change in component cost and, even if tolerance assignment is attempted, will usually be carried out first.

This paper proposes a new radial exploration approach to yield estimation and design centering, an approach in which the information obtained during the yield estimation stage is used, at little additional cost, to improve the existing design. The potential savings of the approach have been demonstrated using test circuits containing up to 58 variable components.

II. Manufacturing Yield

Consider a circuit in which there are m variable components $p_1,p_2,\cdots,p_m$ such that the ith component p_i varies between its upper limit $\bar{p}_i$ and lower limit $\underline{p}_i$ with a probability density function (PDF) $\phi_i(p_i)$.[1] Let $\bar{q}_j$ and $\underline{q}_j$ be the upper and lower specification limits on the jth of the circuit's n responses of interest. The manufacturing yield can then be expressed as

$$Y=\int_{\underline{p}_m}^{\bar{p}_m}\cdots\int_{\underline{p}_1}^{\bar{p}_1} g(p_1,\cdots,p_m)\phi(p_1,\cdots,p_m)\,dp_1\cdots dp_m \tag{1}$$

where $\phi(p_1,\cdots,p_m)$ is the joint *pdf* of the m variable components, and $g(p_1,\cdots,p_m)$ is a testing function which indicates acceptance or rejection of the circuit under consideration, viz:

$$g(p_1,\cdots,p_m)=1, \quad \text{if } \underline{q}_j\leqslant q_j(p)\leqslant\bar{q}_j;\ j=1,\cdots,n$$
$$=0, \quad \text{otherwise.} \tag{2}$$

Alternatively, the yield can be expressed as a mathematical expectation of $g(p_1,\cdots,p_m)$ with respect to the *pdf* $\phi(p_1,\cdots,p_m)$:

$$Y=\langle g(p_1,\cdots,p_m)\rangle \quad \text{w.r.t. } \phi(p_1,\cdots,p_m). \tag{3}$$

Manuscript received February 1, 1979; revised April 30, 1979.
The authors are with Imperial College of Science and Technology, London, England.

[1] ϕ: For nonlinear circuits, a search involving a small number of circuit analyses may be involved. For linear circuits, a very efficient procedure [17], [18] is available.

Reprinted from *IEEE Trans. Circuit Syst.*, vol. CAS-26, no. 9, pp. 768–774, September 1979.

In practice, an estimate $\hat{Y}$ of this expectation can be obtained by simulating a number (N) of the circuits according to the *pdf* $\phi(p_1,\cdots,p_m)$ and then applying the testing function (2), such that

$$\hat{Y}=\frac{1}{N}\sum_{i=1}^{N} g(p_1,\cdots,p_m). \quad (4)$$

The estimate $\hat{Y}$ can also be interpreted as the proportion of circuits which pass the specification test, and is the most commonly understood meaning of the definition of manufacturing yield.

III. Radial Exploration

For notational convenience it is useful to define the various regions of interest in the component space. Let the tolerance region be denoted by T, where

$$T \triangleq \{p \mid \underline{p}_i < p_i < \bar{p}_i;\ i=1,2,\cdots,m\} \quad (5a)$$

and the acceptance region by A, where

$$A \triangleq \{p \mid \underline{q}_j < q_j(p) < \bar{q}_j;\ j=1,2,\cdots,n\}. \quad (5b)$$

The feasible region F is then the intersection of the tolerance and acceptance regions:

$$F=A \cap T. \quad (6)$$

Using these definitions, the manufacturing yield Y as given by (1) can now be expressed as

$$Y=\int \cdots_{F} \int \phi(p_1,\cdots,p_m)\,dp_1\cdots dp_m. \quad (7)$$

The manufacturing yield can also be expressed in terms of the "weighted" volumes of the tolerance and feasible regions. For the simple case of uniform and uncorrelated component distributions, the yield is simply the ratio of the feasible (V_F) and tolerance (V_T) volumes:

$$Y=\frac{V_F}{V_T}. \quad (8)$$

On the basis of extensive experimental evidence it is suggested that an assessment of Y can be made (which may not necessarily be an unbiased estimator of Y, but is nevertheless useful for controlling the yield improvement) in the following way. First, in multiparameter component space, a number of lines having random directions are generated, in each case passing through the nominal design. For each such line, and in both directions away from the nominal design, the distances r_F and r_T to the edges of the feasible and tolerance regions are determined (Fig. 1). Again in both directions, the normalized distance $r_0 (= r_F/r_T)$ to the feasible region boundary is calculated so that two distances, r_0^+ and r_0^-, are associated with each line. The proposed assessment of Y is then

$$\hat{Y}=\frac{1}{2L}\sum_{j=1}^{L}(r_{0_j}^+)^m+(r_{0_j}^-)^m \quad (9)$$

where L is the number of randomly generated lines, and $r_{0_j}^+$ and $r_{0_j}^-$ are the normalized distances associated with the jth line.

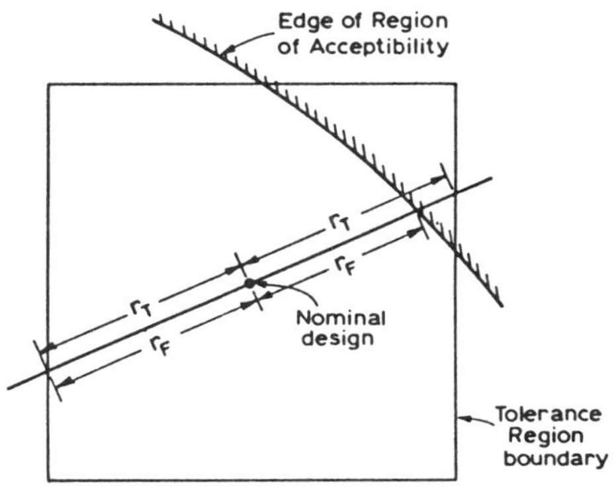

Fig. 1. Illustrating the distances r_F and r_T to the boundaries of the feasible and tolerance regions, respectively, from the nominal design. Two examples are shown.

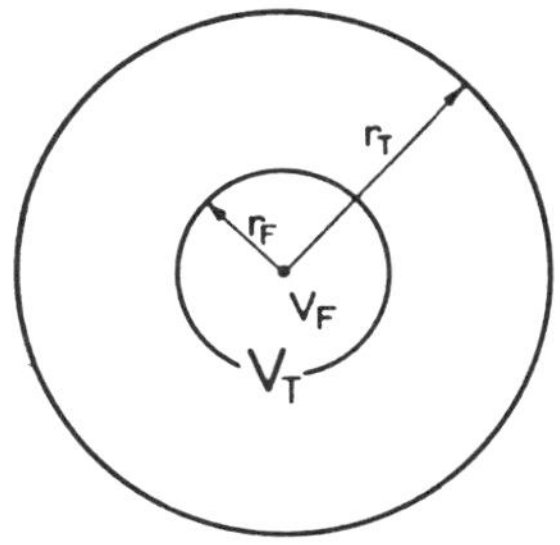

Fig. 2. For circular and concentric volumes V_F and V_T in two dimensions, the ratio of the two volumes is equal to $(r_F/r_T)^2$.

A single contribution (e.g., $(r_{0_j}^+)^m$ for a specific j) to the summation in (9) may be viewed as the result of a one-sample experiment to estimate the ratio of two volumes. For example, for two variable components, and for the special case of circular and concentric volumes V_F and V_T (Fig. 2), the ratio of volumes is $(r_F/r_T)^2$; for m-dimensional spherical volumes the ratio would be $(r_F/r_T)^m$. A succession of such (independent) samples of relative volume is then used, in (9), to provide a better estimate, just as a Monte Carlo analysis employs a series of one-sample experiments of circuit success or failure.

Although the derivation of (9) has not been mathematically rigorous, the algorithm (see Appendix A) for yield estimation based on this expression has been subjected to extensive testing both with circuits ranging in size from 7 to 58 variable components as well as with two- and three-dimensional geometric patterns for which the yield is known. In all cases very satisfactory confirmation was obtained. For the circuits, Table I compares yield estimates with those obtained from a 500-sample Monte Carlo analysis. For the geometric patterns, Table II allows estimated relative volumes to be compared with known exact values.

TABLE I
COMPARISON OF YIELD ESTIMATES OBTAINED BY THE RADIAL EXPLORATION AND MONTE CARLO METHODS FOR VARIOUS DESIGNS OF THREE DIFFERENT CIRCUITS; THE RELATIVE COSTS ARE ALSO INDICATED

Circuit	Yield Estimates (%)				Cost of Radial Exploration Relative to 500 sample Monte Carlo
	Radial Exploration		Monte Carlo		
	Number of lines			95 % Confidence Limits	
Directional filter (Fig 6)	20	54.5	52.4	47.9-56.9	0.07
	50	53.8	"	"	0.14
	100	58.6	"	"	0.25
	200	57.8	"	"	0.47
	500	56.2	"	"	1.13
Active high-pass filter (Fig 9)	50	63.6	61.4	57.0-65.8	0.17
	50	84.5	89.8	87.1-92.5	0.12
	100	62.9	61.4	57.0-65.8	0.34
	100	86.4	89.2	86.4-92.0	0.25
Band-pass filter (Fig 4)	50	44.0	39.8	35.4-44.2	0.18
	50	92.7	91.4	88.8-94.0	0.14
	50	61.1	61.0	56.6-65.4	0.13
	50	99.8	99.6	99.0-100.0	0.11

TABLE II
GEOMETRICAL EXAMPLES FOR ILLUSTRATING THE ACCURACY OF THE RADIAL EXPLORATION METHOD (USING 50 DIRECTIONS) OF VOLUME RATIO COMPUTATION

FIGURE	EXACT RATIO	RADIAL APPROXIMATION
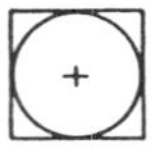 2-dimensional	0.785	0.800
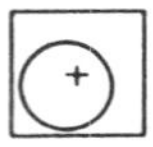 2-dimensional	0.385	0.382
 2-dimensional	0.393	0.400
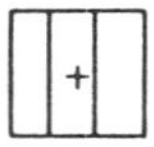 2-dimensional	0.375	0.345
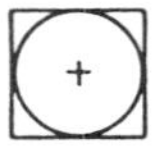 3-dimensional	0.524	0.555

IV. DESIGN CENTERING ALGORITHM

In general, the manufacturing yield is less than the maximum attainable because the tolerance region is not centered within the region of acceptability. However, because the yield estimation algorithm described earlier provides information concerning the *feasible* region, we employ the asymmetry of the feasible region relative to the current tolerance region as the basis of a new design centering procedure.

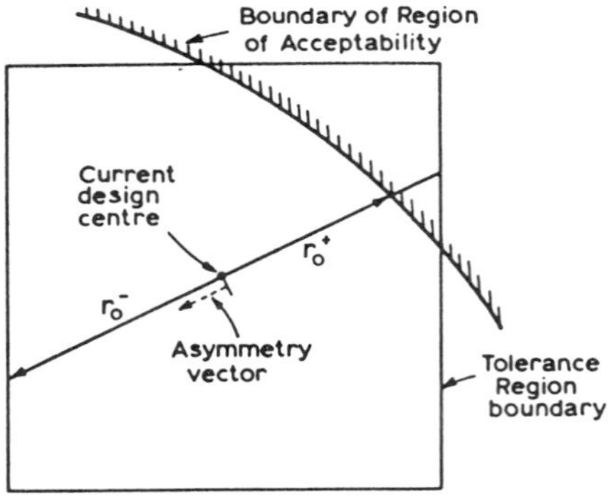

Fig. 3. Definition of the asymmetry vector and asymmetry level for a line in multidimensional component space. Asymmetry level for line shown $= |r_0^+ - r_0^-|$.

A measure of asymmetry—the asymmetry vector—is assigned to each of the lines generated in the yield estimation algorithm, and is the difference between its two feasible lengths r_0^+ and r_0^- (Fig. 3). Thus the asymmetry associated with a line has both magnitude and direction. The *direction* of movement of the design centre (i.e., the nominal component values) is taken to be that of the vector sum of the individual (line) asymmetry vectors. The *magnitude* of movement is chosen somewhat empirically. A choice that has been found suitable in practice is obtained by dividing the magnitude of the vector sum of the asymmetry vectors by the number of "important lines." Initially, a line is regarded as important if its asymmetry level (which is the difference between its two feasible lengths (Fig. 3)) exceeds a given value (e.g., 0.9), but the critical level is gradually reduced as design centering proceeds. The centering scheme terminates either when each of the asymmetry levels has fallen below a minimum value (which, in the examples reported here, is 0.1), or after undergoing a specified number of design centering iterations. The algorithm is described in some detail in Appendix B.

V. EXPERIMENTAL RESULTS

The algorithms for yield estimation and design centering have been implemented in Fortran on a CDC 6500 computer. The program can handle the frequency-domain behavior of linear circuits containing the basic two-terminal passive components as well as independent sources and mutual conductances. Simple linear models of transistors and operational amplifiers have been incorporated.

The program was first applied to the bandpass filter shown in Fig. 4, for which the performance specifications are listed in Table III. The eight parameters so indicated in Fig. 4 were assumed to be subject to variation within specified tolerances. The results associated with two different component tolerances are shown in Fig. 5; in each case, 50 lines at each of five frequencies were used at each

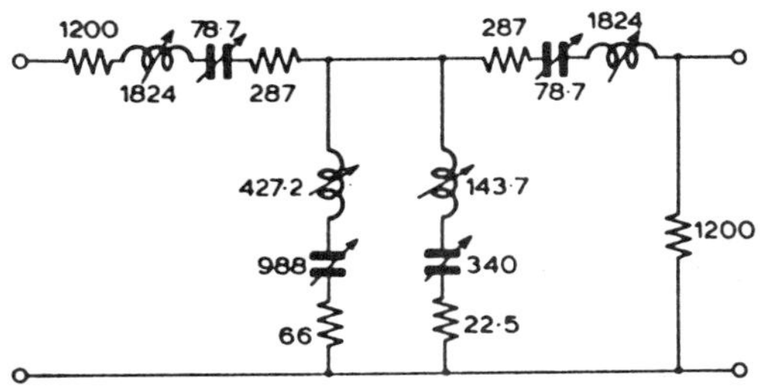

Fig. 4. Bandpass filter. Component values in ohms, picofarads and millihenries.

TABLE III
PERFORMANCE SPECIFICATIONS FOR THE BANDPASS FILTER OF FIG. 4

Frequency range	Loss specification	
	Minimum	Maximum
<240 Hz	L_o+36.5 dB	-
360 Hz to 490 Hz	L_o-0.5 dB	L_o+2.1 dB
>700 Hz	L_o+36.5 dB	-
	L_o = 8.86 dB	

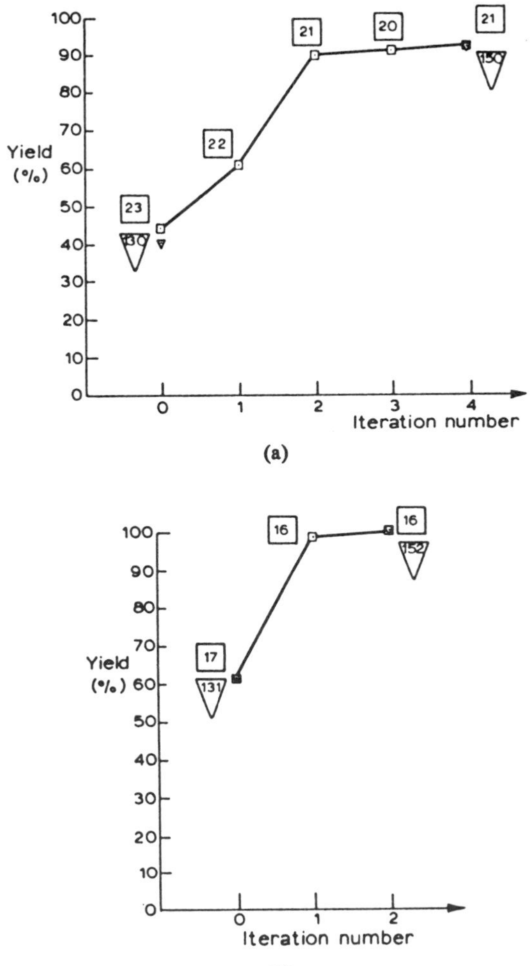

Fig. 5. Yield trajectories and CPU times relating to the bandpass filter of Fig. 4. ▽ 500-sample Monte Carlo estimate □ 50-line radial method estimate. Numbers in squares and triangles are CDC6500 CPU times (in seconds) used for estimating the yield by the radial and Monte Carlo methods, respectively. (a) 5-percent component tolerances (b) 2-percent tolerances.

iteration of the design centering procedure. For the 5 percent component tolerances (Fig. 5(a)) the design is seen to converge from an initial yield estimate of 44 percent to a final estimate of 91 percent. The corresponding 500-sample Monte Carlo estimates are also shown. The CPU-times associated with each iteration are shown in rectangles, and those associated with the 500-sample Monte Carlo estimates in triangles. Fig. 5(b) shows the results obtained for 2-percent component tolerances, but with the same initial design. In general, it is seen that the cost of a complete design centering exercise can be considerably less than that of a 500-sample Monte Carlo analysis of just the initial design.

The program was also used to improve the design of an active filter employing two operational amplifiers, as well as the design of a 58-component filter containing resistances, capacitances, and inductances. As well as to obtain additional experimental evidence in support of the proposed algorithms, the main objective was to test the hypothesis that, since the radial exploration approach is based on a statistical sampling of component space, its computational cost in terms of the number of circuit analyses is largely independent of dimensionality. Fig. 6 shows the form of the filter circuit and Fig. 7 shows the specifications on passband loss and return loss. Fig. 8 shows the results of the application of the design centering procedure for the case when 100 lines were used; for the initial design the result and cost of a 500-sample Monte Carlo analysis is also shown.

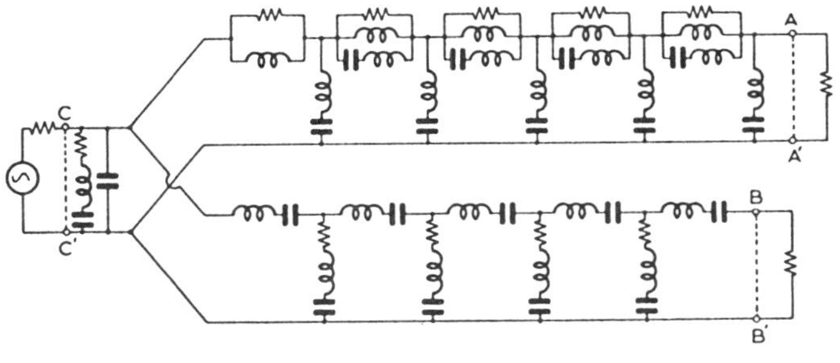

Fig. 6. A directional filter. *CC'-AA'* is the low-band (*LB*) section *CC'-BB'* is the high-band (*HB*) section.

Fig. 9 shows the circuit of the active filter and its gain specifications. All the resistors in this circuit were assigned tolerances of 0.1 percent, and the two capacitors were assigned tolerances of 0.5 percent. Fig. 10 shows the yield trajectories for the cases in which 50 and 100 lines were used. Again, the computational costs are identified.

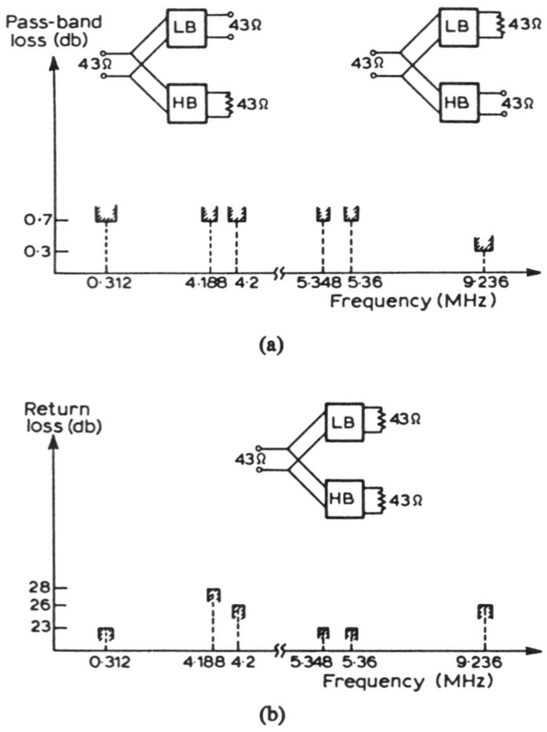

Fig. 7. Passband and return loss specifications for the directional filter of Fig. 6.

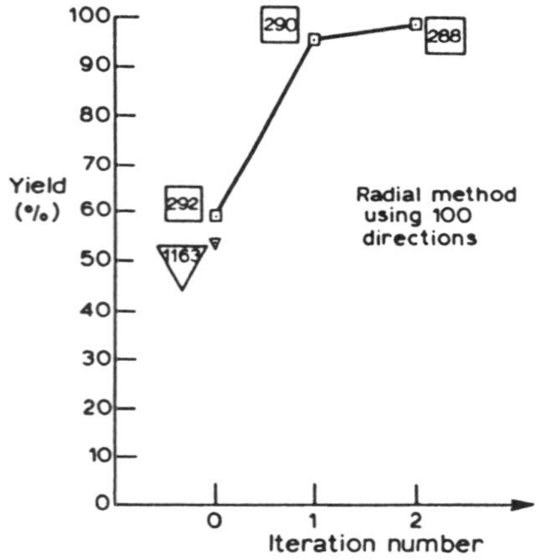

Fig. 8. Yield trajectory and CPU times relating to the directional filter example of Fig. 6.

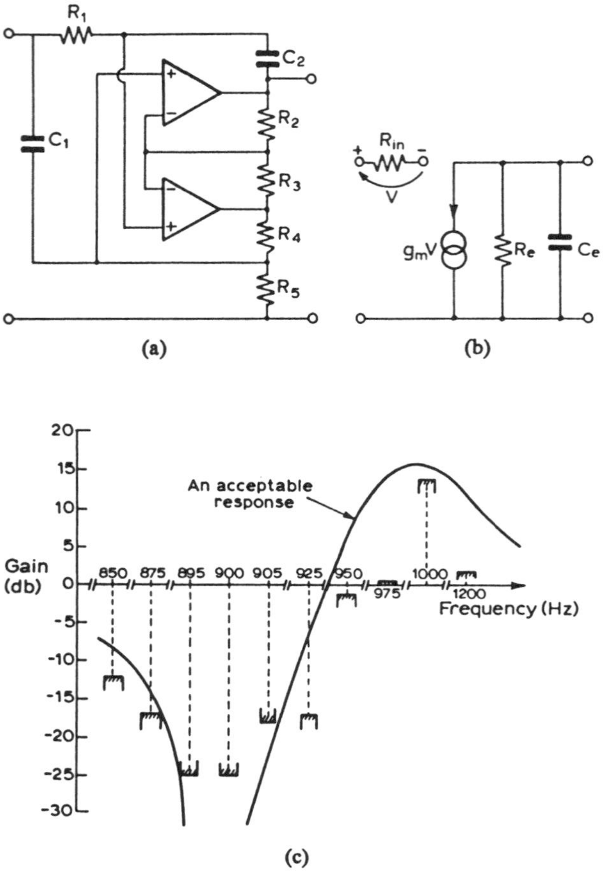

Fig. 9. An active high-pass filter and its gain specifications. (a) Circuit diagram. (b) Equivalent circuit of the operational amplifier. (c) Gain specification.

$C_2 = C_5 = 15$ nF, $R_2 = R_3 = 2$ kΩ	$R_{in} = 1$ MΩ
$R_1 = 1.861$ kΩ, $R_4 = 60.479$ kΩ	$g_m = 10^8$
$R_5 = 318$ kΩ	
$R_1 - R_5$ have 0.1 -percent tolerances	$R_e = 10^3$
C_1, C_2 have 0.5 -percent tolerances.	$C_e = 15.915$ F.

The effect of differing number of radial lines is indicated within Table I for the 58-variable component circuit of Fig. 6. As the number of lines increased from 20 to 500 the yield estimates exhibited little variation and were comparable with the results of a 500-sample Monte Carlo analysis.

Elsewhere [19] we have reported the successful extension of the yield estimation and design centering procedure to the cases in which the component distributions are not uniform, as well as the approach that can be adopted when certain LC pairs in the filter of Fig. 6 are tuned manually to within a given accuracy before assembly.

VI. Conclusions and Extensions

A design-centering scheme based on the radial exploration of multiparameter component space has been described, and its application to circuits of various sizes has shown it to be effective and computationally attractive. In general, the cost of a complete design-centering exercise is considerably less than that of a 500-sample Monte Carlo analysis of the initial design. Although tested exclusively with linear circuits, the scheme is equally applicable to dc and time-domain design, as well as to the design of other physical systems. The results of its application to circuits containing from 7 to 58 variable components tend to support the hypothesis that the number of circuit analyses required is relatively independent of the number of variable parameters in the circuit.

As explained earlier, the basis of the yield estimation algorithm is essentially intuitive. A rigorous mathematical derivation of the algorithm would not only be useful in its own right, but might also provide a relation between the accuracy of the yield estimate and the number of lines.

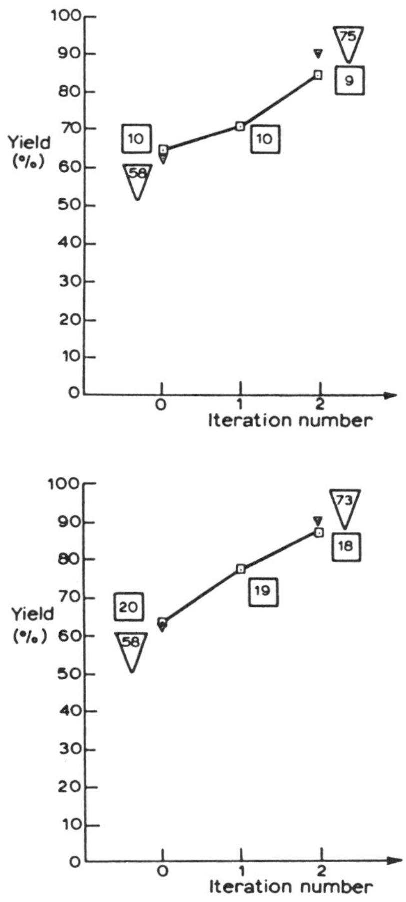

Fig. 10. Yield trajectories and CPU times relating to the active filter example of Fig. 9. (a) Radial method using 50 directions. (b) Radial method using 100 directions. ∇ 500-sample Monte Carlo estimate.

The scheme appears capable of yielding additional information of value to the designer. For example, some form of projection of the points on the feasible region boundary onto the component directions can provide an indicator of yield sensitivity, and may be a suitable means whereby the radial exploration approach can be extended to include tolerance assignment. It is also possible that the information obtained during radial exploration can be used by the designer to examine the tradeoff between specifications and yield; since specifications, normally considered to be fixed, are often open to negotiation.

Appendix A
The Radial Exploration Algorithm for Yield Estimation

1) Compute the impedance matrix at the first frequency. (In practice, LU factors will suffice.)

2) Generate a line in a random direction and find its intersections with the tolerance region.

3) Search in both directions from the nominal design to locate its intersection with the feasible region.

4) Compute the normalized distances r_{0j}^{+} and r_{0j}^{-}.

5) Repeat steps 2 to 4 for each of L lines.

6) Repeat steps 1 to 5 for each of the F frequencies, using the same set of L random directions as for the first frequency, and retain the minimum value of r_0^+ and r_0^- for each line.

7) Use (9) to compute the estimate of yield.

Appendix B
The Radial Exploration Algorithm for Yield Estimation and Design Centering

1) Set the iteration counter I to 1 and apply the yield estimation algorithm described in Appendix A, using I to initiate the random number generator used in that algorithm.

2) Find the SUM of the individual asymmetry vectors (Fig. 3).

3) Set the critical asymmetry level to its initial value $(=0.9)$.

4) Count the number of important lines IL at this asymmetry level.

5) If $IL=0$, reduce the critical asymmetry level by 0.2 and repeat step 4.

6) If IL is still equal to zero when the critical asymmetry level has been reduced below a specified level (e.g., 0.1), no more improvement can occur, so STOP.

7) If $IL \neq 0$, find the new design by updating the existing design by the displacement vector SUM/IL.

8) Increasing I by one and apply the yield estimation algorithm (Appendix A).

9) If the new design is found to be inferior in terms of yield, or is nonfeasible, discard it. Increase I by one to move into a different block of random numbers and reapply the yield estimation algorithm followed by steps 2 to 7.

10) If the new design is better in terms of yield, accept this design and increase I by one. Apply the yield estimation algorithm and repeat steps 2 to 9 until an optimum design results from step 6, or I exceeds the maximum specified number of iterations.

Acknowledgment

The authors wish to acknowledge valuable discussions with R. Soin and R. Tung, and the very helpful advice of R. Brayton. The research reported was supported by the Submarine Systems Division of Standard Telephones and Cables Limited.

References

[1] B. J. Karafin, "The optimum assignment of component tolerances for electrical networks," *BSTJ*, vol. 5, pp. 1225–1242, Apr. 1971.

[2] J. F. Pinel, K. A. Roberts, and K. Singhal, "Tolerance assignment in network design," *IEEE, ISCAS 75*, pp. 317–320.

[3] J. W. Bandler, P. C. Liu, and H. Tromp, "A nonlinear programming approach to optimal design centering, tolerancing and tuning," *IEEE Trans. Circuits Syst.*, vol. CAS-23, pp. 155–165, Mar. 1976.

[4] N. J. Elias "New statistical methods for assigning device tolerances," *IEEE, ISCAS* 75, pp. 329–332.

[5] G. Kjellstrom and L. Taxen, "On the efficient use of stochastic optimization in network design," *IEEE, ISCAS 76*, pp. 714–717.

[6] P. W. Becker and F. Jensen, *Design of Systems and Circuits for Maximum Reliability or Maximum Production Yield.* New York: McGraw-Hill, 1977.

[7] E. M. Butler, E. Cohen, N. J. Elias, J. J. Golembeski and R. G. Olsen, "CAPITOL—Circuit analysis program including tolerancing," *IEEE, ISCAS 77*, pp. 570–574.

[8] B. J. Karafin, "The general component tolerance assignment problem in electrical networks," Ph.D. thesis, Univ. of Pennsylvania, 1974.

[9] S. W. Director and G. D. Hachtel, "The simplicial approximation approach to design centering," *IEEE Trans. Circuit Syst.*, vol. CAS-24, pp. 363–372, July 1977.

[10] J. W. Bandler and H. L. Abdel-Malek, "Optimal centering, tolerancing and yield determination using multidimensional approximations," *IEEE, ISCAS 77*, pp. 219–222.

[11] ——, "Yield estimation for efficient design centering assuming arbitrary statistical distributions," in *Proc. IEEE,* CADEMICS, (Hull, England), pp. 66–71, July 1977.

[12] M. Glesner and K. Haubrichs, "A new efficient statistical tolerance analysis and design procedure for electrical networks," *Proc. IEEE,* CADEMICS, (Hull, England), pp. 59–65, July 1977.

[13] K. H. Leung and R. Spence, "Multiparameter large change sensitivity analysis and systematic exploration," *IEEE Trans. Circuits Syst.*, vol. CAS-22, pp. 796–804, Oct. 1975.

[14] T. R. Scott and T. P. Walker, "Regionalization: A method of generating joint density estimates," *IEEE Trans. Circuits Syst.*, vol. pp. 229–234, Apr. 1976.

[15] T. B. M. Neill, "Variance reduction in the Monte Carlo analysis of electrical networks," *Inst. Elec. Eng. Conf. CAD*, London, pp. 219–224, Apr. 1972.

[16] J. F. Pinel and K. Signhal, "Efficient Monte Carlo computation of circuit yield using importance sampling," *IEEE, ISCAS'77*, pp. 575–578.

[17] K. S. Tahim and R. Spence, "Statistical circuit analysis—A practical algorithm for linear circuits," *IEEE, ISCAS'78*, pp. 180–184.

[18] T. Neumann and D. Agnew, "Tracking sensitivity: a practical algorithm," *Electron. Lett.*, vol. 13, No. 12, pp. 371–372, June 1977.

[19] K. S. Tahim, "Statistical analysis and design of electrical networks," Ph.D. thesis, University of London, 1978.

EFFICIENT MONTE CARLO YIELD PREDICTION USING CONTROL VARIATES

P.J. Rankin and R.S. Soin

Philips Research Laboratories, Redhill, Surrey, England.

ABSTRACT

The Monte Carlo method exhibits generality and insensitivity to the number of stochastic variables, but is expensive for accurate yield prediction of large circuits. We describe a novel application of a "variance reduction" technique, viz. the control variate method, to the improvement of the efficiency of circuit yield estimation. An approximating functional model of the circuit is first analysed in parallel with the actual circuit to estimate accurately their difference in yields. This yield difference is then added to a precise yield estimate of the approximating model to obtain a more accurate estimate of the yield of the actual circuit. For a given accuracy, reductions in the number of full circuit analyses of up to eight times are demonstrated. Methods for constructing the approximate model are suggested, including the use of sensitivities and the method of moments to automate the control variate technique.

1. INTRODUCTION

The Monte Carlo (MC) method is the most reliable technique for the tolerance analysis of electrical circuits. The method is applicable to any type of circuit without requiring simplifying assumptions of the forms of the probability distributions of either component parameter values or circuit responses. Various alternative deterministic methods for yield estimation [1,2] either suffer from a prohibitive proliferation of computational effort with increasing number of toleranced components [1], or approximate the circuit response or its probability distribution [2].

The crude MC procedure is as follows. First a number of sets N of representative values $\underline{P} = p_1, p_2 \ldots p_k$ for the K statistically varying (toleranced) component parameters of the circuit are pseudo-randomly selected according to their joint probability density function (pdf), denoted by $\emptyset(\underline{P})$, For each set of values the M performance functions of the circuit represented by $f_j(\underline{P})$, $j = 1 \ldots M$ can then be calculated, and the circuit tested against Q possible performance requirements of the form:

$$f_j(\underline{P}) \geq L_j \quad \text{or} \quad f_j(\underline{P}) \leq U_j \tag{1}$$

Defining a testing function reflecting a circuit's acceptability or otherwise:

$$g(\underline{P}) \triangleq \begin{cases} 1 \text{ if } \underline{P} \text{ is such that (1) is satisfied} \\ 0 \text{ otherwise} \end{cases} \tag{2}$$

then yield is the expectation of $g(\underline{P})$ with respect to the pdf, $\emptyset(\underline{P})$, i.e.

$$Y = \int_{\underline{P} \in R^K} \cdots \int g(\underline{P}) \emptyset(\underline{P}) d\underline{P} \tag{3}$$

where R^K is the K dimensional space of variation of the component parameters - the "tolerance region". Finally, then the averaged value of g for the N sample circuits is computed as an estimater of (3):

$$\hat{Y} = \frac{1}{N} \sum_{i=1}^{N} g(\underline{P}_i) \tag{4}$$

where $\hat{Y}$ denotes an estimate of Y. Since g may only have the value 1(pass) or 0 (fail), each circuit analysis and test constitutes a Bernoulli trial [3], so the sampling distribution of $\hat{Y}$ is binomial with variance $\sigma_y^2 = Y(1-Y)/N$ which may be approximated by

$$\hat{\sigma}_{\hat{y}}^2 = \frac{\hat{Y}(1 - \hat{Y})}{N}$$

When N and $\hat{Y}$ satisfy the condition $N\hat{Y}(1 - \hat{Y}) \geq 6$, the sampling distribution of $\hat{Y}$ may be taken to be Gaussian giving a 95% confidence interval for the true yield as $\hat{Y} \pm 2\hat{\sigma}_{\hat{Y}}$ (see [4]).

From the above it can be seen that for a yield of 60%, 100 analyses would provide a confidence interval of ±10% with a 95% degree of confidence. Because of the $1/\sqrt{N}$ dependence of the size of the confidence interval, the sample size has to be quadrupled to double the accuracy. Clearly the MC procedure is expensive for large circuits where each analysis may require long computation. Now it is precisely with reference to large circuits that MC analysis finds favour since for a given accuracy the required sample size N is independent of the number of statistically varying parameters.

Several schemes (called variance reduction techniques) for reducing the required sample size for a particular confidence interval have been reported in the general MC literature [5,6,7]. However these techniques are problem dependent

Reprinted from *IEEE Int. Symp. Circuits Syst.*, vol. 1, pp. 143-148, April 1981.

and few have been demonstrated for the problem of yield estimation of electrical circuits. This paper outlines an attempt to apply one method - the control variate (CV) method to the circuit yield problem. Only a very limited application in this context has previously been suggested [8]. After presenting the formal basis for the use of a control variate and the gains in efficiency it offers, reductions in the number of full circuit analyses of up to eight times are demonstrated on a practical circuit example. Concluding, the major advantages and limitations of the CV method are summarized indicating further possible enhancements.

2. THE CONTROL VARIATE TECHNIQUE

The Method

Consider a second system, the control variate or shadow model, for which an analogous set of responses $f'(\underline{P})$ exist for a given $\underline{P}$. We require that the shadow model approximate the performance of the actual circuit for a given $\underline{P}$, but it be computationally much cheaper to analyse.

The shadow model responses are functions of $\underline{P}$ which is subject to the same pdf, i.e. $\emptyset(\underline{P})$ as for the actual circuit. Analogously to (1), we denote shadow model responses and specifications as $f'_j(\underline{P})$, j = 1...M, $f'_j(\underline{P}) \geq L'_j$, $f_j(\underline{P}) \leq U'_j$, etc. Note that specifications L'_j, U'_j may be chosen independently of those for the actual circuit. Analogously to (2) we define a new testing function $g'(\underline{P})$; and write the yield of the shadow model as:

$$Y_c = \int \cdots \int_{\underline{P} \in R^K} g'(\underline{P}) \, \emptyset(\underline{P}) \, d\underline{P} \qquad (7)$$

Denoting the yield of the actual circuit as Y_m, expression (3) becomes:

$$Y = Y_m = \int \cdots \int_{R^K} \left\{ g(\underline{P}) - g'(\underline{P}) \right\} \emptyset(\underline{P}) d\underline{P} + \int \cdots \int_{R^K} g'(\underline{P}) \, \emptyset(\underline{P}) \, d\underline{P} \qquad (8)$$

Denoting the first integral in (8) as ΔY, we get $Y_m = \Delta Y + Y_c$. Clearly ΔY is the difference in yield between the circuit and the shadow model. The CV technique entails separate evaluation of Y_c and ΔY. We chose a shadow model whose yield may be obtained analytically or which is inexpensive to analyse so that Y_c can be estimated from an auxiliary large sample MC run. Thence the main computational cost is in estimating ΔY from a small number of samples n as

$$\Delta\hat{Y} = \frac{1}{n} \sum_{i=1}^{n} g(\underline{P}_i) - g'(\underline{P}_i) \qquad (9)$$

The crucial point about this method is that the variance of the estimate of Y_m is largely due to the variance of the estimate of ΔY, which is diminished because the control variable $g'(\underline{P})$ behaves like $g(\underline{P})$ and hence absorbs some of its variation. The estimation (9), follows the common random numbers or correlated sampling technique [4,9].

Yields $\hat{y}_m = {}^1/n \sum_{i=1}^{n} g(\underline{P}_i)$ and $\hat{y}_c = {}^1/n \sum_{i=1}^{n} g'(\underline{P}_i)$ are individual estimates from the n sample analyses. The variance of $\Delta\hat{Y}$ may be estimated as:

$$\text{var}(\Delta\hat{Y}) = \text{var}(\hat{y}_m) + \text{var}(\hat{y}_c) - 2\,\text{covar}(\hat{y}_m, \hat{y}_c) \qquad (10)$$

where $\text{var}(\hat{y}_m) = \hat{y}_m(1-\hat{y}_m)/n$ and $\text{var}(\hat{y}_c) = \hat{y}_c(1-\hat{y}_c)/n$; $\text{covar}(\hat{y}_m, \hat{y}_c) = \rho\sqrt{\text{var}(\hat{y}_m)\text{var}(\hat{y}_c)}$. ρ is the correlation coefficient between estimates $\hat{y}_m, \hat{y}_c$.

Figure 1 is a schematic of the experiment to estimate ΔY (see (9)). Sample sets of parameter values $\underline{P}_j$, j = 1..n are generated and both systems analysed for each set, to evaluate $g(\underline{P}_j)$ and $g'(\underline{P}_j)$. If n_{11} denotes the number of times both $g(\underline{P}_j)$ and $g'(\underline{P}_j)$ are 1, n_{00} when both are zero and so on for n_{10}, n_{01}; then $\Delta\hat{Y} = ((n_{11} + n_{10})/n) - ((n_{11} + n_{01})/n)$. Also $(n_{11} + n_{10} + n_{00} + n_{01}) = n$. An estimate of ρ is:

$$\rho = \frac{n_{11}n_{00} - n_{01}n_{10}}{\sqrt{(n_{11}+n_{10})(n_{11}+n_{01})(n_{00}+n_{10})(n_{00}+n_{01})}} \qquad (11)$$

Efficiency

Like other variance reduction techniques, the CV method achieves increased accuracy by using knowledge of the system being simulated to replace part of the experimentation by theoretical analysis or simpler experimentation. Here knowledge of the circuit behaviour is exploited to construct the approximating model. The quality of this approximation is expressed in the correlation coefficient ρ.

We consider now the efficiency of the CV technique over the primitive MC technique. Primitive MC analysis comprises n circuit analyses with attendant variance $\text{var}(\hat{y}_m)$. Assuming the shadow model's true yield is available, the variance of the CV estimate of yield of the actual circuit is given by the variance of the estimate of $\Delta\hat{Y}$. Since primitive MC analysis variance is inversely proportional to sample size, neglecting the cost of analysing the shadow model, the efficiency η of the CV estimation over the primitive method is

$$\eta = \frac{\text{var}(\hat{y}_m)}{\text{var}(\Delta\hat{Y})} \qquad (12)$$

Since the two models are expected to have similar yields, i.e. $\hat{y}_m \approx \hat{y}_c$, then $\text{var}(\hat{y}_m) \approx \text{var}(\hat{y}_c)$. Therefore (12) becomes

$$\eta = \frac{1}{2(1-\rho)} \qquad (13)$$

Theoretically ρ can lie in the range $-1 \leq \rho \leq 1$. Although similarities in the two models ensure $\rho>0$, gains $(\eta>1)$ are only realized when $\rho>0.5$. When the yield Y_c of the shadow model cannot be found analytically but must itself be estimated from a large number of sample analyses N in an auxiliary MC analysis, the efficiency in equation (13) is moderated to

$$\eta = \frac{1}{2(1-\rho)+r}, \text{ where } r = n/N \qquad (14)$$

Expression (14) predicts the variance reduction achieved rather than overall cost savings. To reduce r and increase η, a large value of N may be preferred. However, if the cost of analysis of the shadow is not negligible then the cost of a longer auxiliary MC run may not be justified vis-à-vis the costs of achieving the same overall variance by using a higher number of full circuit analyses n. If t_m and t_c are the unit computational costs of analysing the actual circuit and the shadow model, then overall reductions in costs will still be achieved if

$$n.t_m.\eta > n.t_m + (N + n)t_c$$

To maximize computational saving, for a given ρ and given choice of n, the optimum choice of N may be shown to be:

$$N_{opt} = n \sqrt{\frac{t_m}{2t_c(1-\rho)}} \qquad (15)$$

Figure 2 plots expression (14), for r = 0 together with a family of curves of η versus ρ for different parameters t_m/t_c. At each point in the family N_{opt} has been found from (15) and substituted in (14). Therefore the curves show practical obtainable efficiencies for given ρ and tm/t_c, when Y_c cannot be calculated analytically. Clearly a compromise must be made in the choice of the shadow model - a more complex model will tend to give a higher ρ but may cost more to analyse.

One area where the efficiency of the CV technique may be enhanced is in the choice of shadow model specifications L'_j, U'_j. As stated earlier these may be chosen independently of those of the actual circuit L_j,U_j. To increase efficiency L'_j,U'_j should be chosen to maximize ρ. In practice only an estimate of ρ (expression (11)) determined from the n samples is available. For example, if the insertion loss of the shadow model was 1dB lower than that of the circuit, the optimum L'_j,U'_j tend to be 1dB lower than L_j,U_j. Adjustment of L'_j and U_j may be carried out as a post-processing operation after the n circuit analyses of both models are complete.

General Methods for Constructing Control Variates

The CV technique requires a model approximating the performance of the circuit in question. This model may be a circuit, a mathematical model or a combination. Its yield should be accurately obtainable either because it has a mathematical structure amenable to analytic calculation, or because it is simple enough to perform a large sample auxiliary MC run economically. Some general methods of constructing such models are now discussed.

One approach is to use the designer's insight into his circuit's behaviour to build a functional equivalent while the circuit itself is being designed. Sometimes the desired behaviour may be summarized in a simple circuit model whose elements can be expressed as functions of the statistical variables of the actual circuit.

A second approach is via such technique timing simulation, macromodelling [10] or model simplification [11], breaking down the circuit by semi-automatic methods to leave only the effects which dominate its performance variability.

A third approach offering promise is to express each response as a Taylor series in terms of the deviation of the statistical variables from their nominal values [12]. The yield of the Taylor series model could be found generally from an auxiliary MC run. However if the Taylor series were truncated after the first or second derivatives, the variances of the performances could be obtained via the method of propagation of errors [4] (method of moments). The CV method would then combine the speed of the method of moments with the reliability of the MC method, the latter being used only to estimate the departure from linearity of the actual circuit.

As a final enhancement of the CV model, regression equations may be added at the cost of a small number of analyses of both systems. Deviations between the performances of the two systems could be regressed on a subset of the statistical variables, selected as being dominant in producing discrepancies between the two systems. Each evaluation of the full CV model would thus require a calculation of the performance of the basic model and the addition of the prediction of the appropriate regression equation.

3. CIRCUIT EXAMPLE AND RESULTS

Circuit

The CV approach was tested on the gyrator circuit of figure 3. The circuit is of a basic type used at high frequencies for fully integrated active filtering [13]. A resonance near 5MHz is produced by terminating both gyrator ports with integrated capacitors.

To analyse the complete circuit, simplified component models were employed with an idealized statistical description of the integrated circuit process. Electrical models for each component were derived by geometrical scaling rules from mask dimensions and a set of 9 basic process variables such as sheet resistivities, specific capacitances, base transit time, lateral mask overetch, etc. Different components then tracked electrically as a process variable was altered.

For MC work a multinormal pdf was assigned to the 9 process variables, 5 of which were mutually correlated. Each new set of process values then required a dc analysis, followed by ac analysis at 4 frequencies on the full 62 element - 28 node circuit (after expansion of 16 component models) to test the ac voltage transfers G(f), in dB, against 5 arbitrary specifications determining the yield.

Shadow Model

For its application the gyrator must represent as closely as possible a pure inductance. The equivalent LCR circuit in figure 4 was therefore an obvious candidate for a shadow model. Designer's understanding of the gyrator circuit allowed the values of the 5 elements in the shadow to be expressed in terms of some of the electrical elements in the gyrator, and so as functions of the same 9 process variables. For example, R_p included the effects of various Q-controlling mechanisms such as (a) phase shifts from the transistors and from parasitic capacitances in the 590Ω resistors, and (b) damping due to the gyrator port conductances, 67Ω resistor and series resistance of one terminating capacitor. Finally the nominal Q and centre frequency of the LCR model were trimmed by adding a constant capacitance and conductance to C_T and R_p respectively.

Having described the 5 dependencies using an in-house circuit analysis package, PHILPAC [14], ac analysis at 4 frequencies sufficed to evaluate the performance of the 3-node shadow model. Figure 5 illustrates the tracking between the two systems when one process variable was disturbed to its 2σ values.

Experiments

A base of 20000 analyses at 4 frequencies of both the gyrator and its shadow was formed for statistical experimentation. Common random numbers in the two long MC runs were forced by using the same random number seed. The empirical variances of the primitive MC and CV yield estimators could then be compared by subdividing the base to give 200 independent yield estimations (each requiring 100 gyrator analyses) by the two methods.

In the experiments below, N = 20000 analyses of the shadow model were used to estimate Y_c. Therefore with $r = n/N = 0.005$, the variance ratios η were only marginally affected by the variance in finding Y_c. In practice the choice of N would depend on the value of $\hat{\rho}$ achieved in the first n analyses and on t_m/t_c, which depends strongly on the way the analysis package handles expressions, etc. (here $t_m/t_c \approx 14$).

In Table 1 the same 5 specifications were applied to both systems: (i) G(5.428) ⩽ 6; (ii)G(5.428) ⩾ -11; (iii) G(5.676 - G(5.176) ⩽ 10; (iv) G(5.676) - G(5.176) ⩾ -10; (v) G(6.167) ⩽ 0. As well as the combined outcome with all specifications active, partial yields, correlations and efficiencies are also quoted as valid tests of the CV method. The table gives the averages of 200 CV yield estimates $\hat{Y}_m$ and of 200 estimates $\hat{\rho}$. η values are the ratio of the measured variances of the primitive MC yield estimations (100 gyrator analyses only) and CV yield estimations (100 analyses of gyrator and shadow, plus 20000 shadow analyses). Average values of the primitive MC and CV estimates of yield agreed, confirming that the CV estimator is unbiassed [3]. Measured efficiencies agreed closely with theoretical predictions and verified that CV efficiencies depend only on ρ rather than the level of the yield.

Maintaining gyrator specifications as before, Table 2 shows the improvement after maximizing $\hat{\rho}$ by altering the LCR specifications on each lot of n=100 analyses. Figure 6 shows the variability of the overall yield estimates. A $\hat{\rho}$ optimization algorithm was applied individually for the five specifications, each optimal LCR specification then being used both for finding $\Delta\hat{Y}$ over n=100 samples, and $\hat{Y}_c$ over 20000 samples. When specification optimization is performed over a larger number of samples n, the $\hat{\rho}$ achieved is less subject to sampling errors or bias, giving higher efficiencies and a more accurate prediction of the final confidence interval on yield.

4. CONCLUSIONS AND EXTENSIONS TO THE METHOD

In common with the basic MC technique, the CV method can deal with complex component distributions and large numbers of stochastic parameters without requiring simplifying assumptions about the performance distributions. Since our approximating model of the circuit's behaviour is required to function as a control in the statistical simulation, the technique may be particularly appropriate for large IC's performing relatively simple tasks. The CV model may also be used during tolerance synthesis. Only small modifications to existing circuit analysis packages providing MC facilities are required for incorporation of this method.

The main limitations of the method lie in the effort required to construct a model which mimics the circuit adequately yet is simple enough to be cheaply analysed. Additionally, for problems with a small number of stochastic variables and well behaved performance distributions, deterministic methods are likely to offer greater potential.

Current work is directed at automation of the technique using first order sensitivities to construct the approximating model with the addition of regression equations. Experiments are also planned to assess the CV technique when the aim of the simulation is to predict circuit performance variability rather than yield. Finally, the estimation of yield derivatives with respect to statistical design parameters which can be performed in a primitive MC analysis [15,16] is also amenable to the control variate approach.

5. ACKNOWLEDGEMENTS

Searching discussions with K.W. Moulding at Philips Research Laboratories and Dr R. Spence and co-workers at Imperial College, London, are gratefully acknowledged.

6. REFERENCES

[1] S.W. Director, G.D. Hachtel and L.M. Vidigal, "Computationally Efficient Yield Estimation Procedures based upon Simplicial Approximation", I.E.E.E. Trans. Cir. and Sys. CAS-25, 3, March, 1978.

[2] J.W. Bandler and H.L.A. Malek, "Optimal Centering, Tolerancing and Yield Determination via Updated Approximations and Cuts", I.E.E.E. Trans. Cir. and Sys., CAS-25, 10, October,1978.

[3] H.J. Larson, "Introduction to Probability and Statistical Inference", John Wiley & Sons, 1969.

[4] P.W. Becker and F. Jensen, "Design of Systems and Circuits for Maximum Reliability or Maximum Production Yield", McGraw Hill,1977.

[5] E.J. McGrath et al, "Techniques for Efficient Monte Carlo Simulation", Science Applications, Incorporation, March 1973, NTIS No. AD-762-721 (U.S. Dept. of Commerce, Springfield, USA).

[6] J.M. Hammersley and D.C. Handscomb, "Monte Carlo Methods", Methuen & Co. Ltd., London, 1964 (Chapter 5).

[7] J.P.C. Kleijnen, "Statistical Techniques in Simulation", Marcel Dekker, New York, 1974.

[8] T.B.M. Neill, "Variance Reduction in Monte Carlo Analysis of Electrical Networks", I.E.E. Conf. on CAD, April 1972, Soton.,pp 219-224.

[9] P.W. Becker, "Finding the Better of Two Similar Designs by Monte Carlo Techniques", I.E.E.E. Trans. on Reliability R-23, 4, October, 1974.

[10] H.Y. Hsiek, N.B. Rabbat, A.E. Ruehli, "Macromodelling and Macro-simulation Techniques", 1978, I.E.E.E. Int. Symp. Circuits and Systems, New York, pp 336-339.

[11] R. Spence and T. Neumann, "On Model Simplification", Proc. 1978, I.E.E.E. Int. Symp. Circuits and Systems, New York, pp 350-353.

[12] J.F. Pinel and K.A. Roberts, "Tolerance Assignment in Linear Networks using Nonlinear Programming", I.E.E.E. Trans. Circuit Theory CT-19, pp 475-479, Sept. 1972.

[13] K.W. Moulding, J.R. Quartly, P.J. Rankin, R.S. Thompson, G.A. Wilson, "Gyrator Video Filter I.C. with Automatic Tuning", I.E.E.E. Journ. Solid State Ccts., Dec. 1980, SC-15, No. 6, pp 963-968.

[14] C. Niessen, "Computer Aided Design of LSI Circuits", Philips Techn. Rev. 37, pp 278-290, 1977, No. 11/12.

[15] B.V. Batalov, Y.N. Belyokov, F.A. Kurmaev, "Some Methods for Statistical Optimization of Integrated Microcircuits with Statistical Relations among the Parameters of the Components", Soviet Microelectronics (USA), vol. 7, No.4, pp 228-238, 1978.

[16] M.A. Styblinski, "Estimation of Yield and its Derivatives by Monte Carlo Sampling and Numerical Integration in Orthogonal Subspaces", European Conference on Cct. Theory and Design, Sept. 1980, Warsaw, Poland.

Spec. No.	Aver. $\hat{Y}_m$ %	Aver. $\hat{\rho}$	Yield Var. Ratio η
i	99.49	0.6178	0.23
ii	91.15	0.8492	2.89
iii	85.39	0.9121	6.49
iv	88.65	0.8587	4.05
v	85.95	0.9282	6.19
All	51.29	0.7942	2.55

TABLE 1

Spec. No.	Aver. $\hat{Y}_m$ %	Aver. $\hat{\rho}$	Yield Var. Ratio η
i	99.26	0.9279	0.5
ii	91.20	0.9706	5.56
iii	85.31	0.9571	5.71
iv	88.64	0.9542	6.31
v	86.02	0.9729	8.36
All	51.59	0.9336	3.74

TABLE 2

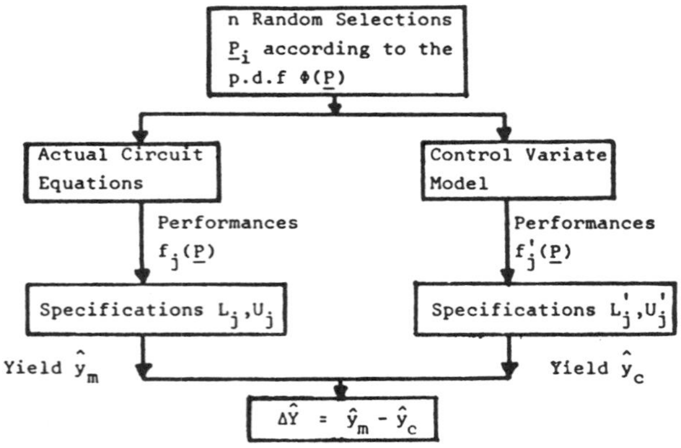

Figure 1: Estimation of ΔY with common random numbers

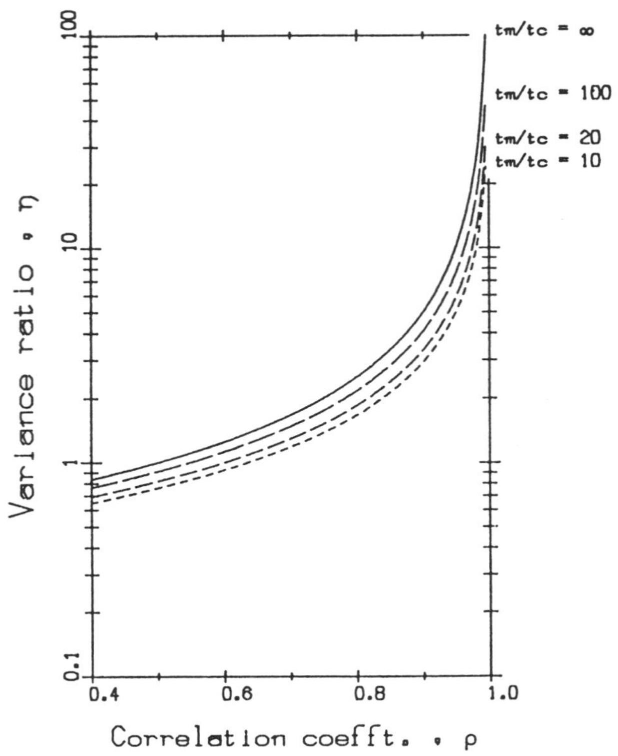

Figure 2: Efficiency vs. Correlation, N optimized

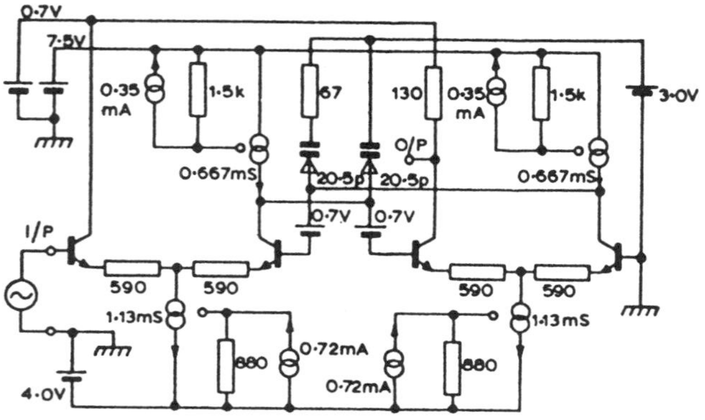

Figure 3: Gyrator resonator circuit with nominal values.

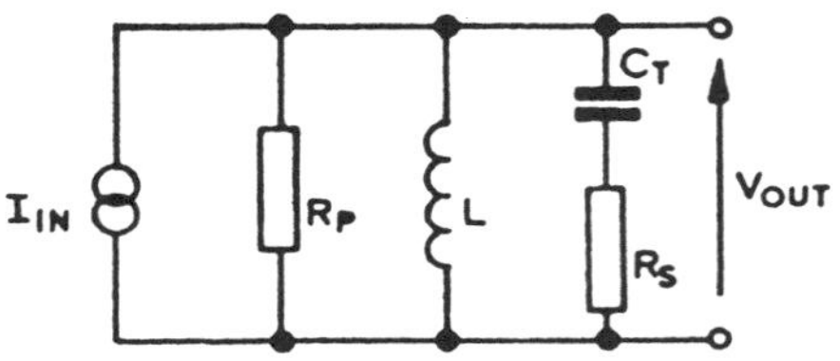

Figure 4: LCR Equivalent Circuit

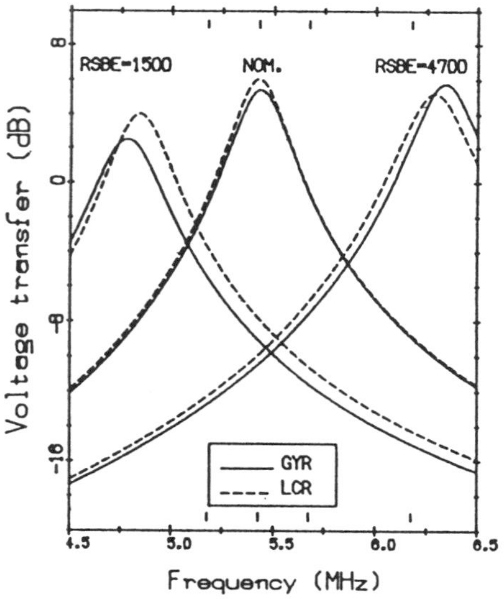

Figure 5: Responses of Gyrator and LCR Circuits

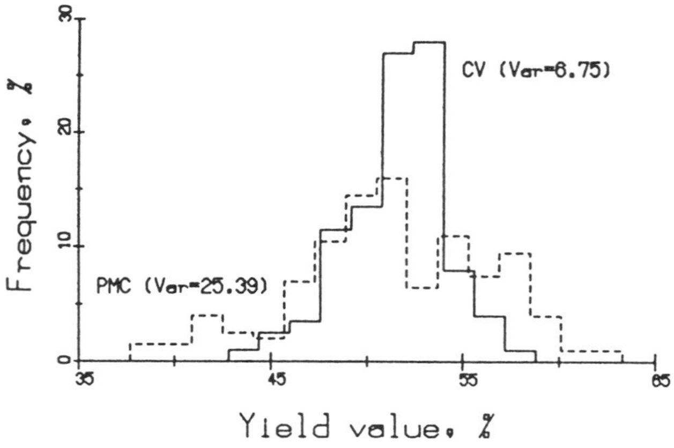

Figure 6: Histograms of Yield Estimates

Statistical Design Centering and Tolerancing Using Parametric Sampling

KISHORE SINGHAL, MEMBER, IEEE, AND J. F. PINEL, SENIOR MEMBER, IEEE

Abstract—A new statistical circuit design centering and tolerancing methodology based on a synthesis of concepts from network analysis, recent optimization methods, sampling theory, and statistical estimation and hypothesis testing is presented. The method permits incorporation of such realistic manufacturing constraints as tuning, correlation, and end-of-life performance specifications. Changes in design specifications and component cost models can be handled with minimal additional computational requirements.

A database containing the results of a few hundred network analyses is first constructed. As the nominal values and tolerances are changed by the optimizer, each new yield and its gradient are evaluated by a new method called *Parametric sampling* without resorting to additional network analyses. Thus the most costly phase of statistical design, statistical simulation, may be carried out only once, which leads to considerable computational efficiency. Equivalent or superior designs for intermediate size networks are obtained with less computational effort than previously published methods. For example, a worst-case design for an eleventh-order Chebychev filter gives a filter cost of 44 units, a centered worst-case design reduces the cost to 18 units and statistical design using Parametric sampling further reduces the cost to 5 units (800 analyses, 75 CPU seconds on an IBM 370/158).

I. Introduction

STATISTICAL design centering and tolerancing of electronic circuits is necessary to produce cost-effective product designs. The last decade has seen the gradual emergence of a more effective design methodology based on optimization techniques to supplement the traditional intuitive statistical design approaches. The problem, simply stated, is to transform a designer's "breadboard" design into a manufacturable product that minimizes manufacturing cost by optimizing both nominal component values and their tolerances. A knowledge of component costs and tolerance distributions (manufacturing, temperature, drift, $\cdots$) is assumed.

Effective use of optimization for statistical design requires efficient techniques to estimate circuit yield and its gradient. It is well known that except for trivial circuits Monte Carlo techniques are required to compute yield. Hence the two fundamental obstacles of statistical design optimization are:

(1) Excessive CPU time required to continually recompute accurate yield as part of the iterative optimization.

(2) Inability to use powerful optimizers because accurate yield gradients are difficult and very costly to compute.

Many authors have provided meaningful insight into the statistical design problem. Analysis of their teachings reveals that direct assault on the fundamental obstacles has been avoided. The result is a number of geometric techniques that assume convexity and suffer from dimensionality problems for circuits having more than a very few variables [1], [2], and statistical techniques that simply consume excessive CPU [3]. Others have limited applicability due to simplifying assumptions like linearity. Only one general purpose algorithm has been successfully demonstrated on large examples [4], [7].

The purpose of this paper is to present an efficient technique for yield gradient computation and to describe a new cost-effective statistical design centering and tolerancing algorithm (incorporating yield gradients) that does not require Monte Carlo yield computation for every iteration. This then is a frontal assault on the above fundamental obstacles. The number of design variables is primarily limited by the selected constrained nonlinear optimization algorithm [12]. Examples demonstrate comparable circuit

Manuscript received February 7, 1980; revised January 5, 1981. This work was supported in part under contract with Bell-Northern Research, and by NSERC, Canada.

K. Singhal is with the University of Waterloo, Waterloo, Ont., Canada.
J. F. Pinel is with Bell-Northern Research, Ottawa, Ont., Canada.

Reprinted from *IEEE Trans. Circuits Syst.,* vol. CAS-28, no. 7, pp. 692–702, July 1981.

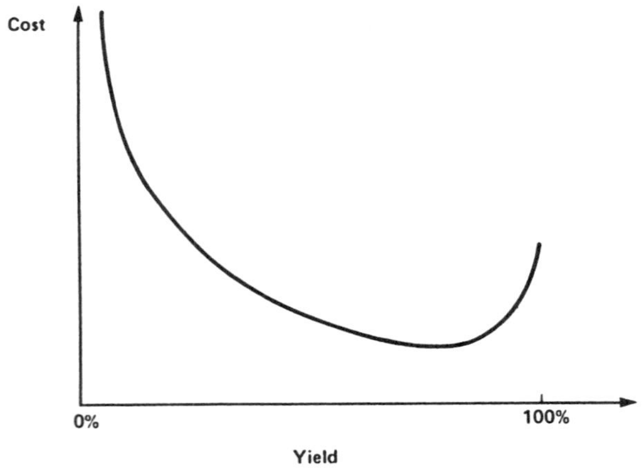

Fig. 1. A typical circuit cost-versus-yield curve.

designs to [4] and superior designs to [3] with less computational effort. It does not appear practical to run examples of such complexity (22 design variables) on [1] or [2].

The entire approach is derived from importance sampling concepts [5], a well-known statistical technique for variance reduction. It is important to note that the mission here is to adapt these concepts to the statistical circuit design environment and not to achieve variance reduction in yield estimates *per se*. This leads to a definition of Parametric sampling, a concept which gives the desired penetration of fundamental obstacles.

Initially, one possible objective function is defined to illustrate the dominant role of yield in cost minimization.

II. Product Cost Minimization

Numerous product cost models based on the strategic concepts of throw-away or repair are possible [6], [7]. A throw-away model is assumed in the following, though repair models are equally applicable. The objective then is to minimize the cost C_s of units satisfying all requirements, where

$$\left.\begin{aligned} C_s &= C_u \frac{\text{Total number of manufactured units}}{\text{Number of passing units}} \\ \text{i.e.,} \\ C_s &= \frac{C_u}{Y} \end{aligned}\right\} \tag{1}$$

where C_u is per unit manufacturing cost (component cost and fixed costs such as printed circuit board, assembly and test labor, etc.) and Y is yield. The statistical design problem is clearly to find a set of nominal component values and their tolerances that represent the minimum on the cost-versus-yield curve shown in Fig. 1.

For discrete circuits, both C_u and Y depend on component values and tolerances.

When circuit manufacture involves a number of process steps such as for hybrid and integrated circuits, each step has a cost C_i and yield Y_i. Then C_s has the form

$$C_s = \left(\cdots \left(\left(\frac{C_1}{Y_1} + C_2 \right) \frac{1}{Y_2} + C_3 \right) \frac{1}{Y_3} + \cdots + C_n \right) \frac{1}{Y_n} = \frac{C_u}{Y_n} \tag{2}$$

where

$$Y_n = \frac{\text{Number of units passing final test}}{\text{Number of units entering final test}} = \text{final test yield}.$$

Except when component trimming is used, process tolerances and geometries control component tolerances. Clearly, statistical design techniques mainly influence final test yield where circuits whose components are within process tolerances fail functional testing. Thus the objective is to minimize cost C_s by adjusting nominal values to maximize Y_n. This is sometimes referred to as the problem of yield maximization through design centering (see, for example, [11]).

III. Basic Yield Estimation

To introduce notation and to provide a quick review, consider the basic Monte Carlo yield estimation technique. Let $f(\boldsymbol{x})$ be the joint probability density function of the components.[1] Then the yield

$$Y = \int_{-\infty}^{\infty} z(\boldsymbol{x}) f(\boldsymbol{x})\, d\boldsymbol{x} \tag{3}$$

is the expected value of the real variable $z(\boldsymbol{x})$ where

$$z(\boldsymbol{x}) = \begin{cases} 1, & \text{when circuit with component values } \boldsymbol{x} \text{ meets all specifications} \\ 0, & \text{when it fails.} \end{cases}$$

The simplest yield estimation technique is based on the Strong Law of Large Numbers (see Appendix). It consists of generating N random samples $\boldsymbol{x}_1, \boldsymbol{x}_2, \cdots, \boldsymbol{x}_N$ from the density $f(\boldsymbol{x})$, analyzing the circuit with each of these samples, evaluating $z(\boldsymbol{x}_i)$ for each and then forming the estimate $\hat{Y}$ of Y as

$$\hat{Y} = \frac{1}{N} \sum_{i=1}^{N} z(\boldsymbol{x}_i) = \frac{1}{N} \sum_{i=1}^{N} z_i = \frac{\text{Number of passing circuits}}{\text{Total number of circuits}}. \tag{4}$$

The variance $V(\hat{Y})$ of $\hat{Y}$ is

$$V(\hat{Y}) = \frac{V(Y)}{N} = \frac{1}{N} \int_{-\infty}^{\infty} [z(\boldsymbol{x}) - Y]^2 f(\boldsymbol{x})\, d\boldsymbol{x} \tag{5}$$

where $V(Y)$ is the variance of Y. The square root of $V(\hat{Y})$ is a measure of the expected error in the estimate $\hat{Y}$. Generally $V(Y)$ is not known and the variance is estimated from the sample response values z_i as (see Appendix)

$$V(\hat{Y}) \cong \frac{1}{N-1} \left[\frac{1}{N} \left\{ \sum_{i=1}^{N} z_i^2 \right\} - \hat{Y}^2 \right] = \frac{\hat{Y}(1-\hat{Y})}{N-1}. \tag{6}$$

It is seen that the expected error of the estimate decreases roughly as $1/\sqrt{N}$.

[1]In many applications the components are *statistically independent* and the joint probability density function is simply the product of the individual densities.

IV. Importance Sampling and Parametric Sampling

Many techniques to reduce the variance of the estimate without a corresponding increase in the computing cost are available in the literature [5]. One of these techniques called *importance sampling* has previously been used to reduce the error in the yield estimate [8], [9]. This technique forms the basis of the proposed design methodology. A brief review follows:

Equation (3) is rewritten as

$$Y=\int_{-\infty}^{\infty}\left\{z(x)\frac{f(x)}{h(x)}\right\}h(x)\,dx \qquad (7)$$

where $h(x)$ is some other density function, called the *sampling density*. Theoretically, $h(x)$ is arbitrary except for the requirement that $h(x)\neq 0$ whenever $z(x)f(x)\neq 0$. The circuit yield is estimated as in (4) with $z(x_i)$ replaced by $z(x_i)f(x_i)/h(x_i)$: $f(x_i)$ and $h(x_i)$ being the values of the density functions $f(x)$ and $h(x)$ evaluated at the point x_i. Thus

$$\hat{Y}_{\text{IS}}=\frac{1}{N}\sum_{i=1}^{N}\left\{z(x_i)\frac{f(x_i)}{h(x_i)}\right\}=\frac{1}{N}\sum_{i=1}^{N}z_i\frac{f_i}{h_i}=\frac{1}{N}\sum_{i=1}^{N}z_iW_i. \qquad (8)$$

The sample points x_i in (8) are generated from the sampling density $h(x)$ rather than the component density $f(x)$. The *weight factors* $W_i=f_i/h_i$ compensate for the use of a different density function. The variance of the estimator $\hat{Y}_{\text{IS}}$ is given as

$$V(\hat{Y}_{\text{IS}})\cong\frac{1}{N-1}\left[\frac{1}{N}\left\{\sum_{i=1}^{N}z_iW_i^2\right\}-\hat{Y}_{\text{IS}}^2\right] \qquad (9)$$

and can be larger or smaller than the variance of the direct sampling estimator depending on the sampling density $h(x)$. Large gains in precision are possible by a suitable choice of h.

Returning to the design problem, note that the density $f(x)$ usually depends on some parameters $\boldsymbol{\theta}$. For example, if the components are assumed to have normal distributions, the parameters $\boldsymbol{\theta}$ represent the various means (nominal values), the standard deviations (tolerances), and, if the components are correlated, the correlation coefficients. The objective of the design is to find a set of parameters $\boldsymbol{\theta}^*$ that minimize circuit cost starting from some specified $\boldsymbol{\theta}^\circ$. During the design process many intermediate values $\boldsymbol{\theta}^1,\boldsymbol{\theta}^2\cdots,\boldsymbol{\theta}^k$ must be considered and the yield and cost computed for each $\boldsymbol{\theta}^k$. While cost computation in (1) presents no difficulty once yield is known, yield estimation based on (4) generally requires a new Monte Carlo simulation for each $\boldsymbol{\theta}^k$ as the sample values x_i depend on $\boldsymbol{\theta}^k$. To make this dependence explicit, the component joint density is written as $f(x;\boldsymbol{\theta})$. Now, recall that in (8) the x_i are picked from $h(x)$, the circuit is analyzed at these values and the z_i computed. The parameters $\boldsymbol{\theta}$ appear only in the weight functions

$$W_i(\boldsymbol{\theta})=f(x_i;\boldsymbol{\theta})/h(x_i)$$

and thus (8) can, in theory, be used for yield computation for *any* value of $\boldsymbol{\theta}$. We call this technique *Parametric sampling*. Thus

$$\hat{Y}_{\text{PS}}(\boldsymbol{\theta})=\frac{1}{N}\sum_{i=1}^{N}z(x_i)\frac{f(x_i;\boldsymbol{\theta})}{h(x_i)}=\frac{1}{N}\sum_{i=1}^{N}z_iW_i(\boldsymbol{\theta})$$

and

$$\frac{\partial\hat{Y}_{\text{PS}}(\boldsymbol{\theta})}{\partial\theta_j}=\frac{1}{N}\sum_{i=1}^{N}\frac{z_i\partial W_i(\boldsymbol{\theta})}{\partial\theta_j}. \qquad (10)$$

The advantage of the parametric approach is that no new circuit analyses are required when $\boldsymbol{\theta}$ is changed,[2] and gradient information is obtained for virtually no cost. As seen above, the functional form $W_i(\boldsymbol{\theta})$ in (10) permits simple *computation of the derivatives* of the Parametric sampling yield estimate thus allowing use of powerful optimization techniques. Although statistical fluctuations in the Monte Carlo sampling process introduce random errors in $\hat{Y}_{\text{PS}}(\boldsymbol{\theta})$, (10) provides accurate gradients for the estimated yield. Note that while importance sampling concepts form the basis of Parametric sampling, it is not being used here in its traditional variance reduction role.

To illustrate the use of (10), consider a single standardized *component* whose density f is normally distributed with mean 0.5 and unit variance. For simplicity, the range of *component* values (0, 1) is considered as "acceptable." In yield estimation using (4), samples x_i are selected from the distribution f and the fraction that falls in the range (0, 1) is found. To use the importance sampling estimate (8) consider the sampling density to be normal with zero mean and standard deviation 2. Sampling will now be from this new density. Values outside the range (0, 1) are still rejected. However, those falling inside the range have weights

$$W_i=\frac{(2\pi)^{-1/2}\exp\left\{-\frac{1}{2}\left(x_i-\frac{1}{2}\right)^2\right\}}{0.5(2\pi)^{-1/2}\exp\left\{-\frac{1}{2}\left(\frac{x_i}{2}\right)^2\right\}}$$
$$=2\exp\left\{-\tfrac{3}{8}x_i^2+\tfrac{1}{2}x_i-\tfrac{1}{8}\right\}.$$

These weights compensate for the distortion introduced by the use of a different density $h(x)$.

Parametric sampling can now be used to evaluate the yield for a family of component densities f that, for simplicity, all have unit variances but different mean values μ, thus the component density $f(x;\theta)$ becomes $f(x;\mu)$. Instead of computing a set of x_i from $f(x;\mu)$ for each case separately, only one set of x_i values is selected from the *sampling density* (normal, zero mean, standard deviation = 2). For those x_i that lie in the range (0, 1) the weights

$$2\exp\left\{-(3/8)x_i^2+\mu x_i-\mu^2/2\right\}$$

are computed for each new value of μ. Fig. 2 shows the

[2] While (8) is theoretically valid for any $\boldsymbol{\theta}$, depending on $h(x)$ and N there will only be a finite range of values for which the answers will be sufficiently accurate. Equation (9) can still be used to test if the probable errors are acceptable.

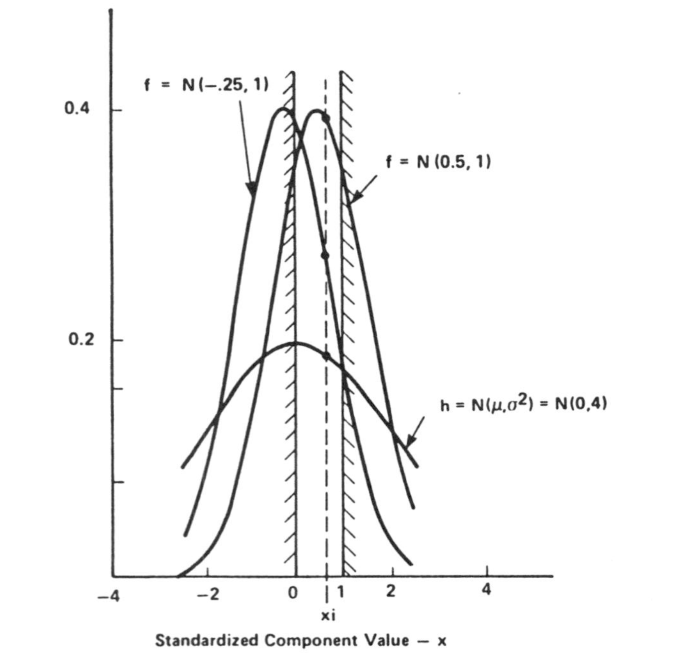

Fig. 2. Sampling density h and two possible component densities f for standardized component x.

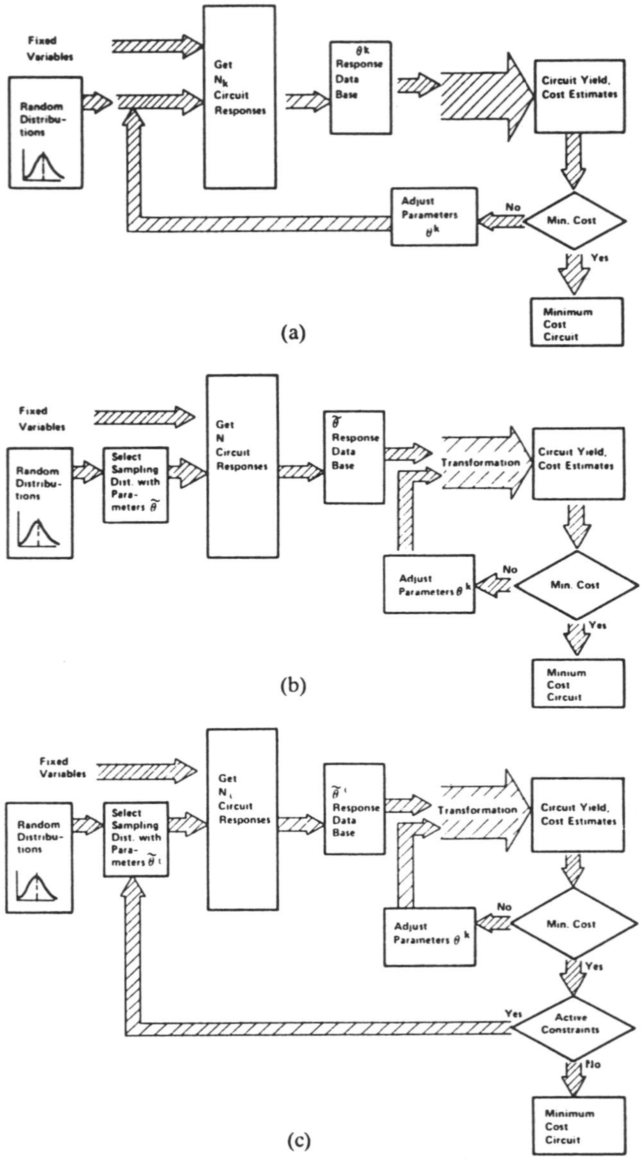

Fig. 3. Design cycle for (a) Elias' and Kjellstrom's methods [3], [4], [7]; (b) Conceptual Parametric sampling; (c) Practical implementation of Parametric sampling.

sampling density h and two possible component densities $f(x;\mu)$ with $\mu=-0.25$ and 0.5. The weight W_i for a given x_i is the ratio of the distributions f and h evaluated at x_i.

Moreover, the derivative of the weight function for each new value of μ is readily computed at the same time using

$$\frac{dW_i}{d\mu}=(x_i-\mu)W_i.$$

V. Statistical Design with Parametric Sampling—Concept

Statistical design methods have been proposed by many authors [1]–[4], [6], [7], [9]–[11]. Some appear to be suitable when there are only a few variables [1], [2] or for linear circuits [10], [11] or for tolerancing without centering [6], [10] or for yield maximization [11]. Two methods that are sufficiently general and whose use has been demonstrated on intermediate size problems are due to Elias [3] and Kjellstrom *et al.* [4], [7]. The Parametric sampling approach is also applicable to the general statistical design problem.

The design cycle in Elias' and Kjellstrom's methods can be depicted, as in Fig. 3(a). Both methods require a form of Monte Carlo analysis *inside* the design loop. The number of circuit responses N required in each iteration in Elias' method is fairly large whereas in Kjellstrom's method, N_0 is of the order of several hundred and $N_k=1,\ k=1,2,3,\cdots$ Elias' method has no "cost" criterion nor "yield gradient" (though a "yield sensitivity" is defined) and Kjellstrom's method does not provide a yield gradient. Designs produced by these methods explicitly require component distributions and constraint specifications. Any change in these specifications requires a new statistical design optimization.

In contrast, designs based on Parametric sampling, Fig. 3(b), have a number of useful features. The Monte Carlo analysis is carried out once only using a sampling distribution $h(\boldsymbol{x})$.[3] The component values $\boldsymbol{x}_i$, the values $h(\boldsymbol{x}_i)$, and the circuit response are stored in a "data base." Yield estimation and design optimization are carried out by using the database but *no new* circuit analyses. Thus Monte Carlo analysis is *outside* the design loop. Changes in specifications, component densities, cost models, etc., are handled by running the design loop again, but the most expensive part of the algorithm—the database construction—is carried out only once. Moreover, efficient optimizers such as [12] can be used as full gradient information, and if necessary higher derivatives of the yield, are available. Due to the finite size of the database, the nomi-

[3] Often the density $h(\boldsymbol{x})$ is selected from the same family as $f(\boldsymbol{x})$ with $\boldsymbol{\theta}$ being assigned some value $\tilde{\boldsymbol{\theta}}$.

nal values and tolerances produced by the algorithm are biased. A verification Monte Carlo analysis is desirable.

VI. Statistical Design with Parametric Sampling—A Practical Implementation

Two problems arise when attempting to implement the above conceptual design algorithm:

(1) Monte Carlo analysis cannot, in general, be totally placed outside the design loop. However, it can be removed from an "inner" design loop, thus allowing repetitive use of a database.

(2) The algorithm must ensure that yield estimates are sufficiently accurate.

The inner design loop strategy (which influences yield accuracy) and formulas for more accurate yield estimates are discussed below.

A. Data Base Updating for Inner Design Loop

Fig. 4 illustrates the need for both inner and outer design loops. The contours show points of equal probability of occurrence in the joint *sampling density* space created by components $\boldsymbol{x}=(x_1, x_2)$. Fig. 4(a) shows conditions for $\boldsymbol{\theta}^{\circ}$ where both component and sampling densities have the same mean. Initially N samples $\boldsymbol{x}_1, \boldsymbol{x}_2, \cdots, \boldsymbol{x}_N$ are generated using sampling densities and z_i is evaluated for each sample. Next yield $\hat{Y}_{\mathrm{PS}}(\boldsymbol{\theta}^{\circ})$ is computed from (10) and new parameters $\boldsymbol{\theta}^1$, are generated by the optimizer. After k iterations, a component density condition $\boldsymbol{\theta}^k$ such as shown in Fig. 4(b) may occur. Now the same points $\boldsymbol{x}_1, \cdots, \boldsymbol{x}_N$ and hence the same z_i are used to compute $\hat{Y}_{\mathrm{PS}}(\boldsymbol{\theta}^k)$, but the weights $W_i(\boldsymbol{\theta}^k)$ are considerably different. For example, in Fig. 4(a) and (b) the point $\boldsymbol{x}_i$ has $z_i=1$, but in Fig. 4(b) the relative weight makes $\boldsymbol{x}_i$'s contribution to the yield summation (10) insignificant. If fact, very few of the original N points now contribute "meaningfully" to the yield estimate since a predictably small percentage of the data base points fall inside the "active" tolerance box shown in Fig. 4(b). Thus the accuracy of the $\hat{Y}_{\mathrm{PS}}(\boldsymbol{\theta}^k)$ estimate is suspect.

Three alternatives for increasing $\hat{Y}_{\mathrm{PS}}(\boldsymbol{\theta}^k)$ accuracy are considered.

(1) For the kth iteration, increase the number of meaningful points to N_k. Generate additional points using sampling distributions, compute $W_i(\boldsymbol{\theta}^k)$ for each, but only determine the value of z_i when $W_i>\epsilon$, otherwise assume $z_i=0$. Continue to generate points until N_k "meaningful" points exist. Then evaluate $\hat{Y}_{\mathrm{PS}}(\boldsymbol{\theta}^k)$ using (10) where N becomes N_k, the total number of meaningful points in the updated data base. This method maintains yield estimate accuracy by causing some additional circuit analyses during each iteration.

(2) Generate a new data base around the $\boldsymbol{\theta}^k$ design center by shifting sampling distributions, as shown in Fig. 3(c). This approach has the same effect as the previous alternative, should be slightly more accurate because the $W_i>\epsilon$ decision is not required, but requires generation of N_l *new* values z_i since the old data base is destroyed.

(3) Generate a new data base as in (2) above, but do not destroy the original data base. Then, form the pooled estimate

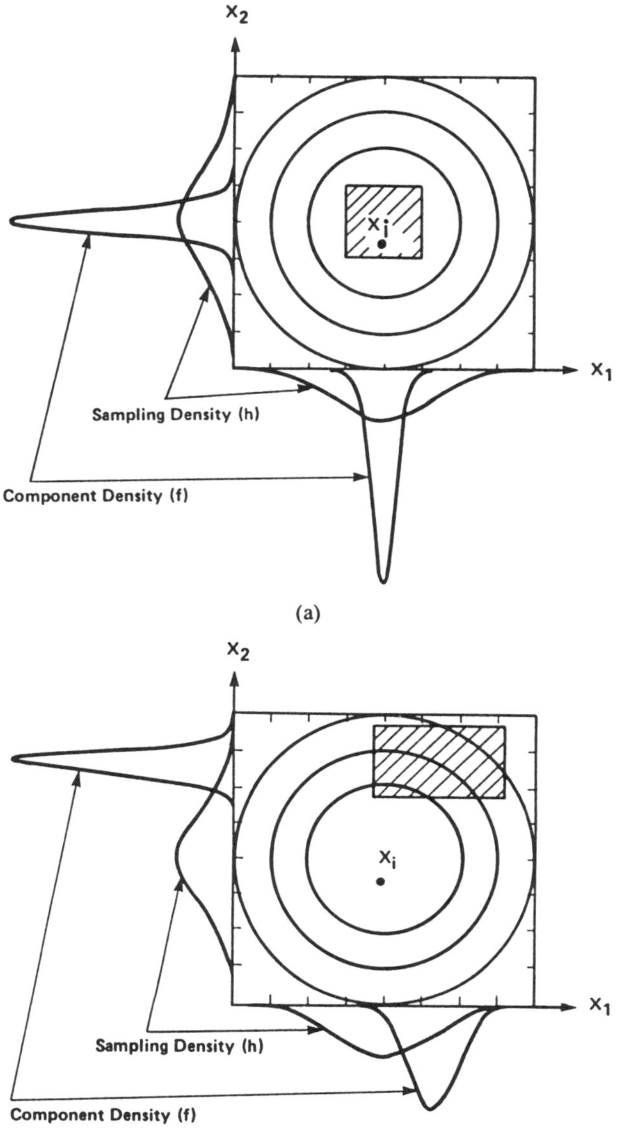

Fig. 4. Relative locations of sampling density and component density (tolerance box). (a) Favorable. (b) Unfavorable.

$$\hat{Y}_{\mathrm{PS}}(\boldsymbol{\theta}^k)=\sum_{j=0}^{l}\phi_j\hat{Y}_{\mathrm{PS}j}(\boldsymbol{\theta}^k) \tag{11}$$

where

$$\sum_{j=0}^{l}\phi_j=1$$

and a strategy for definition of ϕ_j is required. The usual approach is to select ϕ_j to minimize $V(\hat{Y}_{\mathrm{PS}}(\boldsymbol{\theta}^k))$ which gives (see Appendix)

$$\phi_j=\frac{1}{V\left(\hat{Y}_{\mathrm{PS}j}(\boldsymbol{\theta}^k)\right)\left\{\displaystyle\sum_{m=0}^{l}\frac{1}{V\left(\hat{Y}_{\mathrm{PS}m}(\boldsymbol{\theta}^k)\right)}\right\}}. \tag{12}$$

In practice, accurate estimates of $V(\hat{Y}_{PSm}(\theta^k))$ using (9) are suspect after k iterations since the accuracy of $\hat{Y}_{PSm}(\theta^k)$ estimates have forced the data base to be updated. Note that

$$\phi_j = 0, \qquad \text{for } j=1,2,\cdots,(l-1)$$

reverts to case (2).

Alternative (2) is preferred over (1) since experience has shown that a number of iterations within each data base are possible before updating. This gives a very efficient inner design loop where circuit analysis is NOT required. Alternative (3) requires additional bookkeeping and storage but provides some increase in $\hat{Y}_{PS}(\theta^k)$ estimate accuracy.

A number of criteria can be used to make the update decision. The following approach has been used in the examples. Optimality is not claimed, but useful results have been obtained. Normal distributions for components are assumed for convenience only.

Referring to Fig. 3(c):

Step 1): Set $l=0$ and for each component, establish a normal distribution with

$$3\sigma_i^\circ = t_i^* \tag{13}$$

where t_i^* is worst-case or "safe" tolerance. "Safe" may be nothing more than engineering judgement on good tolerance to use, and σ_i is standard deviation of component distribution, and a sampling distribution with

$$\left.\begin{aligned} \tilde{\sigma}_i^\circ &= \gamma^\circ \sigma_i^\circ, \gamma^\circ > 1 \\ \tilde{\mu}_i &= \mu_i \end{aligned}\right\} \tag{14}$$

where γ° is inflation factor and μ_i is component nominal value.

Step 2): Generate lth database using the sampling distributions.

Step 3): Using [12] solve the following constrained minimization problem.

$$\left.\begin{aligned} &\text{minimize} \quad C_u(\theta, \hat{Y}_{PS}(\theta)) \\ &\text{subject to} \\ &\qquad \mu_i + \beta_2\sigma_i \leqslant \tilde{\mu}_i^l + \beta_1\tilde{\sigma}_i^l \\ &\qquad \mu_i - \beta_2\sigma_i \geqslant \tilde{\mu}_i^l - \beta_1\tilde{\sigma}_i^l \\ &\qquad |\gamma^l\sigma_i - \tilde{\sigma}_i^l| \leqslant \delta_{1l} \\ &\text{and} \\ &\qquad \left|\frac{\mu_i - \tilde{\mu}_i^l}{\tilde{\mu}_i^l}\right| \leqslant \delta_{2l} \end{aligned}\right\}. \tag{15}$$

The first two sets of constraints ensure that the $\pm\beta_2\sigma_i$ component density limits do not exceed the $\pm\beta_1\tilde{\sigma}_i^l$ sampling density limits, while the last two restrict the standard deviation and nominal value shifts. This is only one possible constraint set.

Step 4): If any of the constraints in (15) are active (i.e., satisfied by equality) at the end of Step 3), increment l and shift the sampling distribution mean to the new design center, take $\tilde{\sigma}_i^l = \gamma^l\sigma_i$ and repeat Steps 2) and 3). Terminate when Step 3) converges without active constraints.

Experience to date suggests reasonable results are achieved with $\gamma^\circ = 4$ and $\beta_1, \beta_2 = 2$ initially. It is desirable to reduce γ^l, δ_{1l} and δ_{2l} as convergence approaches, since large changes in σ and μ are not expected, and a joint sampling density function requiring such broad limits is no longer required, nor desirable to maintain yield estimate accuracy.

B. Yield Estimator Accuracy

Simple yield estimators (8), (10) have been considered for importance and Parametric sampling, respectively. For the design methodology outlined in the last section to work, it is necessary that the yield estimate be sufficiently accurate for moderate sized samples. One of the unfortunate features of (8) and (10) is that the yield estimate can become larger than 1. This happens fairly often when the yield is high (as would be the case in most statistical designs including final test yields for integrated circuits—see Section II) and the optimizer selects component values to increase the yield estimate even further.

Alternate estimates based on failure densities and on ratios have associated problems. A more precise estimator suited to this application is presented below.

1) The Regression Estimator [13]:

Consider the problem of finding the expected value $\hat{u}$ of some random variable u based on a set of observations $u_1, u_2, \cdots, u_N$. The simple estimator for $\hat{u}$ is the average value $(1/N)\Sigma_{i=1}^N u_i$ with variance $(1/N)V(u)$. In many situations, in addition to the u_i, values of another variable v_i may be available. Suppose that (u_i, v_i) are correlated and that the mean of v, $\bar{v}$ is known. The regression estimator u_R,

$$\left.\begin{aligned} u_R &= \hat{u} + b(\bar{v} - \hat{v}) \\ \text{where} \qquad \hat{v} &= \frac{1}{N}\sum_{i=1}^{N} v_i \end{aligned}\right\} \tag{16}$$

with b fixed is "consistent" (i.e., gives correct results for large N) and has variance

$$V(u_R) = V(u) + b^2V(v) - 2b\text{Cov}(u, v) \tag{17}$$

where Cov(u,v) is the covariance between $\hat{u}$ and $\hat{v}$. The best value of b in (16) can be estimated as

$$b = \frac{\text{Cov}(u, v)}{V(v)} \tag{18}$$

to minimize the variance in (17). Generally, however, the covariance, just like the variance, is not known exactly and must be replaced by its estimate based on the sample (see Appendix).

Substitution of (18) in (16) and (17) gives

$$\hat{u}_R = \hat{u} + \frac{\text{Cov}(\hat{u}, \hat{v})}{V(\hat{v})}(\bar{v} - \hat{v}) \tag{19}$$

and

$$V(\hat{u}_R) \cong V(\hat{u}) - \frac{\text{Cov}^2(\hat{u}, \hat{v})}{V(\bar{v})}. \tag{20}$$

The regression estimator thus always has smaller variance than the simple estimator $\hat{u}$ and the variance reduction can be considerable when $\hat{u}$ and $\hat{v}$ are strongly correlated. The key, of course, is in finding a suitable auxiliary variable v. For importance sampling, an auxiliary variable appears naturally.

Consider the case when all the circuits are considered to be passing irrespective of their response, i.e., there are no specifications the circuits must satisfy. Then yield $\bar{v}$ should clearly be unity. However, use of the importance sampling relation (8) (or the Parametric relation (10)) gives the yield estimate $\hat{v}$ as

$$\hat{v}=\frac{1}{N}\sum_{i=1}^{N} W_i. \tag{21}$$

Simple algebra shows that the regression estimator to be used in importance/Parametric sampling has the form

$$\left.\begin{aligned}\hat{Y}_R &= \frac{1}{N}\sum_{i=1}^{N} z_i W_i + \hat{b}\left(1-\frac{1}{N}\sum_{i=1}^{N} W_i\right)\\ \text{where}\quad \hat{b} &= \frac{\left\{\sum_{i=1}^{N} z_i (W_i)^2\right\}-\frac{1}{N}\left\{\sum_{i=1}^{N} z_i W_i\right\}\cdot\left\{\sum_{i=1}^{N} W_i\right\}}{\left\{\sum_{i=1}^{N} (W_i)^2\right\}-\frac{1}{N}\left\{\sum_{i=1}^{N} W_i\right\}^2}\end{aligned}\right\}. \tag{22}$$

Note that in using (22) for Parametric sampling the values W_i will depend on the parameters $\boldsymbol{\theta}$.

Use of an estimated $\hat{b}$ in (22) makes $\hat{Y}_R$ "biased." Statistical techniques (like jacknife [14]) exist that can be used to reduce the bias and provide more robust estimates of the variance than (20). These techniques have been used to monitor the probable error in the following examples, but are not discussed here.

VII. Results

A. Example 1

Consider the high-pass filter, Fig. 5(a), first discussed in [16]. The specifications have been tightened slightly in that the -0.05-dB attenuation limit has been extended to 630 Hz, and the bandedge to 650 Hz. The complete specifications for the filter are shown in Fig. 5(b). The reference frequency is 990 Hz and 11 other frequencies at 170, 350, 440, 630, 650, 720, 740, 760, 940, 1040, and 1800 Hz are used. The cost model for this example is taken as

$$C_s=\frac{C_A+\sum_{i=1}^{5}\frac{1}{t_i}+2\sum_{i=6}^{7}\frac{1}{t_i}}{Y}$$

where $C_A=2.5$ represents the fixed cost part of C_u as defined in (1).

The components are assumed to be independently normally distributed except that the resistors modelling inductor losses are taken as completely correlated with the

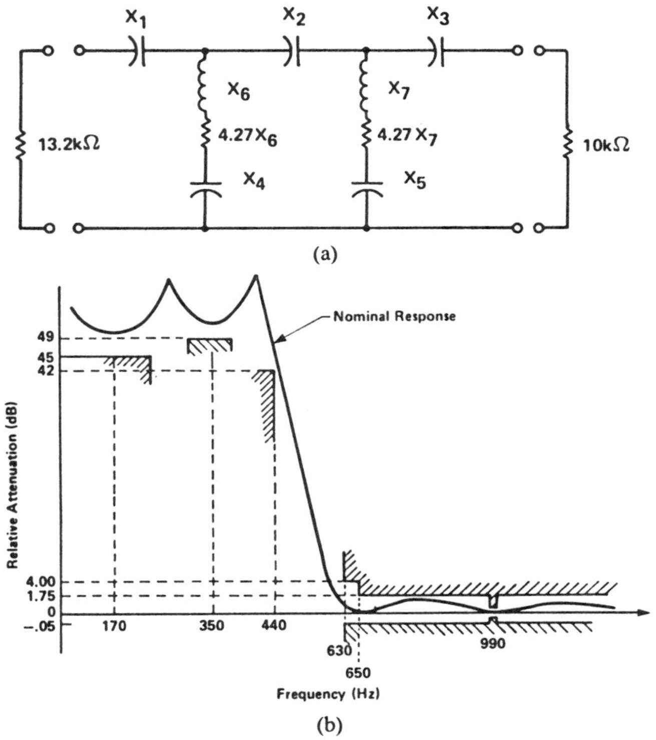

Fig. 5. Example 1 high-pass filter. (a) Schematic. (b) Response specification.

inductors. The nominal component values correspond to the mean values μ_i of the distributions, while the tolerances t_i are related to the standard deviations as

$$t_i=\frac{3\sigma_i}{\mu_i}.$$

The worst-case design with fixed nominal values was given in [16] and had a cost of 5.99 units. This worst-case cost was reduced later [17] to 4.53 by using centering.

Starting from the nominal values used in [16] a Parametric sampling design with three data bases constructed with 100, 200, and 300 sample points (600 points total) and using pooled estimates resulted in costs of 3.88, 3.78, and 3.50, respectively. The total design time (excluding verification Monte Carlo sampling) was 25 s on an IBM 370/158. Statistical design with a 300 point data base starting from the centered worst-case design reduced the cost to 3.39 and required only 8 s of CPU.

For comparison, starting from the nominals of [16] Kjellstrom's algorithm [18] resulted in costs of 3.70, 3.65, and 3.49 after 500, 1000, and 2000 analyses, respectively. All the final designs are summarized in Table I. Elias' method [19] gave a cost of 3.9 after three iterations with 100 samples each starting from the centered worst-case design (as compared with 3.39 for the same effort using Parametric sampling).

B. Example 2

Consider the low-pass Chebyshev filter first discussed by Neirynck [20], as shown in Fig. 6. The nominal design for this filter was produced by a standard filter synthesis

TABLE I
EXAMPLE 1 HIGH-PASS FILTER DESIGN ALTERNATIVES. NOMINAL VALUES OF CAPACITORS IN NANOFARADS AND INDUCTORS IN HENRIES

Component	Starting nominal values μ	Design Cycle with 3 data bases: 100-point data base		additional 200 points		additional 300 points		Statistical design starting from worst-case centered design: worst-case centered design		design with 300-point data base		Kjellstrom's design with 2000 analyses	
		Δμ%	t%	Δμ%	t%	Δμ%	t%	Δμ%	t%	Δμ%	t%	Δμ%	t%
x_1	11.80	-1.80	9.25	-1.01	8.06	1.11	9.33	4.83	5.23	8.46	12.68	5.48	11.04
x_2	8.73	-1.50	7.25	0.92	7.64	1.43	8.00	19.70	3.96	17.90	11.77	11.16	9.57
x_3	10.45	3.83	8.92	6.77	11.05	6.52	9.39	39.43	4.25	33.71	9.09	24.46	9.63
x_4	39.04	-3.39	10.00	-3.22	9.34	-4.20	9.27	-13.09	4.05	-13.56	8.50	2.40	11.25
x_5	90.13	4.35	7.97	3.66	10.11	5.04	14.00	7.77	5.24	5.11	15.00	-2.96	20.31
x_6	3.85	0.40	10.00	2.94	12.50	1.36	8.87	5.38	4.48	4.75	12.90	-5.04	9.30
x_7	3.18	-3.57	10.00	-3.53	10.34	-2.93	11.85	-16.82	4.32	-14.89	11.37	-8.34	9.66
Yield verification based on 300 samples		89.67		84.33		97.33		100		97.33		96.33	
Cost estimate		3.88		3.78		3.50		4.53		3.39		3.49	

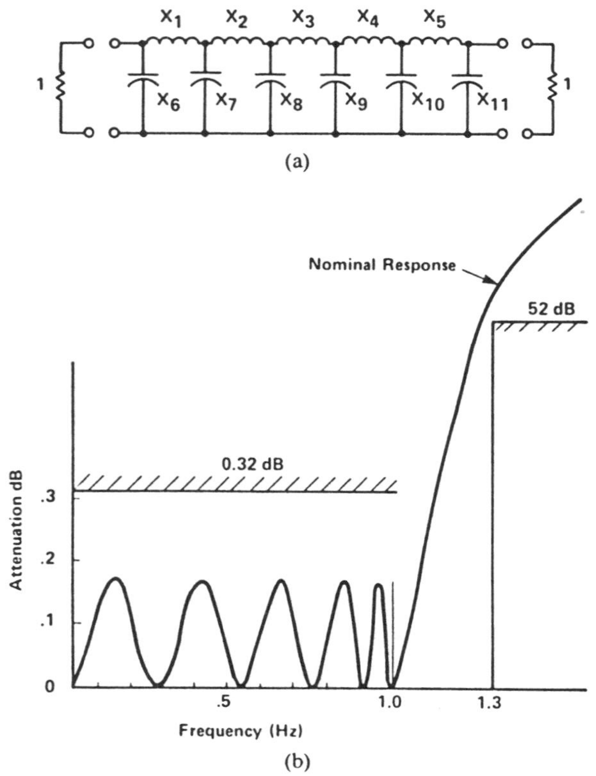

Fig. 6. Example 2 low-pass filter. (a) Schematic. (b) Response specification.

program. A cost model of the form

$$C_s = \frac{\sum_{i=1}^{11} \frac{1}{t_i}}{Y}$$

was set up and constraints were specified at 51 frequencies (0.02, 0.04,$\cdots$,1., and 1.3 Hz). A worst-case design with fixed nominal component values gave a cost of 44.3 which was reduced to 18.68 in a centered worst-case design. For the statistical design the components were taken as independently normally distributed.

The Parametric sampling approach was used and both the initial design produced by the filter synthesis package and the centered worst-case design were taken as starting values. In each case two databases were constructed, the first with 300 points and the second with an additional 500 points, a total of 800 analyses in each case. Starting from the initial nominals the cost was reduced from 44.3 to 7.48 after 300 analyses and to 6.76 after the next 500 analyses. The corresponding experiment with the centered design gave costs of 5.26 and 5.00, respectively. The total CPU time required in each case was 75 s. This includes the 800 analyses and the two 22 variable (μ, t) constrained optimization steps but excludes the cost of the verification Monte Carlo analyses. Table II summarizes the results.

For comparison this example was also run using Kjellstrom's method [21] and after 800 analyses a cost of 7.48 was achieved. The resulting design had a much higher yield than the statistical designs produced through the Parametric sampling approach but the tolerances were considerably smaller.

VIII. CONCLUSIONS

An algorithm for statistical design centering and tolerancing has been presented. The Parametric sampling technique is used to eliminate costly circuit analysis inside the optimization loop. It has been shown that yield derivative information is virtually free, thus allowing the use of

TABLE II
EXAMPLE 2 LOW-PASS FILTER DESIGN ALTERNATIVES

Component	Starting nominal μ values μ	Design cycle with two databases: design with 300-point data base		Design cycle with two databases: additional 500-point data base		Statistical design starting from centered worst-case design: Worst-Case design		Statistical design starting from centered worst-case design: design with 300-point data base		Statistical design starting from centered worst-case design: additional 500-point data base		Kjellstrom's design with 800 analyses	
		Δμ%	t%	Δμ%	t%	Δμ%	t%	Δμ%	t%	Δμ%	t%	Δμ%	t%
x_1	.2251	-.49	2.28	.12	2.49	-5.82	.62	-6.49	2.66	-6.40	2.90	-.04	1.62
x_2	.2494	.56	2.15	-.30	1.97	-6.58	.58	-6.66	3.32	-6.58	3.71	-.20	1.83
x_3	.2523	-.22	2.68	.16	1.84	-6.86	.45	-7.02	2.47	-6.78	2.00	-.67	1.08
x_4	.2494	-.46	2.32	-.34	2.23	-6.58	.58	-6.98	2.86	-7.10	3.40	-.16	1.62
x_5	.2251	.66	2.00	.66	3.00	-5.82	.62	-5.15	2.65	-4.80	2.76	-.67	1.86
x_6	.2149	.43	2.35	.43	3.53	9.35	.85	8.61	3.78	8.61	4.72	.65	1.41
x_7	.3636	.09	2.14	-.70	2.05	6.99	.57	7.92	2.11	7.92	2.63	-.47	1.32
x_8	.3761	.62	2.06	1.47	1.80	6.89	.51	6.83	3.00	5.93	2.50	.90	1.26
x_9	.3761	-.68	2.00	-1.08	2.40	6.89	.51	6.89	3.06	7.92	2.36	-.11	1.32
x_{10}	.3636	-.68	2.00	-1.38	2.34	6.99	.57	6.00	2.03	6.00	2.54	.58	1.26
x_{11}	.2149	.59	2.10	1.09	2.39	9.35	.85	7.82	3.06	7.12	2.83	3.54	3.12
Yield verification based on 300 samples		67.67		71.33		100		76.76		78.67		98.67	
Cost estimate		7.48		6.76		18.68		5.26		5.00		7.48	

powerful optimization techniques. The algorithm has the potential to handle nonlinear circuits and can easily handle practical design requirements like tuning, correlation, limits on manufacturing and end-of-life yield and changes in cost models and specifications. Examples demonstrate that the method is a practical and efficient approach to the statistical design problem, and comparable or superior designs are produced by the algorithm at lower CPU cost than other established techniques.

APPENDIX

In this Appendix some of the basic concepts and results from probability theory and statistics that have been used in the paper are outlined. These results are stated without proof as they can be found in any standard text, e.g., [15]. The equations are given for one dimensional distributions, but extension to multiple dimensions should be obvious.

The Strong Law of Large Numbers

If a sequence of N random variables $x_1, x_2, \cdots, x_N$ is picked from the probability density function $f(x)$ and if a random variable q is defined through

$$\hat{q} = \frac{1}{N}\sum_{i=1}^{N} q(x_i)$$

and the integral

$$\bar{q} = \int_{-\infty}^{\infty} q(x) f(x)\, dx$$

exists, $\hat{q}$ will, almost always, approach $\bar{q}$ as N tends to infinity.

Further, if

$$\overline{q^2} = \int_{-\infty}^{\infty} q^2(x) f(x)\, dx$$

also exists, the amount that $\hat{q}$ deviates from $\bar{q}$ can be estimated. Let

$$V(\bar{q}) = \overline{(q^2)} - (\bar{q})^2$$

be the variance. Then the *Central Limit Theorem* states that, for large N, $(\hat{q}-\bar{q})/\sqrt{V(\bar{q})/N}$ is approximately distributed as standard normal with zero mean and unit variance. This result permits the construction of *confidence intervals* for $\bar{q}$ based on $\hat{q}$. If $K_{\alpha/2}$ is a point such that the integral of the standard normal density from $K_{\alpha/2}$ to infinity is $\alpha/2$ then with probability $1-\alpha$,

$$|\hat{q}-\bar{q}| \leqslant K_{\alpha/2}\sqrt{\frac{V(\bar{q})}{N}}$$

$K_{\alpha/2}$ points are tabulated in mathematical handbooks. Generally the variance $V(\bar{q})$ is not known but must be estimated through

$$V(\bar{q}) \cong s^2 = \frac{1}{N-1}\sum_{i=1}^{N}\left[q(x_i) - \hat{q}\right]^2$$

$$= \frac{1}{N-1}\left[\left\{\sum_{i=1}^{N} q^2(x_i)\right\} - N\hat{q}^2\right]$$

where s^2 is a random variable based on the observed values $q(x_i)$. The use of s^2 (rather than $V(\bar{q})$) in constructing confidence intervals requires that the point $K_{\alpha/2}$ be replaced by $t_{\alpha/2, N-1}$, the corresponding point on a Student t-distribution with $N-1$ degrees of freedom. As N becomes large, the $t_{\alpha/2, N-1}$ points approach the points $K_{\alpha/2}$ and, in most yield estimation problems, $K_{\alpha/2}$ can be used.

Pooled Estimators

If $\hat{Y}_0, \hat{Y}_1, \hat{Y}_2, \cdots, \hat{Y}_l$ are independent estimates of the yield Y then the pooled estimate is given as

$$\hat{Y} = \sum_{j=0}^{l} \phi_j \hat{Y}_j$$

where ϕ_j are nonnegative weights with

$$\sum_{j=0}^{l} \phi_j = 1.$$

To minimize variance of $\hat{Y}$,

$$V(\hat{Y}) = \sum_{j=0}^{l} \phi_j^2 V(\hat{Y}_j)$$

with respect to the weights form the augmented function

$$\lambda\left(1 - \sum_{j=0}^{l} \phi_j\right) + \sum_{j=0}^{l} \phi_j^2 V(\hat{Y}_j)$$

where λ is a Lagrange multiplier and, for the moment, the nonnegativity constraints are ignored. Setting the derivatives of the augmented function to zero results in

$$\sum_{j=0}^{l} \phi_j = 1$$

$$\phi_m = \frac{\lambda}{2V(\hat{Y}_m)}, \qquad m = 0, 1, 2, \cdots, l.$$

Solving for λ and the ϕ_m

$$\lambda = \frac{2}{\sum_{j=0}^{l} \frac{1}{V(\hat{Y}_j)}}$$

$$\phi_m = \frac{1}{V(\hat{Y}_m)} \cdot \frac{1}{\sum_{j=0}^{l} \frac{1}{V(\hat{Y}_j)}}.$$

Note that as the variances are nonnegative, the weights are also nonnegative.

Substituting the values of the weights, the optimal estimator is

$$\hat{Y} = \left\{ \frac{1}{\sum_{j=0}^{l} \frac{1}{V(\hat{Y}_j)}} \right\} \sum_{j=0}^{l} \frac{\hat{Y}_j}{V(\hat{Y}_j)}$$

with variance

$$V(\hat{Y}) = \frac{1}{\sum_{j=0}^{l} \frac{1}{V(\hat{Y}_j)}}.$$

Estimation of Covariance

The covariance between u and v is defined as

$$\mathrm{Cov}(u, v) = \overline{uv} - \bar{u}\bar{v}$$

where

$$\overline{uv} = \int_{-\infty}^{\infty} u(x) v(x) f(x)\, dx.$$

Generally the covariance is not known and must be estimated from the sample values $\{u(x_1), v(x_1)\}, \cdots, \{u(x_N), v(x_N)\}$ as

$$\mathrm{Cov}(\hat{u}, \hat{v}) \cong \frac{1}{N-1} \sum_{i=1}^{N} [u(x_i) - \hat{u}][v(x_i) - \hat{v}]$$

$$= \frac{1}{N-1}\left[\left\{\sum_{i=1}^{N} u(x_i) v(x_i)\right\} - N\hat{u}\hat{v}\right].$$

Acknowledgment

The authors wish to thank D. Agnew of Bell-Northern Research for numerous stimulating discussions and assistance in preparation of the examples, G. Kjellstrom of L.M. Ericsson, Sweden, for his generous assistance in preparation of comparative numerical data and K. Singhal for information on regression and pooled estimators.

References

[1] S. W. Director and G. D. Hachtel, "The simplicial approximation approach to design centering," *IEEE Trans. Circuits Syst.*, vol. CAS-24, pp. 363–372, July 1977.

[2] J. W. Bandler and H. L. Abdel-Malek, "Optimal centering, tolerancing, and yield determination via updated approximation and cuts," *IEEE Trans. Circuits Syst.*, vol. CAS-25, pp. 853–871, Oct. 1978.

[3] N. Elias, "New statistical methods for assigning device tolerances," in *Proc. 1975 IEEE Int. Symp. Circuits and Syst.*, (Newton, MA), pp. 329–332, Apr. 1975.

[4] G. Kjellstrom and L. Taxen, "On the efficient use of stochastic optimization in network design," in *Proc. 1976 IEEE Int. Symp. Circuits and Syst.*, (Munich, Germany), pp. 714–717, Apr. 1976.

[5] H. Kahn, "Use of different Monte Carlo sampling techniques," in *Symp. on Monte Carlo Methods*, (ed. H. A. Mayer), New York: Wiley, 1954, pp. 146–190.

[6] B. J. Karafin, "The general component tolerance assignment problem in electrical networks," Ph.D. dissertation, Univ. of Pennsylvania, Philadelphia, PA, 1974.

[7] G. Kjellstrom, L. Taxen, and L. E. Blomgren, "Optimization methods for statistical network design," in *Proc. 1975 IEEE Int. Symp. Circuits and Syst.*, (Newton, MA), pp. 321–324, Apr. 1975.

[8] J. F. Pinel and K. Singhal, "Efficient Monte Carlo computation of circuit yield using importance sampling," in *Proc. 1977 IEEE Int. Symp. Circuits Syst.*, (Phoenix, AZ), pp. 575–578, Apr. 1977.

[9] P. W. Becker and F. Jensen, *Design of Systems and Circuits for Maximum Reliability or Maximum Production Yield*, New York: McGraw-Hill, 1977.

[10] A. R. Thorbjorsen and S. W. Director, "Computer aided tolerance assignment for linear circuits with correlated elements," *IEEE Trans. Circuit Theory*, vol. CT-20, pp. 518–523, 1973.

[11] K. S. Tahim and R. Spence, "A radial exploration approach to manufacturing yield estimation and design centering," *IEEE Trans. Circuits Syst.*, vol. CAS-26, pp. 768–774, Sept. 1979.

[12] M. J. D. Powell, "Algorithms for nonlinear constraints that use Lagrangian functions," *Math. Prog.*, vol. 14, pp. 224–248, 1978.

[13] W. G. Cochran, *Sampling Techniques*, (3rd ed.), New York, Wiley, 1977.

[14] R. G. Miller, "The jacknife—A review, *Biometrika*, vol. 61, pp. 1–15, 1974.

[15] P. L. Meyer, *Introductory Probability and Statistical Applications*, (2nd ed.), Reading, MA: Addison-Wesley, 1970.

[16] J. F. Pinel and K. A. Roberts, "Tolerance assignment in linear networks using nonlinear programming," *IEEE Trans. Circuit Theory*, vol. CT-19, pp. 475–479, 1972.

[17] J. F. Pinel, K. A. Roberts and K. Singhal, "Tolerance assignment in network design," in *Proc. 1975 Int. Symp. Circuits and Syst.*, (Newton, MA), pp. 317–320, Apr. 1975.

[18] G. Kjellstrom, private communication, Sept. 1979.

[19] K. Gopal, "Efficient statistical analysis and design of systems using regionalization," Ph.D. dissertation Univ. of Waterloo, 1978.

[20] J. Neirynck and L. J. Milic, "Equal ripple tolerance characteristic," *Circuit Theory* and *Appl.*, vol. 4, pp. 99–104, 1976.

[21] G. Kjellstrom, private communication, June 1976.

Design Centering by Yield Prediction

KURT J. ANTREICH, SENIOR MEMBER, IEEE, AND RUDOLF K. KOBLITZ

Abstract —Design centering is an appropriate design tool for all types of electrical circuits to determine the nominal component values by considering the component tolerances. A new approach to design centering will be presented, which starts at the initial nominal values of the circuit parameters and improves these nominal values by maximizing the circuit yield step by step with the aid of a yield prediction formula. Using a variance prediction formula additionally, the yield maximization process can be established with a few iteration steps only, whereby a compromise between the yield improvement and the decrease in statistical certainty must be made in each step. A high-quality interactive optimization method is described which allows a quantitative problem diagnosis. The yield prediction formula is an analytical approximation based on the importance sampling relation. This relation can also be used to reduce the sample size of the necessary Monte Carlo analyses. Finally the efficiency of the presented algorithm will be demonstrated on a switched-capacitor filter.

Manuscript received October 9, 1980; revised June 25, 1981.
The authors are with the Department of Electrical Engineering, Technical University of Munich, D-8000 Munich 2, Germany.

Introduction

FOR A DESIGN of electrical circuits, particulary of semiconductor integrated circuits, the statistical variations of the circuit parameter values must be considered.

These statistical fluctuations due to manufacturing tolerances will involve a variation of the circuit's response. In this case the manufacturing yield—which is the proportion of manufactured circuits fulfilling the desired performance specifications—is a very useful design criterion.

Design centering is the process of defining a set of nominal parameter values to maximize the yield for known but unavoidable statistical fluctuations of the circuit parameters.

During the last years, several proposals for this task have been published which can be classified into geometric and statistical techniques.

Reprinted from *IEEE Trans. on Circuits and Systems,* vol. CAS-29, no. 2, pp. 88–95, February 1982.

The geometric approaches give a meaningful insight into the design centering problems [1], [2], but the proposed algorithms assume some properties (e.g., convexity) of the acceptance region and are difficult to deal with a large number of circuit parameters [3], [4]. In [5] efforts to overcome these difficulties are made.

On the other hand, the significant amount of computer time is a frequently mentioned drawback of the statistical approaches. Nevertheless, only the Monte Carlo method—used in the statistical techniques [6]–[19]—enables the designer to estimate the yield of a given design in a reliable way.

One of the first published approaches concerning statistical network optimization is the method by Kjellström *et al.* [6], [7] which has been successfully demonstrated on large examples.

There are also other approaches applicable for practical examples [10], [11]. Other methods are especially suited for linear circuits [12], [13]. The calculation of yield, its gradient and Hessian matrix based on a single Monte Carlo analysis, and the application to some simple optimization methods are derived in [14] first.

In this context, a design procedure based on "parametric sampling" [15] should be mentioned. This procedure is derived from the importance sampling concept [16]. With this approach, the manufacturing yield and the yield gradient can be computed for different tolerance situations and different nominal values. Thus well-known optimization methods are applicable.

In this paper, a design centering approach is proposed which is also based on the concept of importance sampling and which was first published in [17] and [18]. Here, we derive an analytical yield prediction formula which contains the yield-gradient and the Hessian matrix.

Due to the quadratic form, this yield prediction is valid in an extensive region around the nominal parameter values. In order to control the statistical uncertainty of the yield prediction, we deduce a variance prediction in addition and formulate design centering as a constrained optimization problem. An eigenvalue decomposition is proposed, to solve the optimization problem (in the case of a nonpositive definite system matrix) in an interactive manner.

Problem Formulation

Consider a set of circuit performance functions $\varphi(p)$ which depends on a set of parameter values p. The acceptance of a circuit by a specification test can be expressed as follows:

$$\varphi(p) \in \Phi. \tag{1}$$

Φ is the region of acceptable performance specifications in the performance space. With

$$\varphi(p) \in \Phi \rightarrow p \in R \tag{2}$$

the acceptance region R in the parameter space is defined.

This region R can be described by the acceptance function $\delta(p)$ such that

$$\delta(p) = \begin{cases} 1, & \text{if } \varphi(p) \in \Phi\text{: circuit accepted} \rightarrow p \in R \\ 0, & \text{if } \varphi(p) \in \bar{\Phi}\text{: circuit rejected} \rightarrow p \in \bar{R}. \end{cases} \tag{3}$$

The parameter values p are statistically varying with a probability density function (pdf) $g(p)$. Thus the manufacturing yield Y of a circuit can be formulated by

$$Y = \int_R \cdots \int g(p)\, dp. \tag{4}$$

With the aid of the acceptance function $\delta(p)$, Y can be expressed as an expectation with respect to $g(p)$:

$$Y = \int_{-\infty}^{+\infty} \cdots \int \delta(p) \cdot g(p) \cdot dp = \mathop{E}_{g}\{\delta(p)\}. \tag{5}$$

An unbiased estimator of Y is

$$\hat{Y} = \frac{1}{N} \cdot \sum_{\nu=1}^{N} \delta(p^{(\nu)}) = \frac{n}{N} \tag{6}$$

where n is the number of accepted circuits which pass the specification test and N is the total number of considered circuits.

Hence, the manufacturing yield can be estimated by repeating network analyses and performance specification tests (Monte Carlo analysis).

The proposed approach to design centering starts at the initial nominal values ξ_0 of the circuit parameters. This initial situation is determined by the initial pdf $g_0(p, \xi_0, C_0)$, containing the mean value vector ξ_0 and for instance the variance-covariance matrix C_0, and by the initial circuit yield Y_0

$$\left.\begin{aligned} Y_0 &= \mathop{E}_{g_0}\{\delta(p)\} \\ \xi_0 &= \mathop{E}_{g_0}\{p\} \\ C_0 &= \mathop{E}_{g_0}\{(p-\xi_0)(p-\xi_0)^T\} \end{aligned}\right\} \text{expectations with respect to } g_0(p). \tag{7}$$

For further use, an additional pdf $h(p)$ is introduced, which is obtained from $g_0(p)$ by truncation with $\delta(p)$

$$h(p, \xi_H, C_H) = Y_0^{-1} \cdot \delta(p) \cdot g_0(p, \xi_0, C_0). \tag{8}$$

The corresponding mean value vector and variance-covariance matrix can also be expressed as expectations with respect to $g_0(p)$:

$$\xi_H = \mathop{E}_{g_0}\{Y_0^{-1} \cdot p \cdot \delta(p)\} \tag{9}$$

$$C_H = \mathop{E}_{g_0}\{Y_0^{-1} \cdot (p-\xi_H)(p-\xi_H)^T \cdot \delta(p)\}. \tag{10}$$

Note that the corresponding values $\hat{Y}_0$, $\hat{\xi}_H$, $\hat{C}_H$, estimated on the basis of random samples according to $g_0(p)$, give information about the acceptance region R in the parameter space.

Yield Prediction

We develop a prediction of the circuit yield Y assuming a change of the probability density function from g_0 to g. For this purpose we use the relation known as the method of importance sampling [16] so that the yield Y can be expressed as an expectation with respect to $g_0(p)$

$$Y=\int_{-\infty}^{+\infty}\cdots\int \delta(p)\cdot g_0(p)\frac{g(p)}{g_0(p)}dp$$
$$=\underset{g_0}{E}\left\{\delta(p)\frac{g(p)}{g_0(p)}\right\}. \quad (11)$$

Note that the importance sampling relation—familiar in connection with variance reduction [15]—is used to formulate the expectation with respect to $g_0(p)$.

We now assume multinormal distributions for $g_0(p,\xi_0,C_0)$ and $g(p,\xi,C)$. With regard to practical requirements as well as to an especially clear illustration of the new approach this assumption is very useful. Under this condition we obtain from (11)

$$Y=\underset{g_0}{E}\left\{\delta(p)\cdot\frac{\sqrt{\det C_0}\cdot\exp\left[-\frac{1}{2}(p-\xi)^T\cdot C^{-1}\cdot(p-\xi)\right]}{\sqrt{\det C}\cdot\exp\left[-\frac{1}{2}(p-\xi_0)^T\cdot C_0^{-1}\cdot(p-\xi_0)\right]}\right\}. \quad (12)$$

Design centering implies only a change $\Delta\xi$ from ξ_0 to ξ whereas the initial variance-covariance matrix C_0 remains unchanged:

$$\Delta\xi=\xi-\xi_0 \qquad C=C_0. \quad (13)$$

Hence, (12) can be written in the form

$$Y=\underset{g_0}{E}\left\{\delta(p)\cdot\exp\left[(p-\xi_0)^T\cdot C_0^{-1}\cdot\Delta\xi-\tfrac{1}{2}\Delta\xi^T\cdot C_0^{-1}\cdot\Delta\xi\right]\right\}. \quad (14)$$

Using a Taylor series expansion of the exponential form in (14) about ξ_0 and applying the expectation to the single terms we get

$$Y\approx Y_0\left[1+(\xi_H-\xi_0)^T\cdot C_0^{-1}\Delta\xi-\tfrac{1}{2}\Delta\xi^T\cdot C_0^{-1}\cdot\left(C_0-C_H-(\xi_H-\xi_0)(\xi_H-\xi_0)^T\right)\cdot C_0^{-1}\Delta\xi\right]. \quad (15)$$

Thereby the cubic and higher order terms of $\Delta\xi$ are neglected. In [18] this derivation is explained in detail. This yield prediction formula is an analytical approximation function with respect to the changes $\Delta\xi$ of the nominal parameter values.

In order to get a better global yield-approximation, we propose the following, somewhat heuristical conversion into an exponential form

$$Y_p(\Delta\xi)=Y_0\cdot\exp\left[(\xi_H-\xi_0)^T\cdot C_0^{-1}\cdot\Delta\xi-\tfrac{1}{2}\Delta\xi^T\cdot C_0^{-1}(C_0-C_H)\cdot C_0^{-1}\Delta\xi\right]. \quad (16)$$

Note that the Taylor series expansion of (16) about $\Delta\xi=0$ considering the first- and second-order terms of $\Delta\xi$ is identical with (15).

It is worth mentioning that the step from (15) to (16) is not essential for the yield prediction. But due to this transformation, the system matrix of (16) is "more positive definite" than those of (15). Thus difficulties of the design centering procedure are reduced as pointed out below.

Variance Prediction

An estimation $\hat{Y}_0$ of the initial circuit yield Y_0

$$\hat{Y}_0=\frac{1}{N_0}\sum_{\nu=1}^{N_0}\delta(p^{(\nu)}) \quad (17)$$

is characterized by a binominal random process. For a sufficient large number N_0 of random samples $p^{(1)}\cdots p^{(\nu)}\cdots p^{(N_0)}$ the binominal distribution of $\hat{Y}_0$ can be approximately described by the normal distribution ($N_0\gtrsim 100$). Hence, the variance $V(\hat{Y}_0)$ is a well-known measure of the expected error in the estimate $\hat{Y}_0$,

$$V(\hat{Y}_0)=\frac{1}{N_0}\cdot\underset{g_0}{E}\left\{\left[\delta(p)-\underset{g_0}{E}\{\delta(p)\}\right]^2\right\}. \quad (18)$$

Since the expectation $E_{g_0}\{\delta(p)\}$ is Y_0, (18) leads to

$$V(\hat{Y}_0)=\frac{1}{N_0}\cdot Y_0\cdot(1-Y_0). \quad (19)$$

In order to determine the uncertainty of the yield prediction we must consider the expectation Y in (11). Hence, an estimator of Y with respect to $g_0(p)$ is

$$\hat{Y}=\frac{1}{N_0}\cdot\sum_{\nu=1}^{N_0}\delta(p^{(\nu)})\cdot\frac{g(p^{(\nu)})}{g_0(p^{(\nu)})}. \quad (20)$$

From the considerations above it is evident that the variance $V(\hat{Y})$ of Y is given by

$$V(\hat{Y})=\frac{1}{N_0}\cdot\underset{g_0}{E}\left\{\left[\delta(p)\cdot\frac{g(p)}{g_0(p)}-Y\right]^2\right\}. \quad (21)$$

Under the same conditions as the yield prediction (16) is derived from (11) we can obtain a variance prediction from (21):

$$V_P(\hat{Y})=V(\hat{Y}_0)\left(\frac{Y_P(\Delta\xi)}{Y_0}\right)^2\cdot\left(\frac{\exp\left[\Delta\xi^T\cdot C_0^{-1}\cdot C_H\cdot C_0^{-1}\cdot\Delta\xi\right]-Y_0}{1-Y_0}\right) \quad (22)$$

where $Y_P(\Delta\xi)$ is given by (16).

In this variance prediction formula, derived in [19], the first- and second-order terms of $\Delta\xi$ are considered.

Since in (22) C_H is positive definite, and Y_P is greater than Y_0 for a design centering step, we see that $V_P(\hat{Y})$ is greater than $V(\hat{Y}_0)$ as expected.

An approximation of $\hat{Y}$ in (20) is obtained from (16)

$$\hat{Y}_{P1}(\Delta\xi)=\hat{Y}_0\cdot\exp\left[(\hat{\xi}_H-\xi_0)^T\cdot C_0^{-1}\cdot\Delta\xi-\tfrac{1}{2}\Delta\xi^T\cdot C_0^{-1}\cdot(C_0-\hat{C}_H)\cdot C_0^{-1}\cdot\Delta\xi\right] \quad (23)$$

where $\hat{Y}_0$, $\hat{\xi}_H$, and $\hat{C}_H$ can be estimated via Monte Carlo analysis. Hence, we conclude that the expression in (22) is also a quadratic approximation of the variance $V(\hat{Y}_{P1})$.

Note that the square root of the variance $V_P(\hat{Y}) \approx V_P(\hat{Y}_{P1})$ is a measure of the statistical uncertainty in the yield prediction $\hat{Y}_{P1}$.

In order to improve the statistical certainty of the yield prediction $\hat{Y}_{P1}$ in (23) we use the following yield prediction $\hat{Y}_{P2}$:

$$\hat{Y}_{P2}(\Delta\xi) = \hat{Y}_0 \cdot \exp\Big[(\hat{\xi}_H - \hat{\xi}_0)^T \cdot C_0^{-1} \cdot \Delta\xi - \tfrac{1}{2}\Delta\xi^T \cdot C_0^{-1} \cdot (\hat{C}_0 - \hat{C}_H) \cdot C_0^{-1} \cdot \Delta\xi\Big]. \quad (24)$$

Though, practical experiences and tedious derivations of regression estimates [20] show that normally

$$V(\hat{Y}_{P2}) < V(\hat{Y}_{P1}) \quad (25)$$

the improvement of the statistical certainty is not very significant.

Therefore, we use $V_P(\hat{Y})$ in (22) as a worst-case prediction for $V(\hat{Y}_{P2})$.

Design Centering by Constrained Optimization

The proposed statistical approach to design centering can be characterized by maximizing the circuit yield Y directly with the aid of the yield prediction formula, whose parameters are estimated via Monte Carlo analysis.

$$Y(\Delta\xi) \approx \hat{Y}_{P2}(\Delta\xi) = \hat{Y}_0 \cdot \exp\left[\Gamma_\xi(\Delta\xi)\right] \stackrel{!}{=} \max. \quad (26)$$

Considering $Y_{P2}(\Delta\xi)$ as shown in (24), the generalized problem is to maximize the quadratic function

$$\Gamma_\xi(\Delta\xi) = (\hat{\xi}_H - \hat{\xi}_0)^T \cdot C_0^{-1} \cdot \Delta\xi - \tfrac{1}{2}\Delta\xi^T \cdot C_0^{-1}(\hat{C}_0 - \hat{C}_H) \cdot C_0^{-1} \cdot \Delta\xi \stackrel{!}{=} \max. \quad (27)$$

This optimization problem seems to be very simple. Unfortunately the following difficulties must be considered:

the restricted validity of the yield prediction formula;

the ill conditioning due to nearly linear dependent circuit parameters;

the uncertainties due to the statistical estimation of the elements $\hat{\xi}_0$, $\hat{\xi}_H$, $\hat{C}_0$, and $\hat{C}_H$;

the occasional appearance of saddle-point behavior in the yield prediction formula ($\hat{C}_0 - \hat{C}_H$) is not positive definite.

In order to consider the restricted validity of the objective function and the ill-conditioned maximization problem, the design centering task is to be formulated by constrained optimization which limits the magnitudes of the parameter changes $\Delta\xi$.

As a constraint we, therefore, propose the normalized magnitude $\|\Delta\eta\|$ of $\Delta\xi$.

$$\|\Delta\eta\| = \sqrt{\Delta\eta^T \cdot \Delta\eta} = \sqrt{\Delta\xi^T \cdot C_0^{-1} \cdot \Delta\xi} = \text{const}. \quad (28)$$

This normalization due to the matrix C_0^{-1}, is very useful with regard to the statistical uncertainty of the quadratic form (27), as pointed out below.

The relation between $\Delta\xi$ and $\Delta\eta$

$$\Delta\eta = R_0 \cdot \Sigma_0^{-1} \cdot \Delta\xi \quad (29)$$

is obtained by decomposition of the matrix C_0^{-1}

$$C_0^{-1} = (\Sigma_0 \cdot K_0 \cdot \Sigma_0)^{-1} = \Sigma_0^{-1} \cdot R_0^T \cdot R_0 \cdot \Sigma_0^{-1} \quad (30)$$

with

$$R_0^T \cdot R_0 = K_0^{-1}. \quad (31)$$

Σ_0 is the diagonal matrix containing the standard deviations and K_0 is the correlation matrix involving the correlation coefficients. Since K_0^{-1} is real, symmetric and positive definite, R_0 can easily be calculated, e.g., via Cholesky decomposition and becomes a real upper triangular matrix.

With the aid of these considerations, we get the following constrained maximization problem for each design centering step:

$$\Gamma_\eta(\Delta\eta) = (\hat{\xi}_H - \hat{\xi}_0)^T \cdot \Sigma_0^{-1} \cdot R_0^T \cdot \Delta\eta - \tfrac{1}{2}\Delta\eta^T \cdot R_0 \cdot \Sigma_0^{-1}(\hat{C}_0 - \hat{C}_H) \cdot \Sigma_0^{-1} \cdot R_0^T \cdot \Delta\eta \stackrel{!}{=} \max$$

$$\text{subject to} \quad \|\Delta\eta\| = \sqrt{\Delta\xi^T \cdot C_0^{-1} \cdot \Delta\xi} = \text{const}. \quad (32)$$

Considering the statistical certainty of the yield prediction, it is evident that with increased changes $\Delta\xi$, the yield prediction viz. (27) gives information about a region with a decreased number of random samples. The variance prediction (22) confirms this fact.

Since the random samples were chosen according to $g_0(p)$, the proposed constraint (28) corresponds to elliptical contours of constant densities of the random samples. Using this constraint the statistical certainty of the yield-prediction formula can be controlled by (22) in an appropriate way.

Now we perform an eigenvalue decomposition of the system matrix

$$R_0 \cdot \Sigma_0^{-1}(\hat{C}_0 - \hat{C}_H) \cdot \Sigma_0^{-1} \cdot R_0^T = U \cdot D \cdot U^T \quad (33)$$

where U is an orthogonal matrix, containing all eigenvectors ($U \cdot U^T = I$, I is the unit matrix) and D is a diagonal matrix involving all eigenvalues

$$D = \operatorname{diag}(d_1, \cdots, d_p, \cdots, d_r). \quad (34)$$

Using the linear mappings

$$x = U^T \cdot \Delta\eta; \; b = U^T \cdot R_0 \cdot \Sigma_0^{-1}(\hat{\xi}_H - \hat{\xi}_0) \quad (35)$$

we get the constrained optimization problem

$$\Gamma(x) = b^T \cdot x - \tfrac{1}{2}x^T \cdot D \cdot x \stackrel{!}{=} \max$$

$$\text{subject to } x^T \cdot x = \text{const}. \quad (36)$$

With the aid of the positive parameter λ ($0 < \lambda < \infty$), we transform (36) into the unconstrained optimization problem

$$\Gamma_m(x) = b^T \cdot x - \tfrac{1}{2} x^T \cdot D \cdot x - \tfrac{1}{2}\lambda \cdot x^T \cdot x \stackrel{!}{=} \max. \quad (37)$$

Thus the stationary point of Γ_m is given by

$$x_s = \left[\frac{b_1}{\lambda + d_1}, \cdots, \frac{b_\rho}{\lambda + d_\rho}, \cdots, \frac{b_r}{\lambda + d_r}\right]^T \quad (38)$$

with

$$[b_1, \cdots, b_\rho, \cdots, b_r]^T = b. \quad (39)$$

Due to this eigenvalue decomposition, a saddle-point behavior of the single components of x_s can be recognized by their negative eigenvalues. In order to obtain a proper maximization step, a selection of a well-conditioned subsystem is proposed in [18].

In this paper an alternative treatment of the negative eigenvalues should be mentioned.

Corresponding to a proposal in [21], we change the signs of the negative eigenvalues in (38) and get a somewhat suboptimal solution:

$$x_0 = \left[\frac{b_1}{\lambda + |d_1|}, \cdots, \frac{b_\rho}{\lambda + |d_\rho|}, \cdots, \frac{b_r}{\lambda + |d_r|}\right]. \quad (40)$$

Let us suppose, that λ is determined in an appropriate way. Then the result of maximizing Γ_m yields the desired mean value vector ξ_1 as an improvement of ξ_0 considering (29) and (35)

$$\Delta\xi_0 = \Sigma_0 \cdot R_0^{-1} \cdot U \cdot x_0 \rightarrow \xi_1 = \xi_0 + \Delta\xi_0. \quad (41)$$

The corresponding value of the maximized circuit yield can be derived from (40), (36), and (26):

$$\hat{Y}_{P\max}(\lambda) = \hat{Y}_0 \exp\left[\sum_{\rho=1}^{r} b_\rho^2 \frac{\lambda + |d_\rho| - \frac{1}{2} d_\rho}{(\lambda + |d_\rho|)^2}\right]. \quad (42)$$

Thus the normalized magnitude of $\Delta\xi_0$ is given by

$$\sqrt{\Delta\xi_0^T \cdot C_0^{-1} \cdot \Delta\xi_0} = \|x_0\| = \sqrt{\sum_{\rho=1}^{r} \frac{b_\rho^2}{(\lambda + |d_\rho|)^2}}. \quad (43)$$

The parametric representation with respect to λ in (42) and (43) leads to the maximized yield prediction as a function of $\|x_0\|$:

$$\hat{Y}_{P\max} = f_1(\|x_0\|). \quad (44)$$

From (40), (41), and (43) we obtain the variance prediction according (22) as a function of $\|x_0\|$:

$$V(\hat{Y}_{P\max}) = f_2(\|x_0\|). \quad (45)$$

Note that the simple calculation of x_0 due to the eigenvalue decomposition (see (40)) allows the determination of $\hat{Y}_{P\max}$ and $V(\hat{Y}_{P\max})$ for a large number of λ so that a graphic representation of (44) and (45) can easily be obtained, as shown in Figs. 2 and 3.

For an appropriate interpretation, we prefer the square root of the variance as a measure of the expected error in the estimate $\hat{Y}_{P\max}$. The representation of both figures on a graphic terminal serves as a suitable criterion to control an interactive maximization process. With the aid of these graphic representations, the circuit designer chooses an appropriate normalized magnitude $\|x_0\|$, to get an acceptable compromise between the yield improvement and the decrease in statistical certainty. For this decision, elementary statistical relations can be used, for example, the single-sided confidence level.

If the variance prediction curve indicates an uncertainty which is too large and no compromise can be accepted, additional random samples must be carried out, to get a smaller value of the variance.

Hence, these curves are an appropriate assessment of the necessary sample size so that a minimum number of random samples can be effected in each design centering step. After calculating the improved nominal values according to (41), this design centering step is finished. In order to get a further improvement of the nominal values, the next design centering step with the changed nominal values including the Monte Carlo analysis is to be carried out. The end of the design centering process is reached when the graphic figures indicate that no further yield improvement can be achieved.

Reduction of the Sample Size

In any design centering step the use of the yield prediction and variance prediction formulas (24) and (22) requires the calculation of the estimates $\hat{Y}_k$, $\hat{\xi}_k$, $(\hat{\xi}_H)_k$, $\hat{C}_k$, and $(\hat{C}_H)_k$, where k is the kth step.

For a given accuracy of the estimate $\hat{Y}_k$, the necessary number N_k of random samples is obtained from (19) if the random samples of the kth Monte Carlo analysis are only used.

In this section it is shown that using the relation of importance sampling we can exploit random samples of past steps for the calculation of the estimates in the kth step. Doing this, the number N_k of random samples in the kth Monte Carlo procedure can be reduced while retaining accuracy of estimation [23].

For the kth design centering step with the probability density function $g_k(p)$, the yield estimate using the ith Monte Carlo analysis can be expressed with the aid of the importance sampling relation:

$$\hat{Y}_k^{(i)} = \frac{1}{N_i} \sum_{\nu=1}^{N_i} \delta_i(p^{(\nu)}) \frac{g_k(p^{(\nu)})}{g_i(p^{(\nu)})}. \quad (46)$$

The variance of the estimate $\hat{Y}_k^{(i)}$ is

$$V(\hat{Y}_k^{(i)}) = \frac{1}{N_i - 1}\left[\frac{1}{N_i} \sum_{\nu=1}^{N_i} \delta_i(p^{(\nu)}) \left(\frac{g_k(p^{(\nu)})}{g_i(p^{(\nu)})}\right)^2 - \hat{Y}_k^{(i)^2}\right]. \quad (47)$$

Exploiting all Monte Carlo procedures $(i = 0, 1, \cdots, k)$, the estimate $\hat{Y}_k$ can be formulated as

$$\hat{Y}_k = \sum_{i=0}^{k} \alpha_i \cdot \hat{Y}_k^{(i)}. \quad (48)$$

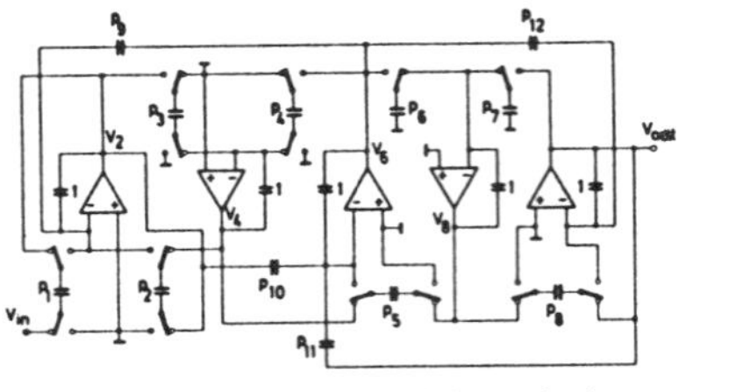

Fig. 1. Switched-capacitor filter with 12 design parameters.

The weights α_i must be chosen so that $V(Y_k)$ is minimized and the constraint

$$\sum_{i=0}^{k} \alpha_i = 1$$

is satisfied. Thus we obtain

$$\alpha_i = \frac{V(\hat{Y}_k)}{V(\hat{Y}_k^{(i)})} \tag{49}$$

where

$$\frac{1}{V(\hat{Y}_k)} = \sum_{i=0}^{k} \frac{1}{V(\hat{Y}_k^{(i)})}. \tag{50}$$

Using (19), the necessary number of random samples in the kth Monte Carlo analysis is

$$N_k = \frac{Y_k(1-Y_k)}{V(\hat{Y}_k^{(k)})} \tag{51}$$

and with (50), we get

$$N_k = Y_k(1-Y_k)\left[\frac{1}{V(\hat{Y}_k)} - \sum_{i=0}^{k-1} \frac{1}{V(\hat{Y}_k^{(i)})}\right]. \tag{52}$$

From (52) it is evident that the exploitation of the past Monte Carlo procedures $(i=0,1,\cdots,k-1)$ always reduces the sample size N_k in the kth design centering step.

The estimates for the past Monte Carlo procedures $(i=0,1,\cdots,k-1)$ are obtained without additional network analyses, because the results of the performance specification tests $[\delta_i(p^{(\nu)})]$ can be stored.

In accordance with (46) and (47) the estimates $\hat{\xi}_k^{(i)}$, $(\hat{\xi}_H)_k^{(i)}$, $\hat{C}_k^{(i)}$, $(\hat{C}_H)_k^{(i)}$ and the corresponding variances must be calculated in the same manner as $\hat{Y}_k^{(i)}$ and $V(\hat{Y}_k^{(i)})$. Note, that in general the calculation of the estimates $\hat{\xi}_k$, $(\hat{\xi}_H)_k$, $\hat{C}_k$, and $(\hat{C}_H)_k$ leads to different values of the corresponding weights α_i.

For determining the sample size N_k, we use only (52) with a desired variance $V(\hat{Y}_k)$ of the circuit yield estimate $\hat{Y}_k$.

Experimental Results

The yield prediction and the design centering algorithms have been implemented using ANS Fortran (approximately 5000 source statements). This program is called TACSY (Tolerance Analysis and Design Centering System). The circuit analysis and the performance specification test must be carried out via subroutine call. TACSY is highly portable and has been tested on several computers.

This program was applied to the switched capacitor filter, shown in Fig. 1. The z-transform of the node voltages can be described with (53):

$$\begin{bmatrix} 1+\frac{p_8}{z-1} & 0 & 0 & p_{12} & \frac{-p_8}{z-1} \\ 0 & 1+\frac{(p_1-p_2)}{z-1} & \frac{p_2}{z-1} & p_9 & 0 \\ 0 & \frac{-p_3 z}{z-1} & 1 & \frac{-p_4 z}{z-1} & 0 \\ p_{11} & p_{10} & \frac{p_5}{z-1} & 1 & \frac{-p_5}{z-1} \\ \frac{p_7 z}{z-1} & 0 & 0 & \frac{p_6 z}{z-1} & 1 \end{bmatrix} \begin{bmatrix} V_{out} \\ V_2 \\ V_4 \\ V_6 \\ V_8 \end{bmatrix} = \begin{bmatrix} 0 \\ \frac{p_1}{z-1} V_{in} \\ 0 \\ 0 \\ 0 \end{bmatrix}. \tag{53}$$

With $z=\exp(j2\pi f/f_c)$ we obtain the frequency-domain behavior where f is the frequency parameter and $f_c=128$ Hz stands for the clock frequency. The insertion loss relative to the loss at $f=800$ Hz is given by

$$a(f) = 20\cdot\log_{10}\frac{V_{out}(800)}{V_{out}(f)}. \tag{54}$$

Equation (54) has to fulfill the PCM performance-qualification listed in Table I. For checking up these specifications, the circuit was tested at the frequencies $f=100$ kHz, $f=10.0$ kHz, and from 1.2 to 4.7 kHz in steps of 100 Hz. In the critical frequency range range from 3.0 to 3.4 kHz, the step size was reduced to 40 Hz.

For solving (53) and checking the performance-specifications, a Fortran subroutine was written and loaded to the TACSY program.

The 12 parameters $p_1 \cdots p_\rho \cdots p_{12}$, as shown in Fig. 1 and in (53), are assumed to have normal probability density functions with mean values ξ_ρ listed in Table II and with standard deviations of $\sigma_\rho = 0.02\cdot\xi_\rho$. Correlation coefficients between the parameters are supposed to be zero.

TABLE I
PERFORMANCE SPECIFICATIONS FOR THE SWITCHED-CAPACITOR FILTER

Frequency range f(KHz)	Loss specification a(f) [dB]	
	Minimum	Maximum
f < 3.0	- 0.25	+ 0.25
3.0 < f < 3.4	- 0.25	+ 0.9
3.4 < f < 3.6	- 0.25	-
3.6 < f < 4.6	+ 14·sin[π·(f-4)/1.2] + 14	-
4.6 < f < 10.0	+ 28	-

TABLE II
MEAN VALUES OF THE CIRCUIT PARAMETERS

circuit parameters	mean values initial design	mean values centered design	mean value changes from the initial to the centered design
p_1	0.177	0.191	7.9 %
p_2	0.0887	0.0911	2.7 %
p_3	0.307	0.320	4.2 %
p_4	0.307	0.305	- 0.65%
p_5	0.0527	0.0537	1.9 %
p_6	0.168	0.163	- 3.0 %
p_7	0.168	0.174	3.6 %
p_8	0.0994	0.103	3.6 %
p_9	0.684	0.623	- 8.9 %
p_{10}	0.406	0.394	- 3.0 %
p_{11}	0.117	0.120	2.6 %
p_{12}	0.220	0.217	- 1.4 %

With the mean values of the initial design a Monte Carlo analysis using 500 random samples was carried out. From this statistical simulation we get the estimates $\hat{Y}_0$, $\hat{\xi}_H - \hat{\xi}_0$, and $\hat{C}_0 - \hat{C}_H$ considering (7), (9), and (10).

Solving the optimization task, we get the graphic representations of the yield prediction and the variance prediction formulas with regard to (44) and (45) for this first step, as shown in Figs. 2 and 3.

With the yield and variance prediction curves the step size of each optimization step can be determined. Thereby a compromise between an increase in yield and a decrease in statistical certainty must be considered. In the two figures it is shown that for small values of $\|x_0\|$ the predictions are almost identical to the "correct" values estimated via Monte Carlo analyses.

For the first maximization step a value of $\|x_0\| = 1.0$ and an increase of yield from 0.24 to 0.36 seems to be an acceptable compromise. With (40) and (41) the new mean values of the circuit parameters were calculated. With these mean values a new Monte Carlo procedure and the next

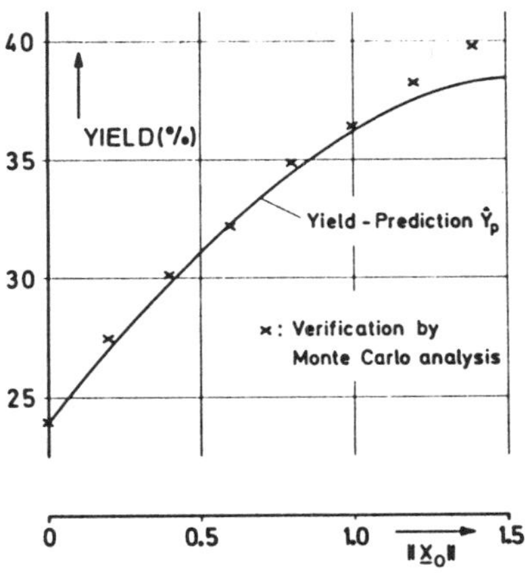

Fig. 2. Yield prediction.

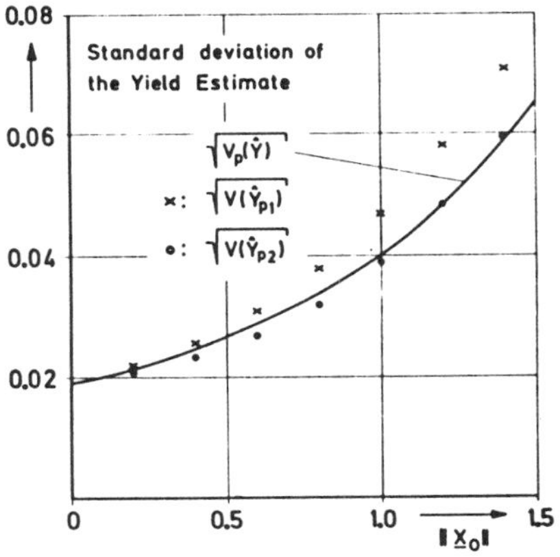

Fig. 3. Variance prediction.

maximization step were performed. After four steps a circuit yield of 0.54 was attainable. Now, near the design center the yield prediction curve becomes very flat and the corresponding variance prediction indicates relative large values in comparison to the increase in yield. To get a certain improvement of the circuit yield near the design center, the variance of the yield estimate must normally be reduced. After several steps with increased sample size the circuit yield of 0.60 was attained. In Table II the changes of the mean values from the initial to the centered design are listed.

Note, that the design center has always to be reached with a circuit yield $\hat{Y} < 1$. For $\hat{Y} = 1$ there is no guarantee that the design center of the acceptance region R with respect to chosen parameter tolerances will be attained. If the situation $\hat{Y} \approx 1$ occurs within the design centering process a change to $\hat{Y} < 1$ must be carried out, either by increasing the parameter tolerances or by restricting the circuit performance specifications.

For this example, in Fig. 4 the circuit's response due to the initial and the centered design are shown. Comparing the insertion loss of the initial design with the centered one and regarding the performance specifications, it is evident that the design centering procedure equalizes the attenuation reserves in the pass and stopband.

For an obvious illustration of the circuit improvement by design centering we propose a presentation according to

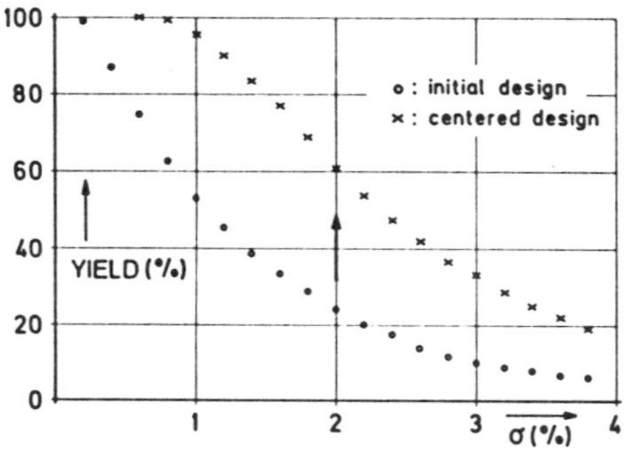

Fig. 4. Performance specifications and insertion loss.

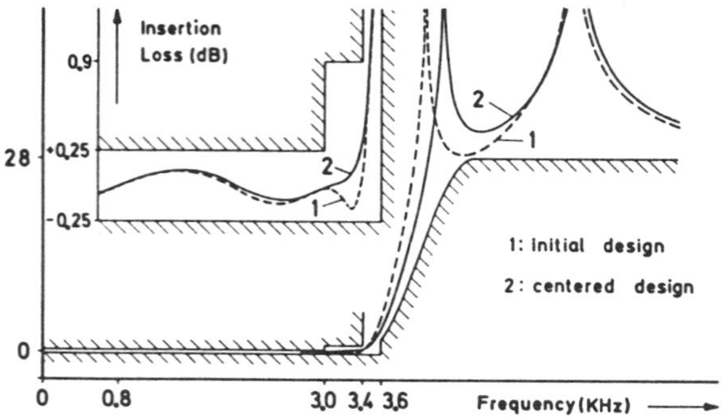

Fig. 5. Circuit yield for different values of σ.

Fig. 5. There, the yield for the initial and the centered circuit design is shown for different values of the standard deviations of the circuit parameters. Note, that the design centering procedure leads to a better design for a wide range of tolerance values, though the maximizing of yield is carried out for the relative standard deviation $\sigma_p = 2$ percent of all parameters ($\sigma_p = \sigma \cdot \xi_p$) only.

It is worth mentioning that these curves can be received with the aid of the previous Monte Carlo simulations, using the importance sampling relation without repeated network analyses.

Conclusions and Extensions

Finally some of the main aspects concerning the proposed approach will be listed.

1) Maximum yield with respect to given specifications is a very practical objective. A circuit, optimized via design centering for a specific set of tolerances will be close the optimum for a wide range of tolerance values (Fig. 5).

2) Note that the yield prediction formula is an analytical approximation based on the importance sampling relation. This relation can also be used to reduce the sample size in each design centering step. The complete Monte Carlo procedure must only be performed for the initial step.

3) With regard to the described difficulties typical for the yield maximization steps, a high-quality interactive optimization method which allows a quantitative problem diagnosis is necessary.

4) The proposed method of yield prediction and interactive optimization can also be used for tolerance assignment.

References

[1] E. Polak and A. Sangiovanni-Vincentelli, "Theoretical and computational aspects of the optimal design centering, tolerancing, and tuning problem," *IEEE Trans. Circuits Syst.*, vol. CAS-26, pp. 795–813, Sept. 1979.

[2] J. W. Bandler, P. C. Liu, and H. Tromp, "A nonlinear programming approach to optimal design centering, tolerancing, and tuning," *IEEE Trans. Circuits. Syst.*, vol. CAS-23, pp. 155–165, Mar. 1976.

[3] S. W. Director and G. D. Hatchtel, "The simplicial approximation approach to design centering," *IEEE Trans. Circuits Syst.*, vol. CAS-24, pp. 363–372, July 1977.

[4] J. W. Bandler and H. L. Abdel-Malek, "Optimal centering, tolerancing, and yield determination via updated approximations and cuts," *IEEE Trans. Circuits Syst.*, vol. CAS-25, pp. 853–871, Oct. 1978.

[5] L. M. Vidigal and S. W. Director, "Design-centering: The quasi-convex, quasi-concave performance function case," in *Proc. IEEE Int. Symp. on Circuits and Systems*, (Houston, TX), pp. 43–46, Apr. 1980.

[6] G. Kjellstroem, "Optimization of electrical networks with respect to tolerance costs," Ericsson Technics no. 3, pp. 157–175, 1970.

[7] G. Kjellstroem and L. Taxen, "On the efficient use of stochastic optimization in network design," in *Proc. IEEE Int. Symp. on Circuits and Systems*, (Munich, Germany) pp. 714–717, Apr. 1976.

[8] J. F. Pinel and K. Singhal, "Efficient Monte Carlo computation of circuit yield using importance sampling," in *Proc. IEEE Int. Symp. on Circuits and Systems*, (Phoenix, AZ), pp. 575–578, Apr. 1977.

[9] A. R. Thorbjornsen and E. R. Armbruster, "Computer-aided tolerance/correlation design of integrated circuits," *IEEE Trans. Circuits Syst.*, vol. CAS-26, pp. 763–767, Sept. 1979.

[10] N. J. Elias, "New statistical methods for assigning device tolerances," in *Proc. IEEE Int. Symp. on Circuits and Systems*, (Newton, MA), pp. 329–332, Apr. 1975.

[11] R. S. Soin and R. Spence, "Statistical design centering for electrical circuits," *Electron. Lett.*, vol. 14, no. 24, pp. 772–774, Nov. 1978.

[12] K. S. Tahim and R. Spence, "A radial exploration approach to manufacturing yield estimation and design centering," *IEEE Trans. Circuits Syst.*, vol. CAS-26, pp. 768–774, Sept. 1979.

[13] K. S. Tahim and R. Spence, "A radial exploration algorithm for the statistical analysis of linear circuits," *IEEE Trans. Circuits Syst.*, vol. CAS-27, pp. 421–425, May 1980.

[14] B. V. Batalov, Yu. N. Belyakov, and F. A. Kurmaev, "Some methods for statistical optimization of integrated microcircuits with statistical relations among the parameters of the components," *Microelectronica* (USA, translated from Russian), pp. 228–238, 1978.

[15] K. Singhal and J. F. Pinel, "Statistical design centering and tolerancing using parametric sampling," in *Proc. IEEE Int. Symp. on Circuits and Systems*, (Houston, TX), pp. 882–885, Apr. 1980.

[16] J. M. Hammersley and D. C. Handscomb, *Monte Carlo Methods.* Norwich, England: Fletcher, 1975.

[17] R. Koblitz, "Design centering and tolerance assignment of electrical circuits with gaussian-distributed parameter values," *Archiv Elek. Übertragung.* vol. 34, pp. 30–37, Jan. 1980.

[18] K. J. Antreich and R. K. Koblitz, "A new approach to design centering based on a multiparameter yield-prediction formula," in *Proc. IEEE Int. Symp. on Circuits and Systems*, (Houston, TX), pp. 886–889, Apr. 1980.

[19] K. J. Antreich and R. K. Koblitz, "An interactive procedure to design centering," in *Proc. IEEE Int. Symp. on Circuits and Systems*, (Chicago, IL), pp. 139–142, Apr. 1981.

[20] W. G. Cochran, *Sampling Techniques.* New York: Wiley, 1977.

[21] J. Greenstadt, "On the relative efficiencies of gradient methods," *Math. Comput.*, vol. 21, pp. 360–367, July 1967.

[22] B. J. Karafin, "The general component tolerance assignment problem in electrical networks," Ph.D. dissertation, Univ. of Pennsylvania, Philadelphia, 1974.

[23] K. Singhal and J. F. Pinel, "Statistical design centering and tolerancing using parametric sampling," *IEEE Trans. Circuits Syst.*, vol. CAS-28, pp. 692–702, July 1981.

FABRICATION BASED STATISTICAL DESIGN OF MONOLITHIC IC's[1]

W.Maly[2], A.J.Strojwas and S.W.Director

Department of Electrical Engineering,
Carnegie-Mellon University
Pittsburgh, PA 15213

Abstract

In this paper we discuss the statement of Yield Maximization problem and the choice of design parameters when considering statistical design of monolithic IC's. Specifically, yield maximization cannot be carried out in terms of the nominal values of the electrical parameters of IC's elements. An extension to the existing methods and a new formulation of the Yield Maximization problem for monolithic IC's is proposed. The necessity of employing a simulator of the manufacturing process , which relates a circuit electrical behavior to the physically designable parameters , is shown.

I. Introduction

A number of statistical design aids for yield estimation and optimization and worst case analysis have been developed (1, 2, 3, 4) to help the circuit designer wherever the random fluctuations inherent in the manufacturing process have to be considered. Specifically, for the case of the statistical design of monolithic IC's, the Yield Maximization (YM) is the central issue. Usually the yield maximization problem was formalized in the following manner.

Let the circuit to be designed be described by a set of algebraic and differential equations:

$$g_k(\mathring{y}, y, t, a, X) = 0 \quad \text{for } k = 1,2,\ldots,n_d \tag{1}$$

where

- y - is a vector of voltages and currents
- t - is time
- a - is a vector of the circuit parameters which can be assumed to be constant
- X - is a vector of random variables[3] representing circuit parameters which are randomly varying due to imperfections and distrubances in the manufacturing process, (typically resistances and capacitances, threshold voltage of MOS transistors etc.)

Let the constraints of the circuit performance be described by a set of inequalities:

$$\Phi_k(x) = \int_0^{t_f} \varphi_k(\mathring{y}, y, t, a, x)\,dt \leq 0 \quad \text{for } k = 1,2,\ldots,n_c \tag{2}$$

where x is a particular value of X. Thus the set

$$\mathbb{R}_X = \{x : \Phi_k(x) \leq 0 \text{ for } k = 1,2,\ldots,n_c\} \tag{3}$$

which is called the acceptability or feasible region, represents the set of realizations of the random variable X (i.e., all particular values of circuit parameters) for which the circuit meets the desired requirements. Assuming that X is described by a joint probability density function[4] (jpdf), $f_X(x, p_X^1, p_X^2)$, where p_X^1 is a vector of mean values of X and p_X^2 is a vector of higher order moments of X, the yield, Y can be defined as:

$$Y = \int_{\mathbb{R}_X} f(x, p_X^1, p_X^2)\,dx \tag{4}$$

In order to state the yield maximization problem and solve it by the existing methods, it is necessary to make the following two assumptions:

A1: It is possible to obtain any desired value of p_X^1 by means of adjusting the process parameters and/or the layout of the IC.

A2: The vector of higher order moments, p_X^2 is independent of the circuit and the process parameters.

Under these assumptions, the yield maximization problem should be stated as the following optimization problem:

$$\max_{p_X^1} \int_{\mathbb{R}_X} f(x, p_X^1, p_X^2)\,dx = Y_{max} \tag{5}$$

Unfortunately, assumptions A1 and A2 are not necessarily valid for monolithic IC applications. In particular:

i) Some components of p_X^1 , which are the optimization variables, are not designable in the sense that they are not necessarily adjustable to any given solution of (5). For example, the mean value of the threshold voltage of an MOS transistor cannot be arbitrarily adjusted to any desired value but can only be determined for a particular process and then only slightly modified.

ii) In general, the components of p_X^1 are not independent of one another. Therefore, it is not always physically possible to obtain a specified combination of values amongst the components of p_X^1. For example, the mean values of β of two n-p-n integrated transistors are strongly related to one another.

iii) The higher order moments, denoted by p_X^2 are not independent of p_X^1. For example, the mean and variance of the resistance of a monolithic resistor are dependent on its geometry.

In this paper we show that by judicious choice of truly independent designable parameters and an alternate formulation of the yield maximization problem we can develop a method which

[1] This research was funded in part by the National Science Foundation under Grant ECS79-23191

[2] On leave from Technical University of Warsaw, Poland

[3] All random variables are denoted by capital letters.

[4] For normally distributed random variables X the jpdf is fully characterized by the first and second order moments, i.e. by vector of mean values m_X and covariance matrix Σ_X

Reprinted from *IEEE Int. Symp. Circuits Syst.*, vol. 1, pp. 135–138, April 1981.

is more suitable for the design of monolithic IC's than those previously reported. Furthermore, we show that a key step in the yield maximization problem is the simulation of the manufacturing process. In particular we introduce the use of a process simulator (5) into the design procedure.

2. Independently Designable Parameters in Yield Maximization of Monolithic IC's

We wish to determine under what conditions the solution of (5), is technologically realizable. Assume that the manufacturing process can be characterized by a vector of deterministic, physically independent and controllable quantities (e.g. temperatures, diffusion times, mask dimensions, etc.),denoted by z and called primary design variables. Further let D denote the set of uncontrollable disturbances, inherent in the manufacturing process. Hence the parameters of the jpdf of X are dependent upon both z and D. Observe that while the values of p_X^1 can be controlled by changing z , the set of technologically realizable combinations is restricted due, in part ,to the existence of D. Thus there may exist the solutions of (5) which cannot be obtained by means of changes of z only. Specifically, the solution of (5) , $p_X^{1\,opt}$, is technologically realizable only if there exists a z^{opt} such that

$$p_X^{1opt} = p_X^1(z^{opt}) \tag{6}$$

It will always be possible to find a z^{opt} which satisfies (6) if the components of the vector $p_X^1 = \{p_{X1}^1, p_{X2}^1, ...\}$ are independent and strictly monotonic functions of components of the vector z . We call the random variable X_i, which is a component of X, whose mean, p_{Xi}^1 is a strictly monothonic function of a primary design variable z, a designable random variable. Furthermore, if we can change the value of p_{Xi}^1 while keeping the other components of p_X^1 constant, then X_i is said to be an independently designable random variable. Hence, a solution of (5) will be technologically realizable if every component of X is an independently designable random variable.

We now wish to illustrate the above general considerations and determine which parameters are the primary design variables and which parameters are the independently designable random variables for the case of monolithic integrated circuits. In order to motivate this discussion consider the following simple example. Let w and l be the width and length of a rectangular window in a photolithographic mask used to fabricate an integrated resistor. They constitute the primary design variables in this example. Due to imperfections in the photolithographic process the dimensions of this window in SiO_2 are described by random variables L and W, respectively. Assume for simplicity that $W = w + \Delta F$ and $L = l + \Delta F$, where ΔF is a zero mean normally distributed random variable with standard deviation $\sigma_{\Delta F}$ which represents the disturbance in the photolithographic process. One can show that if both l and w are much greater than $3\sigma_{\Delta F}$ then $F = L/W$ is a normally distributed random variable with mean $m_F = l/w$ and variance

$$\sigma_F^2 = \left[(w-l)/w^2\right]^2 \sigma_{\Delta F}^2 \tag{7}$$

Furthermore, it can be shown that the actual resistance of the integrated resistor is also a normally distributed random variable R such that $R = F{\cdot}Rs$ with mean m_R and variance σ_R^2 given by

$$m_R = \frac{l}{w} m_{Rs} \tag{8}$$

$$\sigma_R^2 = m_{Rs}^2[(w-l)/w^2]^2\sigma_{\Delta F}^2 + \left(\frac{l}{w}\right)^2\sigma_{Rs}^2 \tag{9}$$

where Rs is a normally distributed random variable with mean m_{Rs} and standard deviation σ_{Rs} which represents the sheet resistance. Thus R is a designable random variable because any specific mean of its resistance can be obtained by adjusting the ratio l/w. Moreover, it is an independently designable random variable because the mean values of the resistance of the different resistors in the circuit can be chosen independently of one another. Hence, it would appear that the nominal resistance values of integrated resistors could be used as optimization variables ,p_X^1, for solving the YM problem (5). However, since m_R and σ_R are dependent on each other this choice of variables would violate assumption A2.

Note that this observation can be extended to all electrical parameters of monolithic elements because both the nominal values and higher order moments of these electrical parameters are related to the mean and variances of the mask dimensions, as well as their ratios or window areas. (The moments of dimension ratios or areas are dependent of each other (e.g. see (7)). Hence none of the moments of electrical parameters of monolithic elements can be used as the optimization variable in the YM problem (5).

Note that for the examples discussed above, we can choose as optimization variables, instead of p_X^1, the means of L and W. (L and W are independently designable random variables because $m_L = l$ and $m_W = w$ and their variances are independent of m_L and m_W, respectively). In general, since any IC design can be described in terms of the mask dimensions, then the mean of the random variable $Z = z + \Delta F$, p_Z^1, representing the dimensions of the IC elements, which are related to the mask dimensions z and disturbances in the photolithographic process ΔF, can be chosen to be the optimization variable in the YM problem (5).

3. Yield Maximization Using the Process Simulation Technique

Since the components of X cannot represent the electrical parameters of the IC, the yield maximization problem (5), for case of monolithic IC's, has to be modified. Towards this end we replace the circuit equations (1) by:

$$g_k(\mathring{y}, y, t, a, x) = 0, \quad k = 1,2,...,n_d \tag{10}$$

$$c_j(x, Z) = 0, \quad j = 1,2,...,n_p \tag{11}$$

where expression (11) models the manufacturing process. Using (10) , (11) and (2) one can define the feasible region $\mathbb{R}_Z$ in the space defined by Z. We can now formally state the yield maximization problem as

$$\max_{p_Z^1} \int_{\mathbb{R}_Z} f_Z(z, p_Z^1, p_Z^2)\,dz \tag{12}$$

where $f_Z(z, p_Z^1, p_Z^2)$, p_Z^1 and p_Z^2 are the jpdf, means and higher order moments of Z, respectively. Thus by adding constraints describing the manufacturing process to the previous statement of the YM problem, and properly choosing elements of Z, the YM problem of monolithic IC's can be solved by the methods proposed previously (1).

Observe that, in general, the dimension of Z is much larger than the dimension of X. (For instance, the number of variables describing the layout coordinates of the zigzag resistor is much larger than the number of its electrical parameters). Thus, the computational expense of approximation to the feasible region $\mathbb{R}_Z$ in the space of independently designable parameters Z could be much larger than the computational expense of approximating $\mathbb{R}_X$ in the space of circuit electrical parameters X. Thus we we propose the following approach .

Assume that we have an approximation, H_X, to the feasible region $\mathbb{R}_X$. Since the random variable X is dependent on Z, which is an independently designable random variable related to the primary design variable z, then at least some moments of X are also dependent on z. Let $\tilde{p}_X^1$ and $\tilde{p}_X^2$ denote those moments of X which are dependent on z, and let $\bar{p}_X^1$ and $\bar{p}_X^2$ denote those moments which are independent of z. Hence $p_X^1 = \{\tilde{p}_X^1, \bar{p}_X^1\}$ and $p_X^2 = \{\tilde{p}_X^2, \bar{p}_X^2\}$ The YM problem can now be stated in the following way:

$$\max_{z} \int_{H_X} f_X\left[x,\tilde{p}_X^1,\bar{p}_X^1,\tilde{p}_X^2,\bar{p}_X^2\right] dx \qquad (13)$$

Note that any solution of (13) is technologically realizable and because we are in a lower dimensional space, the cost of the solution of (13) should be less than cost of the solution of (12). Of course, a key step in being able to solve (13) is the simulation of $\tilde{p}_X^1$ and $\tilde{p}_X^2$.

We now consider the process simulation technique we need for generating $\tilde{p}_X^1$ and $\tilde{p}_X^2$. As we stated earlier

$$X = P(z,D) \qquad (14)$$

where P ( ) is a model of the manufacturing process relating the elctrical circuit parameters X to z and D. The disturbance, D, is most easily simulated using an appropriate random number generator. The advantage of such an approach is that the disturbances of the process are characteristic of a given technology and manufacturing facility, but are independent of the particular circuit to be designed. Hence, the jpdf describing D, $f_D(d)$, needs to be identified only once for each process. Hence, employing (14) we can estimate moments of X in terms of z and therefore solve (13). In the next sections we describe the process simulator which has the capabilities we need.

4. Examples

In our investigation the statistical process simulator FABRICS (FABRication Process of Bipolar Integrated Circuits Simulator) was employed for simulating (14). A detailed description of this simulator is beyond the scope of this paper (see (5, 6, 7)). Suffice it to say that FABRICS was used to generate the data samples composed of electrical parameters of typical bipolar IC elements (i.e., samples of random variable X).

In this section we examine three examples which serve to illustrate the following observations. The first example illustrates the dependence of p_X^1 and p_X^2 on one another. The second example exhibits the fact that even if the nominal values, p_X^1, are constant, the higher moments, p_X^2, can be modified by means of layout changes . The third example demonstrates the computational efficiency of the proposed process simulation technique for solving the YM problem (13).

In each of the examples we have assumed that all of the process parameters are constant and the design variables are mask dimensions. Note that the results reported below were obtained using FABRICS tuned to a real manufacturing process, thus the data presented is physically meaningful.

Example 1.

Consider the following elements of a bipolar integrated circuit: three base diffusion resistors, R_1, R_2, R_3, and one n-p-n transistor. We assumed a fixed layout for the two resistors R_1 and R_2 and for the n-p-n transistor. The length of resistor R_3 was also assumed to be constant while its width, w_3, was treated as a primary design variable,z. The question was whether the changes of w_3 would affect all of the parameters of the jpdf describing the random variables R_1, R_2, R_3 and β of the transistor. Using FABRICS we generated 900 samples (each sample was composed of values for R_1, R_2, R_3 and β) for three values of w_3. The projections of four dimensional scatterplots onto the planes ($R_1 \times R_3$), ($R_2 x R_3$) and (β x R_3) are shown in Fig. 1. We can conclude that the changes of w_3 not only affect m_{R3} and σ_{R3} (according to formulas (7),(8) and (9)) but also the correlation factors of R_3 with the other resistors and β).

Example 2.

We now consider an IC which contains among other two resistors, R_1 and R_2 and an n-p-n bipolar transistor. The primary design variable,z, in this example is the layout scaling factor,λ , which is a quantity by which all element dimensions are multiplied. In Fig. 2 the scatterplots of R_1 and R_2, and R_1 and β, for λ = 1.0 are shown. Similar plots for λ = .5 are shown in Fig. 3. Comparing these two figures we observe that the means of R_1, R_2 and β remain almost unchanged while standard deviations increase and correlation factors decrease if λ decreases. Thus, we see that yield can be affected by holding the nominal values of the designable parameters constant and change the higher order moments which depend upon the scaling factor λ.

Example 3.

We wish to determine the dependence of the yield of the Motorola MC1530 operational amplifier (8) on the layout scaling factor λ. Towards this end we define acceptable performance in terms of the following inequalities:

$$-2mV \leq V_{in\ off} \leq 2mV\,, \quad I_{BIAS} \leq 6\mu A\,, \quad A_d \geq 8{,}000$$

where $V_{in\ off}$ is the dc input offset voltage, I_{BIAS} is the input bias current, and A_d is the differential mode gain.

The yield estimators for several values of λ (λ = .2, .3, .5, .8, 1.0, 1.5, and 2.0) were found by means of the Monte Carlo method ,using FABRICS and BIAS-D (9), to evaluate $V_{in\ off}$,I_{BIAS} and A_d and then computing the yield. The relation obtained between yield and λ is shown in fig. 4. We found that the yield drop for $\lambda < 1$ was due to an increase of σ_{Ad} which was caused by an increase of variances of element parameters. Observe that for increased values of λ, the yield never reaches 100%. This means that the Op Amp yield is determined not only by the designable part of X. Thus, as we have pointed out in the YM problem (13), the designable part of X should be distinguished from its undesignable part. We found also that the CPU time required for process simulation was less than 10% of the time required for circuit simulation.

5. Conclusions

In this paper we have discussed two formulations of the YM problem suitable for the design of monolithic IC's. We have defined the conditions under which the method proposed in (1) can be used for monolithic IC's. We have also proposed a different statement of the YM problem which seems to be suitable when large number of the IC's elements must be taken into account.

We have found that,in general, in order to solve the YM problem for monolithic IC's it is necessary:

i) to define the set of independently designable variables

ii)to use a process simulator to relate the layout of the circuit and process parameters to the circuit parameters.

We have also found that the optimization technique employed for solving the YM problem for the monolithic IC's has to take into account the fact that the space of randomly varying parameters determining circuit yield contains elements which cannot be designed and must be treated as a disturbance only.

References

(1) S.W.Director and G.D.Hachtel, "The Simplicial Approximation Approach to Design Centering", *IEEE Trans. on CAS*, Vol. 24 No. 7, July, 1977, pages 363-371.

(2) J.W.Bandler and H.L.Abdel-Malek, "Optimal Centering ,Tolerancing, and Yield Determination via Updated Approximation and Cuts", *IEEE Trans. on CAS*, Vol. 25 No. 10, Oct., 1978, pages 853-871.

(3) K.Singhal and J.F.Pinel, "Statistical Design Centering and Tolerancing Using Parametric Sampling", *Proceedings of ISCAS* , Houston, April 1980, pages 882-884.

(4) K.S.Tahim and R.Spence, "A Radial Exploration Approach to Manufacturing Yield Estimation and Design Centering", *IEEE on CAS*, Vol. 26 No. 9, Sept., 1979, pages 768-774.

(5) W.Maly and A.J.Strojwas, "Simulation of Bipolar Elements for Statistical Circuit Design", *Proceedings of ISCAS 79*, Tokyo, 1979, pages 788-791.

(6) W.Maly and T.Gutt, "Base and Emitter Diffusion Simulation Models", *Proceedings of European Conference on Electronic Design Automation*, University of Sussex, 1979, pages 88-91.

(7) W.Maly and A.J.Strojwas,"Simulation of Random Properties of Monolithic IC Manufacturing Process". To be published.

(8) D.K.Lynn et al, *Analysis and Design of IC*, Mc Graw-Hill, 1962.

(9) B.L.Biehl, BIASD-Reference Manual. 1978.

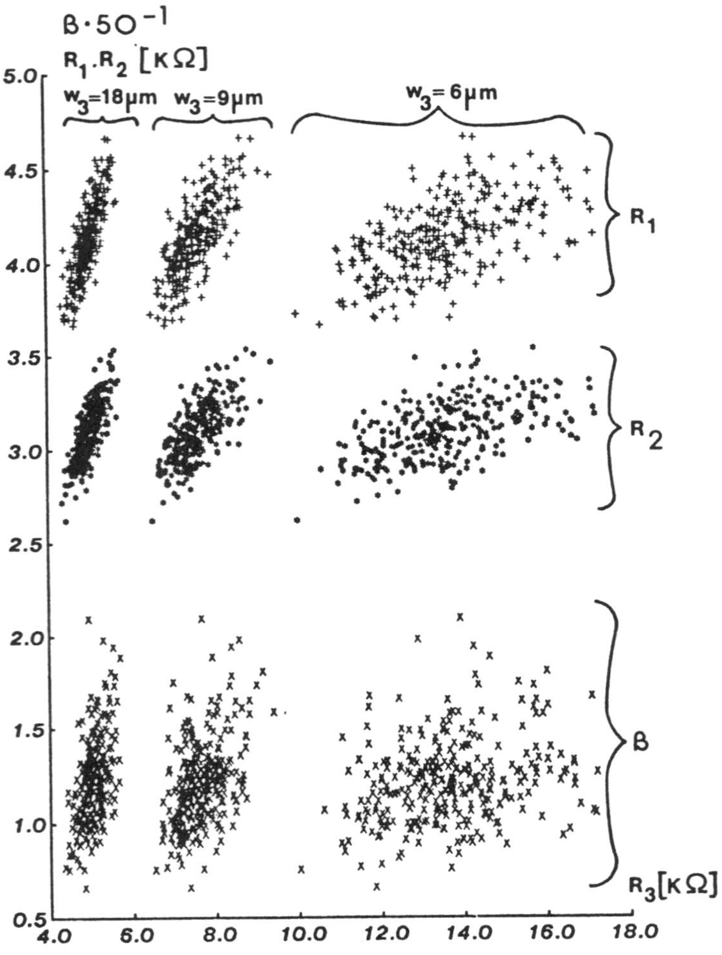

Fig.1. Scatterplots of R_1, R_2 and ß vs. R_3

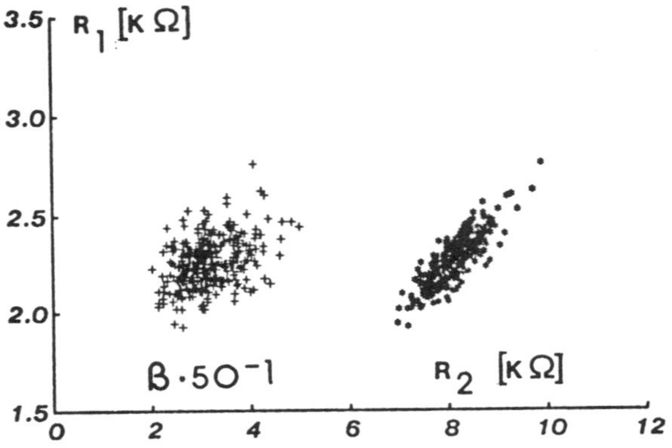

Fig.2. Scatterplots of R_1 vs. ß and R_2 for $\lambda = 1$.

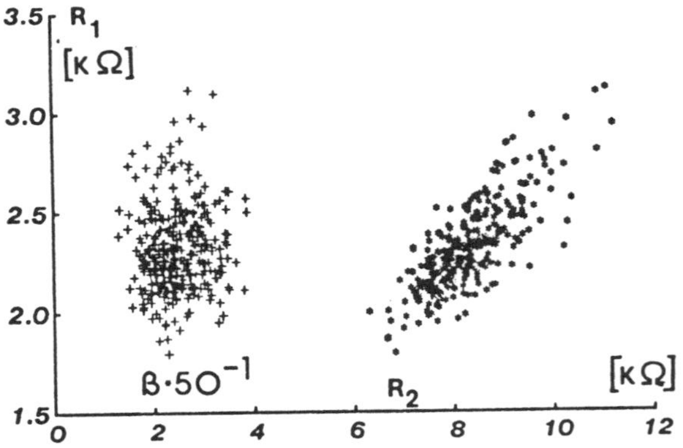

Fig.3. Scatterplots of R_1 vs. ß and R_2 for $\lambda = .5$.

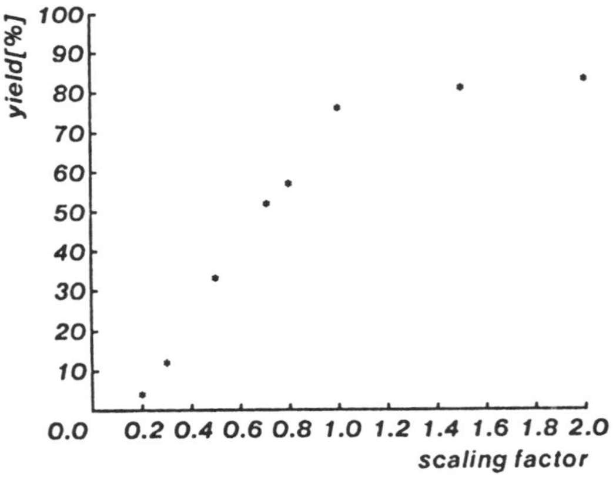

Fig.4. Yield vs. Layout Scaling Factor λ.

Statistical Simulation of the IC Manufacturing Process

WOJCIECH MALY, MEMBER, IEEE, AND ANDRZEJ J. STROJWAS

Abstract—**Information about the random behavior of the IC manufacturing process can be applied for IC and process design tasks. In this paper a methodology for modeling random fluctuations of IC manufacturing process is proposed. A simulator of a complete bipolar manufacturing process called FABRICS, is described. A few applications illustrating advantages of the proposed statistical process modeling method are discussed.**

I. Introduction

ACHIEVING AN acceptable manufacturing yield is one of the principal goals for a designer of a monolithic integrated circuit (IC). Usually, this goal can be accomplished by choosing an appropriate set of nominal values for the mask geometries and a nominal characteristic of a manufacturing process. Therefore, a majority of CAD tools are capable of evaluating only a nominal performance of the circuit in terms of nominal values of certain physical quantities. However, when fluctuations of the circuit elements must be considered, for example, to determine manufacturing yield, such traditional approaches are insufficient and a statistically based approach must be used. Recently there has been increased activity in developing statistically based design methods, (e.g., [12], [26], and [5]). However, for the most part, these approaches can not be directly applied to IC design. This is so because the proposed algorithms are based on optimizing yield in terms of the electrical parameters of circuit elements, rather than the true design parameters such as layout dimensions and some process parameters. However, some of these approaches can be modified for monolithic IC design, if a simulator of the manufacturing process capable of modeling random fluctuations of IC elements [20] is used.

Manuscript received July 21, 1981; revised December 1, 1981. This work was supported by the National Science Foundation under Grant ESC-7923191.

W. Maly is with the Department of Electrical Engineering, Carnegie-Mellon University, Pittsburgh, Schenley Park, PA 15213, on leave from the Technical University of Warsaw, Warsaw, Poland.

A. J. Strojwas is with the Department of Electrical Engineering, University of Carnegie-Mellon, Pittsburgh, Schenley Park, PA 15213.

While several process simulators have been reported (e.g., [3], [4], [29], and [30]), which are very useful for development of new processes, the computational cost of using them for statistical investigations can be prohibitive. In the next section we propose an approach to modeling of the IC fabrication process which can be employed for the statistical design of monolithic IC's. Specifically, we point out some specific features of typical statistical investigations in order to establish the general rules for statistical modeling of IC manufacturing processes. In Section III we present a statistical simulator of the bipolar manufacturing process, called FABRICS, which illustrates our philosophy of statistical modeling. In Section IV we describe examples of the application of FABRICS to various practical design problems in order to show the advantages of the statistical approach to process modeling.

II. Methodology Behind the Statistical Simulation of IC Manufacturing Process

The traditional approach to process simulation, (such as that presented in [30] and [4]), has been to generate the nominal profiles of impurities in silicon. Since the areas of application of a statistical simulator are different, additional information is required. Specifically, in addition to impurity profiles, a statistical simulator of the manufacturing process should generate values for the parameters of IC elements and the so called in-line measurements, i.e., measurements typically made during the fabrication process. Such values are useful for circuit simulation or diagnosis of the manufacturing process.

Reprinted from *IEEE Trans. Computer-Aided Design of Integrated Circuit Systems,* vol. CAD-1, no. 3, pp. 120–131, July 1982.

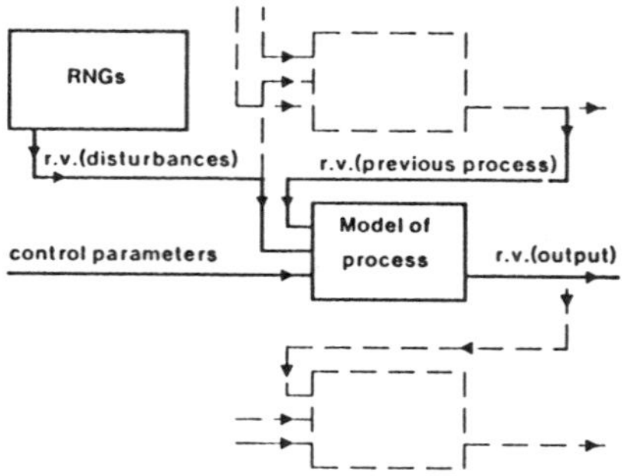

Fig. 1. Model of a single manufacturing step or element parameter (rv denotes random variable).

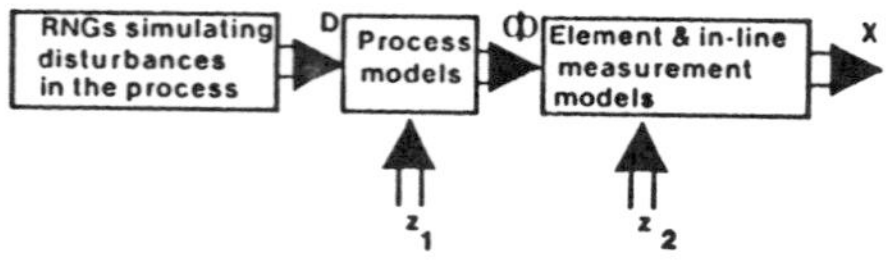

Fig. 2. Structure of a statistical process simulator.

Specifically, parameters of IC elements such as resistances, capacitances, transconductances, etc., are needed for circuit simulation while values of junction depths, sheet resistances or oxide thicknesses are needed for process diagnosis. Furthermore, these values should be described statistically. In other words, since the quantities listed above are in reality random variables, the simulator should generate data which can be used to determine a joint probability density function (JPDF) which characterizes these random variables.

A. Methodology of Modeling

The basic philosophy behind the simulator we developed is to recognize that parameters of an IC element depend upon both layout and process parameters. Furthermore, each process step can be modeled individually; the outcome of each process step depends upon a set of control parameters, a set of disturbances, and the outcome of previous process steps (see Fig. 1). In order to formally describe the simulation procedure we need to define a number of quantities. Let X, a vector of random variables, denote the *parameters of the IC elements* or values of the *in-line measurements.*[1] The vector z_1 represents the *process control parameters*, e.g., temperatures, times, gas fluxes, etc., while vector z_2 represents the *mask dimensions.* Finally, D is a vector of random variables representing uncontrollable *disturbances* in the process. We can simulate these disturbances using appropriately defined random number generators (RNG's), as described more fully in Section II-B.

Symbolically the simulator can be characterized by the flow diagram shown in Fig. 2. While it is possible to develop a detailed, and accurate, mathematical characterization of each step in the manufacturing process (see [3] as an example of such a set of *numerical models*), the use of such a model in a statistically based simulator would be computationally prohibitive. This results from the fact that in order to evaluate the output parameters one has to solve a system of nonlinear partial differential equations which describe the model. Hence, we are motivated to consider less complicated models. Specifically, we establish a simplified physical characterization for both process steps and circuit elements, which can be easily tuned to the particular process being modeled. We propose to use analytical models of each manufacturing step and circuit element which explicitly characterize the dependence of the output parameters on the input parameters. Towards this end we consider models of the fabrication steps of the form

$$\Phi_j = g_j(\hat{\Phi}^j, z_1^j, D^j), \qquad j = 1, \cdots, m \tag{1}$$

where

- Φ_j is a component of the m-vector Φ of *physical parameters* which describes the outcome of a fabrication step (e.g., oxide thickness, parameters of impurity distribution, misalignment, etc.),
- $\hat{\Phi}^j$ is a vector of physical parameters obtained from previous fabrication steps (e.g., in the base diffusion process, impurity concentration in the epitaxial layer is a component of $\hat{\Phi}^j$),
- z_1^j, D^j are the vectors containing those components of z_1 and D, affecting the jth physical parameter.

Note that we will use the vector Φ to denote the outcomes of all process steps, so that Φ_j and $\hat{\Phi}^j$ are subsets of components of Φ. Models of circuit elements are assumed to be of the following form:

$$X_i = h_i(\hat{\Phi}^i, z_2^i), \qquad i = 1, \cdots, n \tag{2}$$

where X_i is an electrical parameter associated with a circuit element (e.g., β of n-p-n transistor), $\hat{\Phi}^i$ and z_2^i are subsets of components of Φ and z_2 which affect this element.

Thus simulation of the random variable X consists of generating samples of D, evaluating components of Φ for subsequent manufacturing steps, and calculating realizations of X using element models. The analytical functions g_j and h_i can be thought of as approximations to the solutions of the (partial) differential equations in the numerical models. The cost of model parameter evaluation can be drastically reduced by using analytical models.

The parameters of the probability distribution of the samples of X generated by the statistical process simulator must be in good agreement with the parameters estimated from the measurements of element parameters, $\tilde{X}$, made in the real process. Note that the process control parameters z_1, and element layout parameters z_2, are known. Thus in order to achieve good agreement between the simulated and measured data, we have to find the parameters of the probability distribution of process disturbances, i.e., to perform an identification of $f_D(d)$ parameters. To accomplish this goal we use the data collected from in-line and test pattern measurements. (Test patterns contain individual elements, e.g., different resistors and transistors, with different sensitivities to the various types of process disturbances.) Simple statistical optimization techniques can be applied to determine param-

[1] In what follows, if a capital letter represents a random variable, the corresponding lower case letter represents a particular value, or realization, from the corresponding distribution.

eters of $f_D(d)$ providing a good fit between $f_X(x)$ and $f_{\tilde{X}}(x)$. If the models of the process steps and circuit elements are complex enough to account for the most significant physical phenomena, one can achieve good accuracy, as will be demonstrated in the next section. Note that identification results are valid for a particular process only (i.e., for a specific z_1). However, once the identification procedure is completed and the probability distribution is determined, we can use the simulator instead of the fabrication process, to optimize the layout of an IC or to fine tune the process control parameters for the particular IC being designed. Also note that while numerical models provide acceptable accuracy over a wide range of process conditions, an extensive identification procedure is required. Sometimes parameters of these models are extremely difficult to identify (e.g., stress-induced bandgap narrowing in SUPREM).

Observe that it is important to clearly identify the model parameters which represent the process disturbances. This is a difficult problem for two reasons. First, most of the physical causes which disturb manufacturing operations can not be explicitly described by parameters of the fabrication step being modeled. For instance, defects in the silicon crystal, which affect the diffusion process can be introduced during any of a number of steps including crystal growth, polishing the surface of the silicon wafer, or any treatment causing thermal or mechanical stresses, etc. It is obvious that models of the diffusion process do not include parameters capable of expressing all possible events affecting these operations. Moreover, in most cases, the dependence of model parameters on a particular class of random events which disturb the process is not explicitly known. However, it is possible to choose a single parameter to characterize physically different random events, if they affect output parameters of the processing step in a similar manner. Thus the many causes of fluctuations in the density of defects in the silicon crystal, which ultimately contribute to fluctuations in the impurity profiles, can be simulated by a single random variable, which describes the diffusivity of the impurity in silicon.

The second difficulty associated with the choice of parameters which represent disturbances, is to ensure their statistical independence. This criterion is important because it allows us to identify, and to generate, each component of D, independently.

B. Structure of RNG's

The goal of the process simulator is to obtain an accurate statistical characterization of the parameters of a circuit element not only within one chip but also on a wafer and in a lot. This task can not be accomplished if the random variables are independently generated by simple RNG's.

In order to illustrate the problem let us assume that the simulator is used to generate data which describes a number of IC chips. In a real process each of these chips is affected by disturbances in a different way, because they are placed in different locations within a wafer, lot, etc. On the other hand, different elements of a single IC chip are located close together, and therefore, are affected by disturbances in a similar way. Thus one can assume that conditions of the

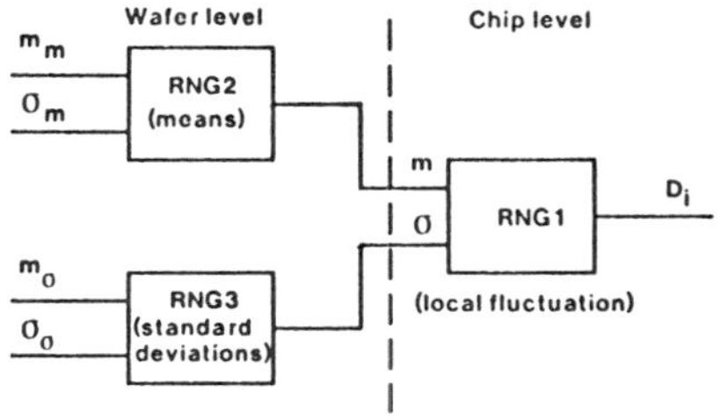

Fig. 3. Two level generator which simulates local (within a chip) and global (within a wafer) fluctuations of process conditions.

process for all elements within one particular chip, may be described by the same set of parameters which characterize disturbances in the process for this particular location of the chip. For some applications, however, such a simplification may be invalid, and, therefore, it is necessary to assume that there is a random factor, unique for each element of the IC, describing small local differences in the condition of the process among elements within a chip. Hence, a single disturbance D_i affecting one particular chip can be characterized by a certain mean value mD_i describing "average" conditions for all elements of the chip and value σD_i representing local (i.e., within a chip) fluctuations in the process conditions. Thus the random variable D_i, affecting a single chip, can be simulated by a RNG, which generates data with a mean equal to mD_i and standard deviation equal to σD_i. Note now that in order to use this RNG for simulating disturbance for another chip new values mD_i and σD_i must be provided. Since the statistical simulator should generate data samples independently of the location of the chips[2] then we have to assume that the location of the chip to be simulated is random. Thus we have to assume that values mD_i and σD_i are changing randomly from one simulated chip to another. Hence, each disturbance in the process must be simulated by at least a two-level RNG, generating local (within a chip) and global fluctuations.

In Fig. 3 we show a two level structure composed of three RNG's. This structure is capable of generating disturbance data for a wafer. RNG1 simulates a disturbance within a chip. Specifically, RNG1 generates a normally distributed random variable D_i. Generators RNG2 and RNG3 provide RNG1 with the mean and standard deviation of D_i for a chip. RNG2 is controlled by two parameters: the mean of means (m_m) of chips in the wafer and standard deviation of means (σ_m) in the wafer. The generator RNG3 is responsible for the mean of standard deviations (m_σ) for chips in the wafer and standard deviation of standard deviations for chips in the wafer (σ_σ).

The structure of the generating procedure presented above can be extended to the lot level, such that differences among wafers in a lot can be simulated as well. This situation results, for instance, when the simulator is used for investigating relationships between in-line measurements and IC parameters. The additional level of RNG's increases the number of parameters which must be identified. Note that while this next level of random number generation may be skipped, the other two levels of generation must be included. This can be for-

[2]The design of an IC cannot be developed for a particular area of the wafer or lot.

mally proven and is evident when we note that the correlation among parameters of IC elements are mainly determined by local differences in the conditions of the process. The variances of these parameters are determined by both local and global variances of the process disturbances. Hence, omission of one of the two levels of generation may cause significant inaccuracy in the simulation.

III. A Statistical Simulator of Bipolar IC Manufacturing Process

In this section we describe a statistical simulator, based on the ideas presented above, for a typical bipolar manufacturing process with boron base and phosphorus emitter diffusions. The statistical simulator, FABRICS (FABRication of Integrated Circuits Simulator), described here, models the following fabrication steps: epitaxy, oxidation, base and emitter diffusions, photolithography, surface treatment processes, and contact deposition and sintering. The simulator contains models of standard in-line measurement parameters (i.e., sheet resistances and junction depths after subsequent diffusions) and models of the dc parameters of typical bipolar IC elements: resistors, n-p-n and p-n-p transistors and diodes.

A. Models Implemented in the Simulator

In order to illustrate the modeling philosophy employed in the process simulator, consider a base diffusion step. Diffusion of impurities in silicon is described by Fick's first and second laws [10] which lead to a diffusion model in the form of a partial differential equation of the parabolic type. Due to the fact that flux of impurities in silicon is dependent on concentrations of vacancies in silicon (e.g., [13]), which may be induced by the diffusion of other impurities [14] or thermal stresses in the silicon substrate, as well as conditions of oxidation of the silicon surface [2], the diffusion model and boundary conditions, which describe the movement of the Si-SiO_2 interface, are nonlinear. Models including all these features are considered in [4] and [29]. An algorithm developed to solve this type of model has been implemented in SUPREM [3]. However, the computational complexity of the algorithms, used for solving the nonlinear parabolic equations with the moving boundary conditions, is too high to compute the large number of different impurity profiles required for a statistical characterization of this step within a reasonable computation time. In addition to the complex, but accurate, models of the diffusion process, as presented in [30] and [4], there are a number of simplified models which have been reported. Specifically, these models are either analytical approximations to the numerical solution of the complex model, or an analytical solution of a simplified description of the process. Commonly known examples of such *analytical models* of the diffusion process are the erfc or Smith's function (e.g., see [28]). While these models are inexpensive in simulation and identification, they can be very inaccurate for process conditions different from those for which the models were developed and identified.

We have chosen to implement in FABRICS the model proposed by Cave [8]. This model is based on a very useful approximation to the base diffusion profile arising from boron

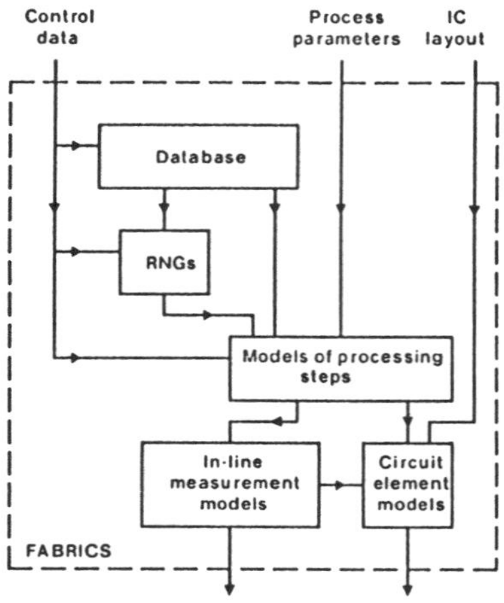

Fig. 4. Structure of FABRICS.

redistribution in the oxide. Cave derived an analytical solution to the diffusion equation under simplified, but practically important, conditions. The result is a fairly complex analytical expression which relates the impurity concentration $N_B(y)$ to the process control parameters (such as times and temperatures) and process disturbances (such as diffusivities of impurities or segregation coefficient). This model has been implemented in [23] and the results proved to be quite accurate. The above model incorporates the effects of growing oxide (moving Si-SiO_2 boundary) and boron redistribution between Si and SiO_2 during the drive-in step. The emitter dip-effect is modeled by the enhancement of the base diffusivity under the emitter region.

Lack of space does not allow us to carefully describe the other models implemented in FABRICS. These models are briefly described in Appendix A in terms of input and output parameters of the model and literature references.

A brief description of in-line parameter models and circuit element models is given in Appendix B. A more detailed model description can be found in [32].

B. Structure of FABRICS

The data-flow diagram of the simulator is shown in Fig. 4. Three types of data are entered to FABRICS: process parameters, IC layout dimensions, and run control parameters which are used to activate the RNG's and model routines in an appropriate sequence. Parameters controlling RNG's and physical constants are stored in a database.

IV. Computational Examples

In the first part of this section we briefly illustrate the methods used to identify, from measured data, the input parameters required by FABRICS. Then we present the results of three examples to illustrate the various applications of FABRICS. Additional examples have been reported in the literature, as noted at the end of this section.

A. Identification and Tuning of FABRICS Parameters

We have performed the identification procedure for two different manufacturing facilities. We describe here the results obtained for one of these two cases; the second case is described in [19]. The identification was performed using in-line measurements and test patterns composed of a set of typical resistors, as well as n-p-n and p-n-p substrate and lateral transistors. Lots with test patterns, and additional test wafers (without IC structures) were manufactured and measured using standard industrial equipment. The goal of the identification procedure is to find the parameters of the probability distribution $f_D(d)$ which minimize the difference between the probability distribution parameters of the simulated (i.e., by FABRICS) and measured data. Note that we have chosen process disturbances that are statistically independent, i.e., the correlation coefficients among components of D are equal to zero. Hence, we can formally state the identification problem as follows:

$$\min_{mD,\sigma D} \ [f_X(x) - f_{\tilde{X}}(x)]^2. \tag{3}$$

This is a highly dimensional statistical optimization problem. However, we can reduce the dimensionality by making several simplifying assumptions.

First, we assume that each in-line or element parameter is only affected by a small number of disturbances. This allows us to consider several subproblems in which we are using the measurement results of a particular parameter, X_i, to identify the disturbances which significantly affect the values of X_i. Furthermore, we can establish a sequence of identification steps such that we consider the parameters with the simplest dependence on process disturbances first and identify a subset of disturbances. Then we can use these to identify the remaining disturbances using the measurement results of the parameters in which the dependence on the process disturbances is more complex. In our case, the sequence of identification steps is the following. From the measurements made in the fabrication process, the parameters of the epitaxial layer (i.e., thickness and resistivity) and oxidation parameters (i.e., linear and parabolic oxide growth coefficients) are identified. Then, based upon sheet resistance and junction depth measurements, the base diffusion parameters (i.e., initial surface concentration, boron diffusivity and impurity segregation coefficient) are identified. A similar procedure is then repeated for the emitter diffusion parameters. Next, using the data obtained for a square resistor (independent of photolithography variations) we can attempt to identify the contact resistivity. By the simultaneous analysis of square and narrow (i.e., with a large ratio of length to width) resistors we can extract disturbances due to the photolithographic processes. Finally, the surface recombination parameters are identified based upon the data collected from the transistor characteristics of base current versus emitter-base voltage [27].

Secondly, we decompose each identification step into two parts. First, we determine the mean value of the disturbances on the wafer, then we identify the variance on the wafer (global variance). In the models in which the dependence of X_i on the disturbances is explicitly given (e.g., resistance dependence on the contact resistivity or photolithography variations), least-squares fitting techniques are used to identify the mean values. In other cases, such as, for instance, base diffusion in which the dependence between the sheet resistance and diffusion disturbances is not explicity, we use nongradient optimization methods to identify the mean values of the input parameters. The standard deviations are identified afterwards by using simple statistical optimization techniques in which a standard deviation (estimated from the simulated data sample) is determined by using one-dimensional line-search algorithms [23].

After determining the mean values and standard deviations on the wafer level, we can attempt to decompose the variance of the disturbances between the wafer (global) and chip (local) levels. This tuning is performed interactively until sufficient agreement in correlation coefficients between the simulated and measured data is reached.

The summary of the identification results is shown in Tables I-III. In Table I we show a comparison between in-line measurements simulated by FABRICS and collected from the test wafers. In Table II we compare means and standard deviations of test pattern element parameters. In Table III correlation factors are presented. All of the estimates were computed using samples composed of 200 measurements or simulated parameters. Note that there is good agreement between parameters of the JPDF's of the simulated and measured data, which indicates that both the models employed in the simulator and the identification methods are accurate enough. Identification of FABRICS input parameters is a computationally intensive task but afterwards the results of identification can be used in the simulation of the manufacturing process for different IC's because they are independent of the layout of a particular IC.

B. Examples of FABRICS Application

The current version of FABRICS is highly interactive program which was developed at Carnegie-Mellon University. FABRICS is written in DEC-Fortran (about 6000 lines of code) and runs on PDP10 and DEC20 computers. To give some idea of computational efficiency, the average time of computing a set of three electrical parameters (saturation current, current gain coefficient and Early voltage) for 100 n-p-n transistors is less than 3 s of CPU time, which includes simulation of impurity profiles, photolithography, and surface recombination parameters for each transistor separately.

Currently, a new PASCAL version of FABRICS is being developed, in which other fabrication steps such as ion implantation, annealing and new models of the emitter diffusion are included. The new program will be capable of generating the complete sets of electrical parameters for IC's which are manufactured in bipolar, NMOS or CMOS processes.

Example I

This example illustrates the use of FABRICS for designing an element of a bipolar IC. Consider a bipolar technology in which the width of the base of the n-p-n transistor can be controlled by varying the emitter drive-in time. Furthermore, assume that from the circuit analysis we know that the β's of the n-p-n transistors in this IC should be as high as possible. On the other hand it is known that the collector emitter breakdown voltage, V_{CEO}, is unacceptable (is too low) if β exceeds a certain value, β_{max}. The objective is to determine the nominal value of the base width which will be used for con-

TABLE I

Comparison of the In-Line Measurement Parameters, Measured in a Real Process and Simulated by FABRICS

Parameter	Unit		Measured	Simulated
R_{sB1}	Ω	m	196	192
		σ	19	18
R_{sB2}	Ω	m	190	186
		σ	12.4	17
R_{sE}	Ω	m	4.3	4.4
		σ	.12	.08
X_{CB2}	μm	m	2.33	2.29
		σ	.05	.043
X_{CB3}	μm	m	2.5	2.52
		σ	.07	.04
X_{EB}	μm	m	1.33	1.3 2
		σ	.07	.04
W_B	μm	m	1.17	1.2
		σ	.08	.06

R_{sB1} and R_{sB2} stand for base sheet resistances after base and emitter drive-ins, respectively. X_{CB2} and X_{CB3} denote the collector junction depths outside and under the emitter region, respectively, and X_{EB} is the emitter junction depth. R_{sE} is the emitter layer sheet resistance and W_B is a base width after emitter drive-in (m denotes a mean value and σ standard deviation).

TABLE II

Comparison of the Means and Standard Deviations of the Following Test Pattern Parameters Measured and Simulated by FABRICS

Parameter	Unit		Measured	Simulated
R_{EPI}	KΩ	m	5.4	5.5
		σ	.62	.53
R_B	Ω	m	178	186
		σ	20	18
R_E	Ω	m	97	96
		σ	5.4	5.1
R_{BE}	KΩ	m	1.4	1.48
		σ	.31	.28
β_1		m	54	56
		σ	16	20
β_2		m	55	57
		σ	16	21
β_S		m	58	57
		σ	15	15.7
β_L		m	11	10.4
		σ	3.6	2.8

Epitaxial resistor (R_{EPI}, 2 squares), base resistor (R_B, 1 square), pinch resistor (R_{BE}, 0.5 square), emitter resistor (R_E, 20 squares), β's of n-p-n transistor for different collector currents (β_1 and β_2), β of substrate transistor (β_S) and β of the lateral p-n-p transistor (β_L).

TABLE III

Comparison of Correlation Factors Estimated from the Data Simulated by FABRICS and Measured in the Process

Parameters	Measurements	Simulation
$R_{EPI} - \beta_S$	.52	.49
$R_B - R_{rE}$	.71	.51
$R_B - \beta_2$	.65	.51
$R_B - \beta_S$	-.32	-.32
$R_B - \beta_L$	-.38	-.19
$R_{BE} - \beta_1$	.81	.85
$R_{BE} - \beta_S$	-.40	-.15

trolling the emitter drive-in time. Hence, we have to find the thickness of the base layer such that the mean (nominal) value of β's of n-p-n transistor in the fabricated IC's is maximal under the condition that the percentage of the IC chips in which β is greater than β_{max} is low (for instance 2 percent).

This problem can be solved in the following way. Using FABRICS, tuned as described in the previous subsection, we determine the distribution of the β's of n-p-n transistors and the mean of the thickness of the base layer, w_B, in a wafer for six different times of emitter drive-in. The mean and standard deviation of the β's within the wafer for each drive-in time are then estimated. The results are shown in Fig. 5 where the dots represent the means of β's for each drive-in time. The vertical segments illustrate the spread[3] of β's within the wafer. Assuming now, for example, that $\beta_{max} = 300$, one finds that for $w_B = 0.9$ μm the mean of β's in the wafer is the highest possi-

[3]These segments contain 98 percent of the generated samples.

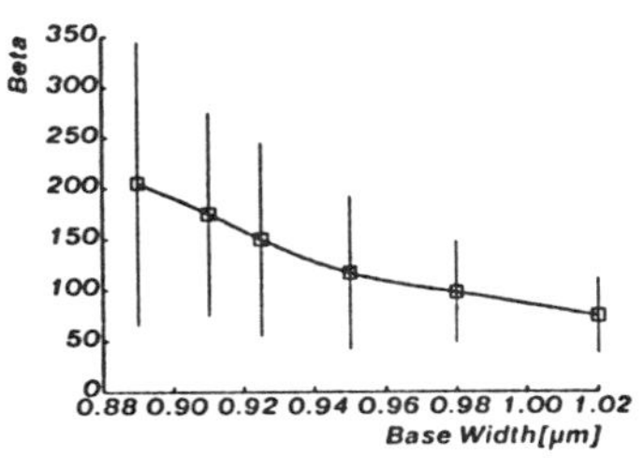

Fig. 5. Distribution (the mean values and the spread) of β of the n-p-n transistor versus base width for 6 values of the emitter drive-in time.

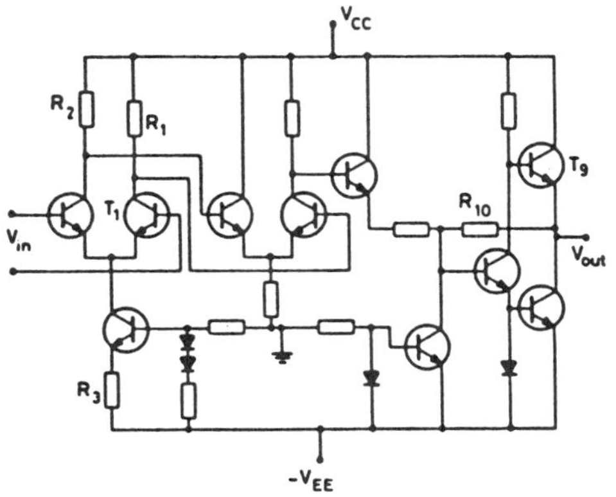

Fig. 6. Schematic of the Motorola operational amplifier MC1530.

ble value for which 98 percent of transistors in the wafer have β less than 300. Hence, in this case it was possible to solve the design problem because FABRICS was able to provide information about the nominal as well as maximal values of interest.

Example II

This example illustrates how FABRICS can be combined with a circuit simulator to optimize the technological parameters of a manufacturing process. Consider the simple operational amplifier (op amp), MC1530, produced by Motorola. (See diagram in Fig. 6.) Suppose we wish to change the process in order to decrease the surface concentration of phosphorus in the emitter layer.[4] To simplify the discussion we assume that the only parameter which can be changed in the process is the surface concentration of phosphorus during predeposition of the emitter layer.

Knowing that the distribution of impurities in the emitter has a great effect on the electrical parameters of the transistors in the circuit, and that the surface concentration can be controlled only by measuring the sheet resistance of the emitter layer R_{SE}, we first investigated the relation between the manufacturing yield and the mean of R_{SE}. We computed this relation using the Monte Carlo method. With FABRICS and a circuit simulator (in this case BIASD [6]), we estimated the manufacturing yield of the op amp for six different phosphorus surface concentrations. For each concentration we generated data simulating 100 chips and computed the mean of R_{SE}. The performance of the op amp was evaluated in terms of the differential gain A_d, input offset voltage $V_{\text{in off}}$, and input bias current I_{bias}. Specifically, performance was accepted if: $A_d > 8000$, $-1.5 < V_{\text{in off}} < 1.5$ mV, and

[4]This kind of the problem could arise, for instance, when the noise characteristic of the op amp must be improved.

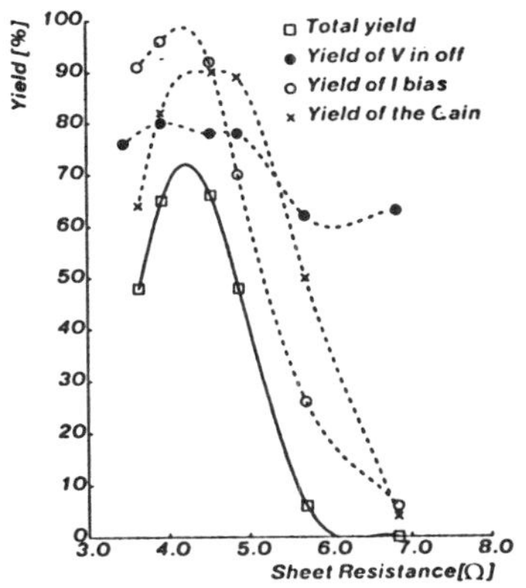

Fig. 7. Total and partial yields of op amp calculated for varying values of the emitter sheet resistance. FABRICS was tuned to the process with the emitter sheet resistance equal to 4.3 Ω/□.

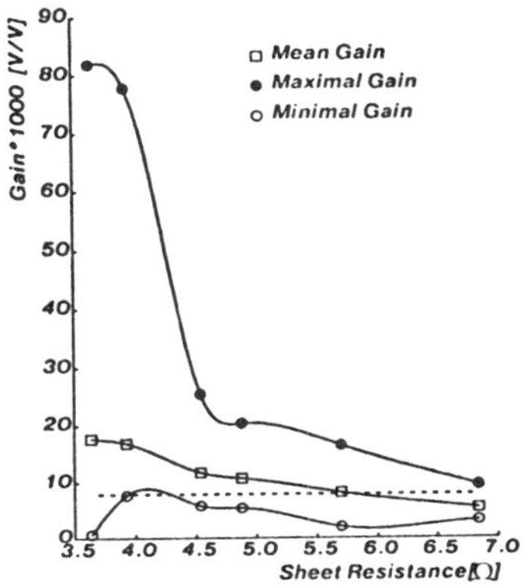

Fig. 8. Maximal, minimal, and mean voltage gains of the op amp versus the emitter sheet resistance.

$-1.2 < I_{\text{bias}} < 1.2$ μA. The results obtained, shown in Fig. 7, are in good agreement with expectations. The best yield is obtained if R_{SE} is in the range close to 4 Ω/□, (i.e., close to the condition in a standard process), and can not be increased very much without a significant drop in yield. A reasonable question at this stage of investigation is: what are the causes of the drop in yield for sheet resistances lower than 4 Ω/□ and higher than 5 Ω/□? To answer this question we compute the partial yields, i.e., yields calculated with respect to each circuit performance parameter separately. (See dashed curves in Fig. 7.) It is seen that for low values of R_{SE} the total yield is low due to the low yield of A_d. For high values of R_{SE} yield is low because of low yields of A_d and I_{bias}. To find the physical interpretation of these relations we computed the mean, maximal and minimal values of A_d versus R_{SE}, which are plotted in Fig. 8. Note that the total yield drops for high values of R_{SE} because of the increase in mean of A_d. Analyzing the dependence of A_d on R_{SE}, one can find very easily that A_d decreases due to the decrease of the mean of the β's. In Fig. 9, both A_d and mean of β, versus R_{SE} are shown. The decrease of the yield A_d for R_{SE} less than 4 Ω/□ is due to the drastic increase of the spread of the actual values of A_d in the wafer, which can be attributed to the increase in spread of β's, caused by a narrower base. (This effect was illustrated in the previous example.) Thus the fraction of op amps with A_d less

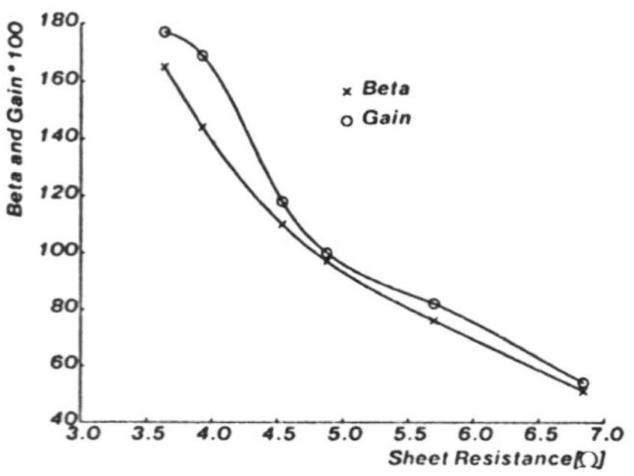

Fig. 9. Mean voltage gain and mean value of β of n-p-n transistors simulated for the op amp MC1530.

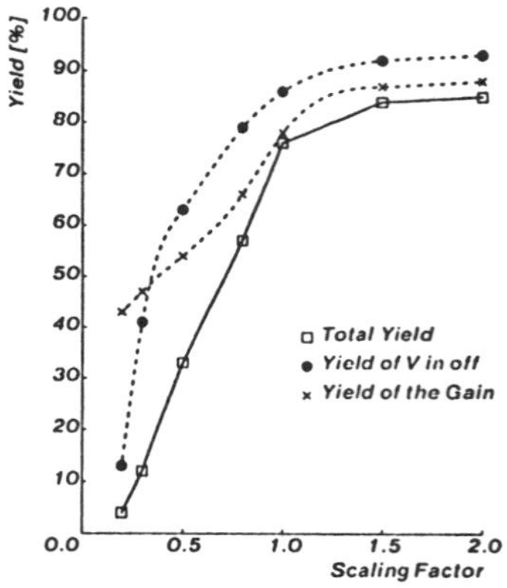

Fig. 10. The total and partial yields of op amp MC1530 versus layout scaling factor λ.

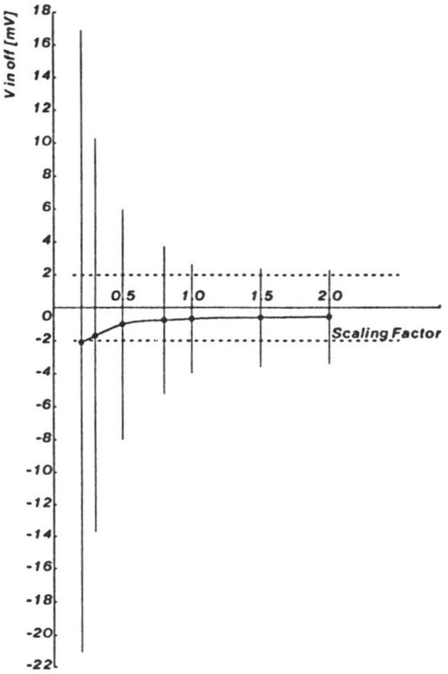

Fig. 11. Mean value and spread of the input offset voltage of the op amp versus layout scaling factor.

than 8000 becomes higher even though the mean of A_d remains high. (See the plot of minimal gain for low values of R_{SE} in Fig. 8.)

From the above discussion we conclude that the surface concentration can not be changed without sacrificing yield. One possible way to allow a decrease in the surface concentration of phosphorus is to introduce an additional amplification stage into the amplifier.[5]

Example III

The layout design rules, i.e., the set of minimal distances between edges of the mask windows, are determined, among other things, by the inaccuracy of the photolithographic processes. These rules are established in order to keep the fluctuations of the IC element parameters low,[6] and thus to assure acceptable yield. Since the sensitivity of IC performance to element parameters vary with elements, then some elements contribute to the fluctuations of IC performance more than others, and only these should be designed using stricter design rules. Thus it is possible to scale down dimensions of some IC elements without increasing the variability of circuit performance (i.e., it is possible to save IC area by relaxing the design rules for some IC elements). The above possibility illustrates one of many layout design problems which should be investigated, and are due to the random fluctuations of the physical dimensions of IC elements. FABRICS can be used to explore the problems involving different aspects of the relation between layout design and the random behavior of the manufacturing process.

Consider the same IC as in Example II, assuming the layout of the op amp as reported in [25]. We state the problem by asking: which parameters of the op amp elements contribute the most to changes in the yield, if the entire layout reported in [25] is scaled.[7] We investigated this problem in the following way. Using FABRICS and BIASD [6], we computed the performance of the op amp and JPDF of β, I_s, and V_A for $\lambda = 0.1$, 0.3, 0.5, 0.8, 1.0, 1.5, and 2.0. Then, assuming $A_d > 8000$, $I_{\text{bias}} < 6\ \mu A$, and $-2 < V_{\text{in off}} < 2$ mV, we estimated the total yield and partial yields versus λ. (See Fig. 10.)

We found that the yield decreases due to: decrease of mean of A_d, increase of the variances of A_d and $V_{\text{in off}}$. We discuss in more detail only the third effect because the first two can be easily explained by the decrease of the mean β (by scaling we change the ratio of emitter area to emitter perimeter), and increase of the variances of the resistors in the circuit, especially resistor R_{10}, which serves as a feedback resistor, fluctuations of which can directly affect A_d. The obtained relation $V_{\text{in off}}(\lambda)$ is shown in Fig. 11 (confidence segments are three standard deviations long). It is evident there that for low λ, i.e., $\lambda = 0.2$ and 0.3, yield is low because of the extremely high variance of $V_{\text{in off}}$. This suggests a mismatch among elements of the input amplifier. We confirmed this supposition by analyzing the JPDF describing the op amps elements. We found, for instance, that the correlation factor between β's of the input transistors becomes low when the dimensions of these elements are decreased. (See Fig. 12.) Hence, a drop in yield, associated with a decrease of the area of the op amp, is

[5]This solution was implemented, for instance, in the Motorola op amp MC1531.

[6]It can be formally shown for the simple monolithic devices [20] that the smaller dimensions of the mask, the higher variance of the parameters of the monolithic device.

[7]By scaling of the layout we mean multiplication of all dimensions in the mask by a factor λ which we call *Scaling Factor*.

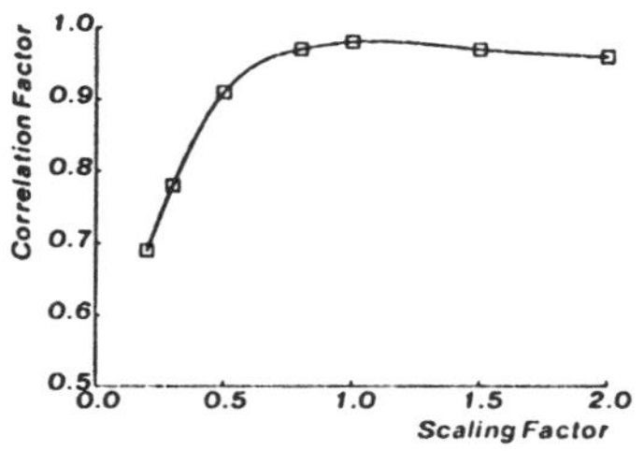

Fig. 12. Correlation coefficient between β's of the op amp input transistors versus layout scaling factor.

caused mainly by changes in the statistical characterization of the elements of the input amplifier and some resistors of op amp. It is now clear that scaling down all transistors and resistors of the op amp, except elements of the input differential amplifier and, for instance, resistor R_{10}, such that the ratio of emitter area to emitter perimeter is constant, one can probably find a better layout of the op amp, i.e., a layout which occupies less area, while still assuring sufficient manufacturing yield.

Other Applications

The examples described above illustrate only the simplest engineering applications of FABRICS. We have found FABRICS to be very useful for several other purposes.

In-line tests, such as junction depth and sheet resistance measurements, are used for controlling the process. These measurements are performed on test wafers accompanying production lots. If some of the manufacturing operations were performed for a certain production lot inaccurately, because of disturbances in the process, then one can expect low yield in this faulty lot. In order to minimize production cost, faulty lots should be rejected from the production stream before all manufacturing steps are accomplished. Thus the problem arises of constructing an algorithm which can be applied for rejecting faulty lots using in-line measurements. The solution of this problem was discussed in [24].

The disturbances in the manufacturing process change with time, causing drops in the manufacturing yield. The diagnosis of physical reasons causing these changes in the yield can be used by process engineers for more efficient process monitoring. FABRICS has been used for developing the diagnostic algorithms providing engineers with information about disturbances in the manufacturing process [22].

FABRICS has been applied as part of a statistical design system using the Simplicial Approximation algorithm [12] and the circuit simulator BIASD. The result of computations obtained by means of this system cited in [21] proved the high computational efficiency of FABRICS. For the Design Centering problem described there, the CPU time of process simulation was only 2 percent of the CPU time used for circuit simulation and the design centering algorithm.

FABRICS has been used for teaching purposes in the IC design course given at Carnegie-Mellon University in Fall 1980. It was used for two tasks. First, as an inexpensive tool familiarizing students with the behavior of the manufacturing process, expressed in terms of the process parameters. Second, as a source of data providing the circuit simulator SPICE with element parameters. This capability of FABRICS was very useful when designing the layout of several TTL IC's which were given to students as projects.

V. Conclusions

In this paper we have described a new approach to the process modeling problem. Analyzing specific features of CAD involving statistical investigations, we have proposed a methodology for the statistical modeling of the IC manufacturing process. The main concept of the method introduced consists of building simplified physical models which can be easily tuned to the random behavior of the real process. The performance of the described process simulator FABRICS shows that the suggested methodology of modeling may be a good solution to the trade-off between accuracy and cost of statistical process simulation. Examples of applications of FABRICS have illustrated a variety of problems which can be solved by means of statistical process simulation. Hence, the method of statistical simulation is found to be a useful, as well as efficient, CAD tool for solving important design tasks. The experience with FABRICS encouraged us to extend the process simulation technique described in this paper to MOS VLSI circuits by implementing additional process and element models.

Appendix A
Process Models

In this Appendix we describe process models implemented in FABRICS. We specify the type of model, literature references, and the input data which are classified into three categories: process control data, random disturbances, and parameters from previous processes. A more detailed model description can be found in [32].

Base Diffusion

Assumption: two-step process consisting of predeposition and drive-in of boron (drive-in consists of three operations: base drive-in and redistribution during emitter predeposition and drive-in).

Model type: analytical function describing boron profile versus depth ([23], [8]).

Input parameters:

process parameters: times and temperatures of diffusion operations;

disturbances: diffusivity of boron impurities in silicon outside emitter region, diffusivity of boron impurities in silicon under the emitter region, surface concentration of boron impurities during predeposition, segregation coefficient of boron impurities on the Si-SiO_2 boundary at equilibrium;

parameters from previous processes: impurity concentration in the epitaxial layer, thickness of the oxide layer after a given step of the fabrication process.

Emitter Diffusion

Assumption: two-step process consisting of predeposition and drive-in of phosphorus impurities.

Model type: emitter impurity profile after the emitter drive-in process is modeled by an analytical approximation [15] to Smith's function [28].

Input parameters:

process parameters: times and temperatures of emitter predeposition and drive-in;

disturbances: diffusivity of phosphorus impurities in silicon and surface concentration of phosphorus impurities during predeposition;

parameters from previous processes: base impurity profile.

Thermal Oxidation

Model type: oxide growth is modeled by a parabolic equation [11]. Oxide thickness after each oxidation step is the output parameter.

Input parameters:

process parameters: times and temperatures of alternate dry O_2 and wet O_2 oxidations;

disturbances: coefficients of parabolic and linear growth in wet and dry O_2 ambients;

parameters from previous processes: initial oxide thickness.

Epitaxy

Assumption: The impurity concentration in the epitaxial layer is uniform.

Model type: concentration of impurities in the epitaxial layer is computed from the resistivity using Irvin's approximation [17].

disturbances: thickness and resistivity of the epitaxial layer.

Photolithography

Model type: statistical model describing photolithographic inaccuracy caused by misalignment, and by the fluctuations of parameters in optical and etching operations [32].

Output parameters: deviations of element dimensions after the process.

Input parameters:

disturbances: misalignment and etching inaccuracy introduced by subsequent (i.e., isolation, base, and emitter) steps.

Surface Treatment Processes

Assumption: By this name we mean all operations that influence the quality of the silicon surface, i.e., surface recombination properties (surface potential Φ_S and surface recombination velocity s_0).

Model type: These processes are modeled in the simulator by the surface recombination component of the base current in a bipolar transistor [27].

Input parameters:

disturbances: surface recombination parameters specified in [27].

Contact Deposition and Sintering

Model type: The influence of contact deposition and sintering is modeled in the process simulator by the contact resistance for the diffusion resistors [1].

Input variables:

disturbances: the specific contact resistance between metal and the diffusion layers.

parameters from other models: R_S sheet resistivity of the diffusion layer.

Appendix B
In-Line Measurement and Circuit Element Models

In-Line Measurement Models:

sheet resistivity of diffusion layers after each diffusion operation;

junction depths of diffusion layers after each diffusion operation.

These quantities are calculated in FABRICS as follows:

The *base diffusion-epitaxial layer junction depths* X_{JC} after each diffusion operation, are computed by solving the equation

$$N_B(y) - N_{\text{EPI}} = 0. \tag{1}$$

The equation is solved numerically using a modified Regula-Falsi algorithm with $N_B(y)$ representing the impurity profile of boron outside or under the emitter layer, and N_{EPI} representing the impurity concentration in the epitaxial layer.

The *sheet resistances of the base diffusion layer* after each diffusion operation are computed as follows:

$$R_{SB} = \left[\int_0^{X_{JC}} q\mu_p(N_B)\, N_B(y)\, dy\right]^{-1}. \tag{2}$$

The integral value is computed numerically by a modified Simpson algorithm (with extrapolation of integral value after each iteration). The dependence of hole mobility μ_p on base impurity concentration is approximated by the formula given in [7].

In-line parameters of the emitter diffusion layer are computed in a similar way.

Circuit Element Models

Resistors

Four types of resistors, i.e., base, emitter, and pinch as well as epitaxial resistor models are implemented in FABRICS. The resistance of these resistors is calculated from the expression given below:

$$R = R_S L_{\text{eq}}/W_{\text{eq}} + 2R_C \tag{3}$$

where

R_S denotes the sheet resistance (or resistance per square) for a given resistor type computed in a similar way to R_{SB};

L_{eq} denotes the equivalent length of the resistor window in silicon oxide;

W_{eq} denotes the equivalent width with the sidewall conductances taken into account (see [16] and [18]);

R_C denotes the contact resistance for a given resistor type.

n-p-n Transistor

The saturation current I_S, Early voltage V_A, and current gain coefficient h_{FE} are simulated as follows:

The value of *collector saturation current* for n-p-n transistors is calculated from Gummel's expression

$$I_s = qn_i^2 A_E\left[\int_{X_{BE}}^{X_{BC}} \{[N_B(y) - N_E(y)]/D_{nB}\}\, dy\right]^{-1} \tag{4}$$

where X_{BC} and X_{BE} are the depths of the emitter and collector edges of the electrical base. D_{nB} denotes the diffusion coefficient of electrons in the base region, A_E is the actual emitter area, and n_i is an intrinsic carrier concentration.

It can be shown that the *Early voltage* for the n-p-n transistor can be expressed by the following formula:

$$V_A = -\frac{Q_B^*}{\delta Q_B^*/\delta V_{CE}} - V_{CE} \tag{5}$$

where

$$Q_B^* = \int_{X_{BE}}^{X_{BC}} \{[N_B(y) - N_E(y)]/D_{nB}\}\, dy. \quad (6)$$

The derivative in the denominator of (5) is approximated by finite differences.

Current Gain Coefficient: Models implemented in the simulator allow for the accurate calculation of h_{FE} over a wide operating range of I_C and V_{CE}. Three dominant components of the base current are taken into account:

back injection to emitter region component;

surface recombination component;

bulk recombination component in the current-induced base at high levels of collector current.

The influence of heavy doping in the emitter region on the emitter efficiency is incorporated in the model. At high impurity concentrations, a bandgap narrowing effect occurs, which can be modeled by an increase in the effective intrinsic carrier concentration [31]. The surface recombination component is modeled as in [27]. The high-injection level component is approximated by relations proposed in [33].

p-n-p Transistors

Collector saturation current, Early voltage and h_{FE} for p-n-p substrate and lateral transistors are calculated.

Collector saturation currents and *Early voltages* are calculated from the expressions similar to n-p-n transistor which are simplified because of the uniform base.

Current gain coefficients are computed over a wide range of I_C and V_{CE}. Two main components of the base current are included:

back injection to the emitter region;

bulk recombination component in the neutral base region.

Degradation of current gain due to high injection level of carriers in the base region is included in the model [18]. Current gain of the lateral transistor is modeled as in [9]. The effects of lateral (active) and vertical (parasitic) carrier flow are included.

Acknowledgment

The authors are very grateful to those who cooperated in the above research by making available presented data and valuable suggestions concerning physics of the manufacturing process. We are especially grateful to S. W. Director who encouraged us to continue this work providing us with aid via numerous valuable discussions, critiques, and remarks.

References

[1] G. D'Andrea and H. Murrman, "Correction terms for contacts of diffused resistors," *IEEE Trans. Electron Devices*, vol. ED-17, pp. 484-485, June 1970.

[2] D. A. Antoniadis *et al.*, "Boron near-intrinsic ⟨100⟩ and ⟨111⟩ silicon under inert and oxidizing ambients–diffusion and segregation," *J. Electrochem. Soc.*, vol. 125, no. 5, pp. 813-819, May 1978.

[3] D. A. Antoniadis, S. E. Hansen, and R. W. Dutton, "SUPREM II–A program for IC process modeling and simulation," Tech. Rep. 5019-2, Stanford Electronic Lab., June 1978.

[4] D. A. Antoniadis and R. W. Dutton, "Models for computer simulation of complete IC fabrication process," *IEEE J. Solid-State Circuits*, vol. SC-14, pp. 412-422, Apr. 1979.

[5] J. W. Bandler *et al.*, "A nonlinear programming approach to optimal design centering and tuning," *IEEE Trans. Circuit Syst.*, vol. CAS-23, pp. 155-165, Mar. 1976.

[6] B. L. Biehl, BIASD–Reference manual.

[7] D. M. Caughey and R. E. Thomas, "Carrier mobilities in silicon empirically related to doping and field," *Proc. IEEE*, vol. 55, pp. 2192-2193, Dec. 1967.

[8] K. Cave, "The base diffusion profile arising from boron redistribution in the oxide–A useful approximation," *Solid State Electronics*, vol. 8, no. 12, pp. 991-993, 1965.

[9] S. Chou, "An investigation of lateral transistor–D.C. characteristics," *Solid State Electronics*, vol. 14, no. 9, pp. 811-826, 1971.

[10] R. A. Colclaser, *Microelectronics: Processing and Device Design.* New York: Wiley, 1980.

[11] B. E. Deal and A. S. Grove, "General relationship for the thermal oxidation of silicon," *J. Appl. Phys.*, vol. 36, no. 12, pp. 3770-3778, Dec. 1965.

[12] S. W. Director and G. D. Hachtel, "The simplicial approximation approach to design centering and tolerance assignment," *IEEE Trans. Circuit Syst.*, vol. CAS-24, pp. 363-371, July 1977.

[13] R. B. Fair, "Boron diffusion in silicon-concentration and orientation dependence, background effects, and profile estimation," *J. Electrochem. Soc.*, vol. 122, no. 6, pp. 800-805, June 1975.

[14] —, "A Quantitative model for the diffusion of phosphorus in silicon and emitter dip effect," *J. Electrochem. Soc.*, vol. 124, no. 7, pp. 1107-1117, July 1977.

[15] J. Gempel, Tech. Univ. of Warsaw, private communication.

[16] D. J. Hamilton and W. G. Howard, *Basic Integrated Circuit Engineering.* New York: McGraw Hill, 1975.

[17] J. C. Irvin, "Resistivity of bulk silicon and of diffused layers in silicon," *Bell Syst. Tech. J.*, vol. 41, pp. 387-410, Mar. 1962.

[18] W. Kuzmicz, Tech. Univ. Warsaw, unpublished manuscript.

[19] W. Maly and A. J. Strojwas, "Simulation of bipolar elements for statistical circuit design," in *Proc. of ISCAS*, pp. 788-791, 1979.

[20] W. Maly, A. J. Strojwas, and S. W. Director, "Fabrication based statistical design of monolithic IC's," in *Proc. of ISCAS*, pp. 135-138, 1981.

[21] W. Maly and S. W. Director, "Dimension reduction procedure for simplicial approximation approach to design centering," *IEE Proc.* 127(6) Proc. Inst. Elect. Eng., vol. 127, no. 6, pp. 255-259, Dec. 1980.

[22] —, "Characterization of random behaviour of the IC manufacturing process," in *Proc. of European Conf. Electronic Design Automation*, Univ. of Sussex, Sept. 1981.

[23] W. Maly and T. Gutt, "Base and emitter simulation models," in *Proc. of European Conference on Electronic Design Automation*, pp. 88-91, Univ. Sussex, 1979.

[24] —, "Simulation of quality control system applied in silicon IC technology," in *Simulation of Control Systems*, I. Troch (ed.) The Netherlands: North Holland, 1978, pp. 285-287.

[25] D. K. Lynn *et al.*, *Analysis and design of IC.* New York: McGraw-Hill, 1962.

[26] J. F. Pinel and K. Singhal, "Efficient Monte Carlo computation of circuit yield using importance sampling," in *Proc. of ISCAS*, pp. 363-371, 1977.

[27] G. Rey and J. P. Bailbe, "Some aspects of current gain variation in bipolar transistor," *Solid State Electronics*, vol. 17, no. 9, pp. 1045-1057, Oct. 1974.

[28] W. R. Runyan, *Silicon Semiconductor Technology.* New York: McGraw-Hill, 1965.

[29] P. L. Shah and W. H. Schroen, "A process model for sequential diffusion and redistribution in silicon LSI technology," in *Proc. of Electrochem. Soc. Meeting*, pp. 400-403, May 1975. Extended Abstr. No. 170.

[30] W. H. Schroen, "Process modeling," in *NATO Study Institute on Process and Device Modeling for Integrated Circuit Design*, pp. 767-793, July 1977.

[31] J. Slotboom and H. C. De Graf, "Measurements of bandgap narrowing in Si bipolar transistors," *Solid State Electronics*, vol. 19, no. 1, pp. 857-862, 1976.

[32] A. J. Strojwas and S. R. Nassif, *FABRICS Manual*, Tech. Rep. DRC-18-30-81, Dep. Elect. Eng. and Design Res. Center, Carnegie-Mellon University, Pittsburgh, PA, June 1981.

[33] R. J. Whittier and D. A. Tremere, "Current gain and cutoff frequency falloff at high currents," *IEEE Trans. Electron Devices*, vol. ED-16, pp. 39-57, Jan. 1969.

Statistical Modeling for Efficient Parametric Yield Estimation of MOS VLSI Circuits

PAUL COX, PING YANG, MEMBER, IEEE, SHIVALING S. MAHANT-SHETTI, MEMBER, IEEE, AND PALLAB CHATTERJEE, SENIOR MEMBER, IEEE

Abstract —Large statistical variations are often found in the performance of VLSI circuits; as a result, only a fraction of the circuits manufactured may meet performance goals. An automated system has been developed to obtain the process statistical variations and extract SPICE model parameters for a large number of MOS devices. Device length and width, oxide capacitance, and flat-band voltage are shown to be the principal process factors responsible for the statistical variation of device characteristics. Intradie variations are much smaller than the interdie variations, therefore, only the interdie variations are responsible for variations in circuit performance. This accurate and simple statistical modeling approach uses only four statistical variables, and thus enables the development of a very computationally efficient statistical parametric yield estimator (SPYE). A linear approximation for the yield body boundary is used to make an accurate prediction of parametric yield. With the addition of temperature and supply voltage as operating condition variables, a maximum of seven simulations are required; only slightly more than the three to five required for "worst case analysis." The method has also been adapted statistical parametric specification of standard cells; performance ranges of circuit building blocks can be characterized once the statistical variations of process-dependent parameters are known.

Predicted performance variations from SPYE have been compared with measured variations in delay and power consumption for a 7000-gate n-MOS inverter chain. Agreement with the mean delay and power are better than 5 percent where SPICE model parameters were obtained from the same slice used for circuit characterization. Excellent agreement was obtained in the predicted spread in the circuit delay and power consumption using measured variations in the statistical variables.

I. Introduction

THE DEMAND for more complex integrated circuits has continued to force a decrease in device geometries and minimum feature size. This scaling of feature size has been more aggressive than process scaling tolerance. The fractional change in the device properties has increased significantly in scaling geometries to submicrometer dimensions. As a consequence, at the micrometer and submicrometer geometries used in current VLSI circuit design, the statistical variation of device characteristics can be very significant. A successful VLSI circuit design for volume production must guarantee acceptable parametric yield with known process-induced fluctuations. In standard cell design methodology, the performance specification of a cell over process variation, temperature, and voltage ranges must be established for high confidence in custom design. It is imperative that an efficient method be available for the accurate estimation of parametric yield. Current methods are not accurate or computationally efficient enough for routine use in integrated circuit designs.

The statistical variations can be classified as intradie and interdie fluctuations. The major design constraint for VLSI stems from interdie variation including variation from lot to lot. We shall assume that the intradie variation is small compared to the interdie variation and present an integrated approach to the generation of statistical device models. These models are aimed towards two classes of application in the transient circuit analysis. The first is a analysis of the circuit performance range (e.g., access time and power) for a given process spread. The second application is to use the statistical models representing process fluctuations to find an optimum design for yield and performance. The requirements for the model extraction for these two applications are described and a procedure for generating both classes of statistical models is presented.

The traditional approach of modeling process statistics in circuit simulation has been concerned with the measurement of variations in the component parameters such as the threshold, gain, body effect, etc., and requiring the circuit to be operational over the variations of these parameters taken in an arbitrary combination. This method requires the selection of the component random variables and a formation of the joint distribution of this set of random variables under the assumption of no correlation between the random variables. These assumptions are not necessary since the joint distribution of all process variables is easily measured as the distribution of the current–voltage characteristics of the MOSFET. The representation of the measured joint distribution in a parameterized form for dc circuit simulation purposes is straightforward if one recognizes that the device current is a function of the correlated random variables. However, two additional factors must be considered in the application of this joint distribution to the transient circuit analysis. First, the model must represent the device capacitance

Manuscript received July 15, 1984; revised September 17, 1984.
The authors are with the Semiconductor Process & Design Center, Texas Instruments, Inc., Dallas, TX 75265.

Reprinted from *IEEE Trans. Electron Devices,* vol. ED-32, no. 2, pp. 471–478, February 1985.

distribution as well as the current distribution in order to provide an accurate transient analysis. Secondly, the correlation of the major geometrical parameters to the intrinsic device parameters is nonlinear. Thus linear correlation matrices are not suitable for the analysis of the model parameters, and transformation to an eigenvector of uncorrelated parameters is not desirable unless both the current and the capacitance can be represented and the geometrical parameters of the transistor can remain invariant to the transformation. It is difficult to satisfy these constraints in a generalized approach. We have, therefore, based our work on an approach similar to the principal factor analysis which allows nonlinear correlation based on quasi-physical relationships between the model parameters. The significant result of the analysis presented here is that the geometrical parameters (length, width, and oxide thickness) and the electrical parameter (flat-band voltage) can be used as the principal factors for the statistical variations of the current and capacitance distribution.

Methods for parametric yield estimation fall into two categories: algorithms which utilize Monte Carlo techniques and geometrically based algorithms. The Monte Carlo techniques have been used extensively for yield estimation and are well characterized statistically. The degree of confidence in the yield estimate is determined by the accuracy of the statistical model and the number of samples taken for the estimate, but is independent of the number of independent statistical variables. However, each sample in the Monte Carlo estimate requires a circuit simulation; several hundred simulations may be required to obtain a reasonable yield estimate. Because of the computational requirements, the Monte Carlo-based methods have limited utility for routine circuit design, however, their use is valuable in the evaluation of other algorithms.

Geometrically based algorithms [3]–[5] offer considerable computational savings if the number of statistically varying parameters can be kept small. These methods utilize search techniques to approximate the boundary of the yield body and numerical integration techniques to estimate the yield. While these methods are attractive for small circuits, the computational requirements can increase exponentially with the number of statistical variables.

II. SPICE Model and Automated Parameter Extraction

The constraints placed on circuit simulation models for VLSI circuit simulation are primarily based on a tradeoff of computational efficiency and the ability to represent the device characteristics. This requirement of computational efficiency is based on the necessity of simulating 10000 element circuits for critical paths of VLSI circuits which have over 100000 transistors. The analytical models developed for this purpose fall into two categories. The first is a model derived from quasi two dimensional analysis which tries to represent all the geometry and non-uniform doping-related variations. The second category of models are quasi-physical and quantized length and width. In general, digital integrated circuits require quantized channel length from process control and speed considerations. The width quantization is not desirable and can be removed without significant computational penalty. However, the channel length affects all parameters and a significantly higher computation penalty is associated with removing the length quantization. It is however, possible to use quasi-physical models which are length quantized, and derive an interpolation formula which represents the perturbation of the model parameters around the extracted length. We consider this approach to statistical circuit simulation models most appropriate, particularly incases where statistical optimization is to be attempted.

The $I-V$ characteristics of the devices represent the joint distribution of all the process variations. The $I-V$ data from MOSFET's with various nominal width and length are easily measured. An optimum parameter extraction program is written to extract the parameters from a measured data set spanning the entire $I-V$ space. A subset of this data for the linear region of five different W/L is used to extract the electrical length and width from the same data base [6], [7]. Thus the entire model parameter extraction is obtained from the measured $I-V$ distribution. The observed distribution of the extracted parameters was best represented by Gaussians.

III. Principal Factors

Simulation of device characteristics for the full range of process variations requires an accurate model for both variations in $I-V$ characteristics and capacitance distribution. The automatic data acquisition system described in the previous section can be used to model the variations in $I-V$ characteristics. However, it is not practical to measure the capacitance distribution and extract the additional parameters required to describe it. It has been shown [8], [9] that current and capacitance of MOSFET's are both sensitive to:

- length;
- width;
- gate oxide thickness; and
- flat-band control.

The sensitivity to other parameters, such as changes in the doping profile, are calculated to be at least an order of magnitude smaller. It is reasonable to expect that these physical quantities are the principal factors which dominate the statistical distribution of both current and capacitance. The current model parameters can be used to derive the capacitance model as well.

Since length, width, oxide thickness, and flat-band voltage are determined by different steps in the manufacturing process, they should be uncorrelated independent statistical variables. The statistical independence of length and width is easy to verify from parameters extracted from the automatic data acquisition system and extraction program. The distribution of W/L and $W \cdot L$ can be extracted at the same time (see Fig. 1). If W and L are uncorrelated

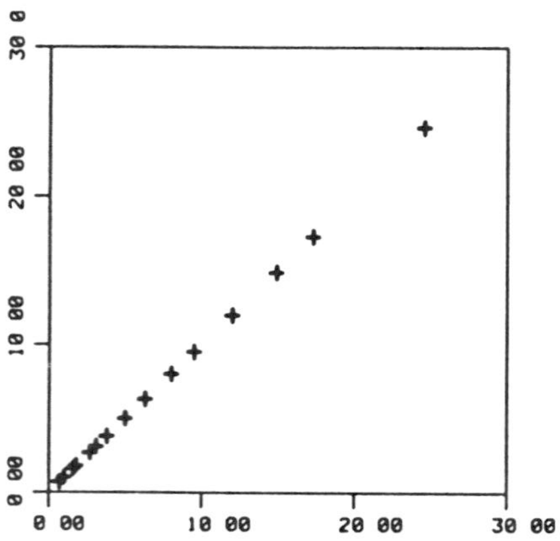

Fig. 1. Comparison of the mean of (W/L) and $W(\text{mean})/L(\text{mean})$.

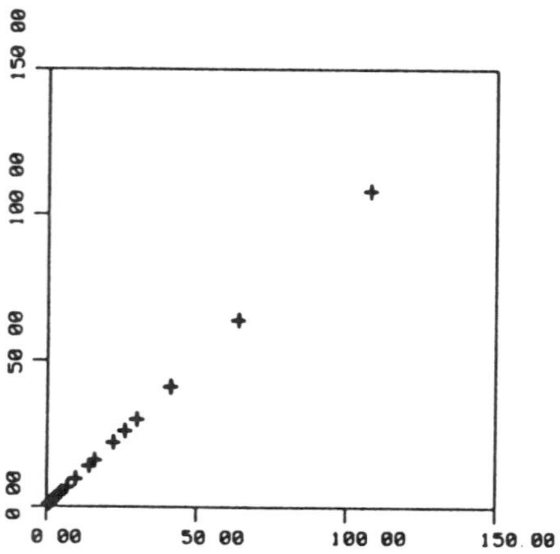

Fig. 2. Comparison of the mean of $(W * L)$ and $W(\text{mean}) \cdot L(\text{mean})$.

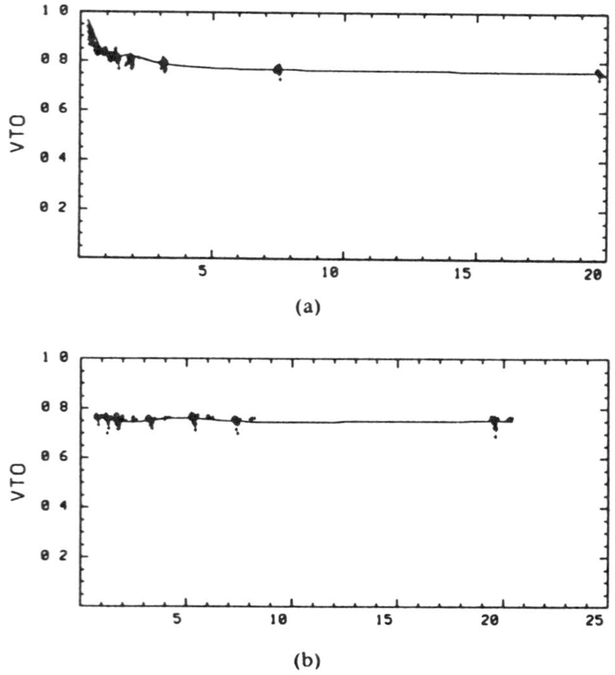

Fig. 3. (a) Length (l) and (b) width dependence of $VT0$.

random variables, then the mean of the joint distribution of W/L and that of $W \cdot L$ should be equal to $W(\text{mean})/L(\text{mean})$ and $W(\text{mean}) \cdot L(\text{mean})$ [10]. Figs. 2 and 3 show the extracted values of these quantities and this data verify that there is no correlation between width and length. Thus the assumption that factors are Gaussian uncorrelated random variables is justified.

IV. Statistical Model For Yield and Performance Optimization

Integrated circuits designed for volume production must meet design specifications over the range of process variations. Otherwise, only a fraction of the circuits manufactured will be acceptable. Classically, device model parameters for circuit simulation are extracted from the $I-V$ curves of devices whose lengths and widths are representative of the devices contemplated for production. It is seldom practical to obtain accurate models for all possible device length and width combinations used in a VLSI circuit; it would be impossible to obtain accurate models from representative devices over the volume of the statistical parameter space. In order to obtain accurate circuit simulation, it is necessary to have valid models, not only for the devices with the specific lengths and widths and process-dependent parameters from which they were derived, but also for other device dimensions and other process variants. The functional relationship between the device model parameters and length, width, oxide capitance, and flat-band voltage must be understood and incorporated in the model. Furthermore, circuit optimization and yield estimation require multiple transient simulations for each VLSI circuit. An efficient method must be derived to express the functional relationship between the principal factors and device characteristics, otherwise the computational cost will be prohibitive. The width and length dependence of our model is derived based on the charge-sharing concept, and has a simple $1/W$ or $1/L$ dependence. The width and length model not only permits accurate simulations in design optimization where the effect of changes in device dimensions must be investigated, but it also reduces the complexity of the statistical optimization problem. The width and length dependence account for a significant part of the statistical variations in device performance. Since the intradie width and length reduction variations can be assumed to be negligible, the total number of statistical variables is greatly reduced and the total computational task becomes manageable for VLSI circuits.

For a short-channel MOSFET, the bulk charge is shared among gate, drain, and source, and hence the threshold voltage and the body effect are reduced. This reduction of the threshold voltage and the body effect can be represented by either a $1/L$ dependence [11] or an exponential dependence on length [12], [13]. Since our experimental results have demonstrated that either model can adequately describe the length dependence, the $1/L$ relationship was chosen because of its computational efficiency. Similarly, for a narrow-width MOSFET, the bulk charge controlled

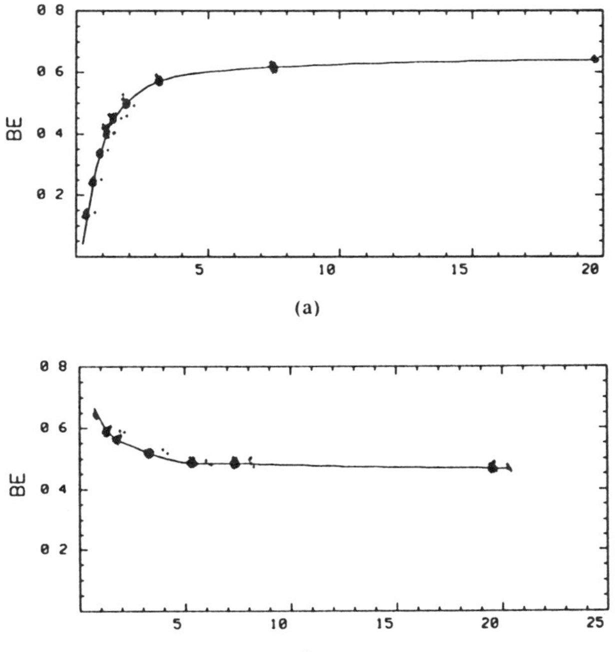

(a)

(b)

Fig. 4. (a) Length (l) and (b) width dependence of *BE*.

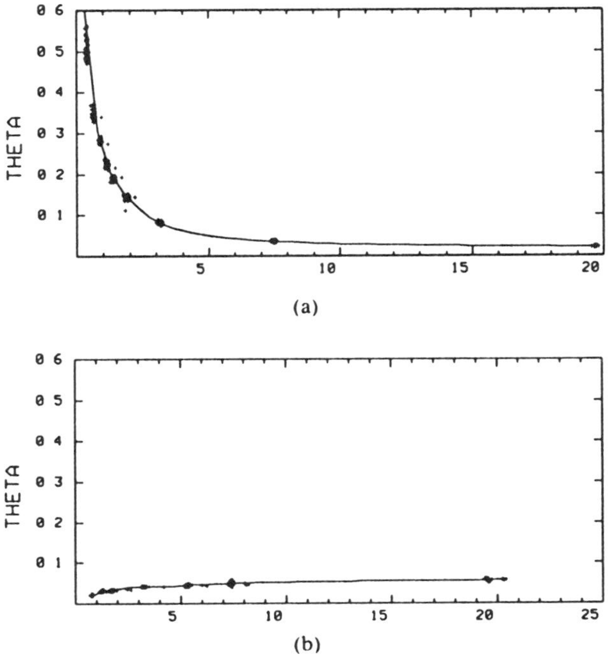

(a)

(b)

Fig. 5. (a) Length (l) and (b) width dependence of *THETA*.

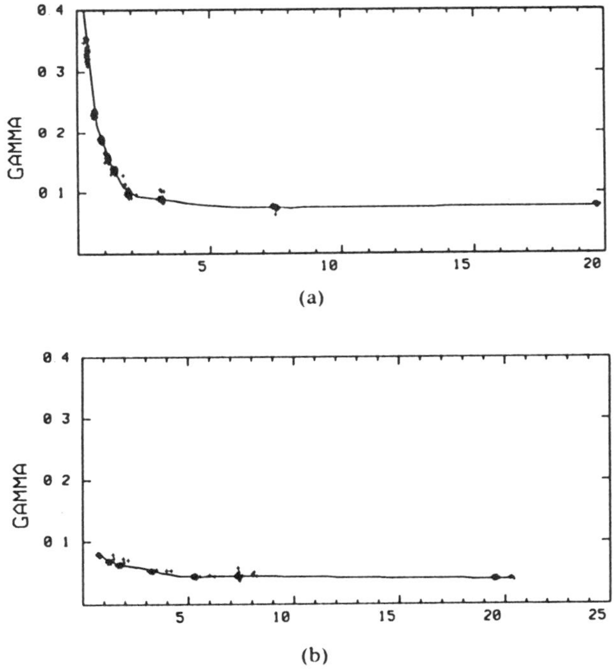

(a)

(b)

Fig. 6. (a) Length (l) and (b) width dependence of *GAMMA*.

by the gate spreads laterally into the transition region between the channel stop and the channel and causes an increase of the threshold voltage and the body effect. It can be shown that this increase of the body effect and the threshold voltage is proportional to $1/W$ [14], [15]. A comparison of experimental values of $VT0$ and BE and the model are shown in Figs. 3 and 4. It can be seen that the model accurately predicts the width and length dependence. In most cases, these and other device model parameters (see the Appendix) can be adequately described as linear functions of $1/L$ and $1/W$. However, the precise form of this dependence is difficult to predict for each of the parameters in the quasi-physical model used for SPICE simulations. Rather using a specific functional relation between the parameters and $1/L$ and $1/W$, the measured parameters are fit to a cubic SPLINE function. The use of the SPLINE interpolation permits an exact fit to the mean of the parameters for each nominal width and length. Furthermore, the sensitivity of each of the parameters to $1/L$ and $1/W$ can be used as boundary conditions for determination of the SPLINE parameters and guarantee that the function is well behaved outside the range of measured geometries.

The reduction of the effective mobility due to the channel vertical electrical field, the surface roughness scattering, and the source–drain resistance induced feedback can be modeled by a parameter THETA [15]. The dependence of THETA on L and W is shown in Fig. 5(a) and (b), and can also be described by the simple $1/W$ or $1/L$ relationship.

The nonlinear velocity saturation characteristics due to the lateral electric field for short-channel devices has been modeled by a parameter GAMMA [15]. It can be shown that GAMMA should be a strong function of $1/L$ and a weak function of $1/W$. The experimental results are given in Fig. 6(a) and (b). This parameter GAMMA provides an accurate representation for the saturation voltage and the device saturation characteristics. Similarly, the nonlinear correlation between the model parameters and the other two principal factors, gate oxide thickness and flat-band voltage, can be determined, and a nonlinear interpolative formula can be derived.

For the parameters in the quasi-physical MOS model used for SPICE circuit simulation, the mean, variance, and

TABLE I
REDUCTION OF VARIANCE

PARAMETER	MEAN	VARIANCE	RESIDUAL
Theta	0.29	3.6e-2	1.7e-4
VTO	0.26	3.0e-3	2.6e-4
BE	0.37	1.6e-2	8.9e-5
Theta	0.29	3.6e-2	1.7e-4
DE	0.06	4.8e-3	7.5e-5
Gamma	0.12	2.9e-3	3.5e-5
Lamda	0.05	3.3e-4	4.6e-6

the residual, after extraction of the length and width dependence, are shown in Table I.

With the exception of *VT*0, less than 2 percent of the variance remains after extraction of the length and width dependence. The large residual in *VT*0 can be attributed to variations in flat-band voltage. With the inclusion of oxide capitance and flat-band voltage dependence, these coefficients extracted from *I*–*V* data of test transistors can be used to generate accurate models for circuits at any arbitrary point in the four-dimensional parameter space. The statistical simulation consists of generating samples by assuming the principal factors are independent normal random variables, calculating the corresponding model parameters by using the nonlinear correlation formula, and using the calculated models in the circuit simulator for yield or performance evaluation.

For VLSI circuits, high correlation exists between the observed *I*–*V* and transient responses of devices within a given die. This high degree of correlation in not only found between devices of the same type, but is also observed between devices of different implant types. The mean of the extracted values of *Lr*, *Wr*, and *Vfb* differ with implant type, but no significant differences are observed in the die-to-die variations about the mean. Since intradie variations of each of principal factors are much less that die-to-die or lot-to-lot differences, only four statistical variables need be considered to model the statistical performance variations of an MOS integrated circuit. The number of statistical parameters is much smaller than that of the circuit design parameters.

V. Statistical Parametric Yield Estimation

Since only four statistical variables are responsible for the most of the process-induced variations in circuit performance, the parametric yield can be estimated efficiently by using geometric approximation for the yield body boundary. For MOS digital circuits, our study indicates that the yield body can be approximated by a convex polygon in the *N*-dimensional statistical parameter space. The performance constraints which define the yield body boundaries can be approximated by linear functions of the statistical variables. While the constraint boundary is certainly not linear for large changes in *Lr*, *Wr*, *Cox* and *Vfb*, process variations make relatively small perturbations in each of these variables. These approximations could result in relatively large errors in the yield estimation if the constraint boundaries pass near the design center. However, for circuits designed for volume production, the yield boundary must be far enough from the design center to obtain acceptable yield. Small errors in the location of the yield boundary in the wings of the probability distribution function will result in insignificant errors in the estimation of the yield.

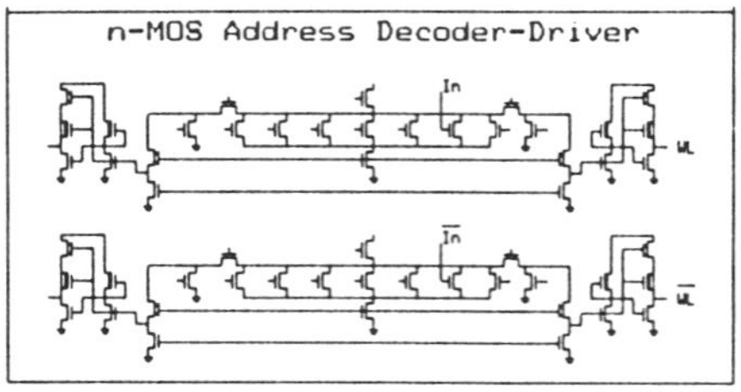

Fig. 7. Circuit diagram of an n-MOS address decoder/driver circuit used for SPYE yield estimation.

Input to the statistical parametric yield estimation (SPYE) program consists of a circuit description, performance constraints or specifications, and statistical information on variation of the principal factors and the functional relationship of these factors to the SPICE model parameters. The circuit description required by the program is a standard SPICE input file; this is used by SPYE for automatic generation of the modified SPICE input files required for simulation at selected points in the statistical parameter space. The circuit performance constraints define the specifications for the circuit to be used in determining the yield. A wide variety of parameters derived from either dc or transient analysis of the circuit may be used to define these constraints. The n-MOS address decoder–driver circuit shown in Fig. 7 has been studied extensively using the SPYE algorithm. Performance constraints for the n-MOS address decoder–driver are:

time (WL)	< 11.8 ns	(1a)
power (V3V)	< 2.2 mW	(1b)
volt cross (WL, WL)	< 1.2 V.	(1c)

These constraints contain limits on the circuit delay time for the word line output (1a), the maximum power consumption (1b), and the voltage crossing point of the complementary word line outputs (1c). The statistical variations in *Lr*, *Wr*, *Cox*, and *Vfb* and the dependence of the other SPICE model parameters were obtained from *I*–*V* data on a large number of test transistors produced by the same process as that intended for use with this circuit.

Information on the number and range of each of the statistical variables is used by SPYE to generate the SPICE input files necessary to perform the circuit simulations for yield estimation. A total of $N+1$ simulations must be performed where N is the number of variables. The location of these points in the *N*-dimensional parameter space is arbitrary as long as all are noncolinear and sufficiently separated to provide an accurate approximation of the yield body. For convenience SPYE selects the vertices of a regular simplex in the *N*-dimensional parameter for simulation; the size of the simplex is chosen so that each of the

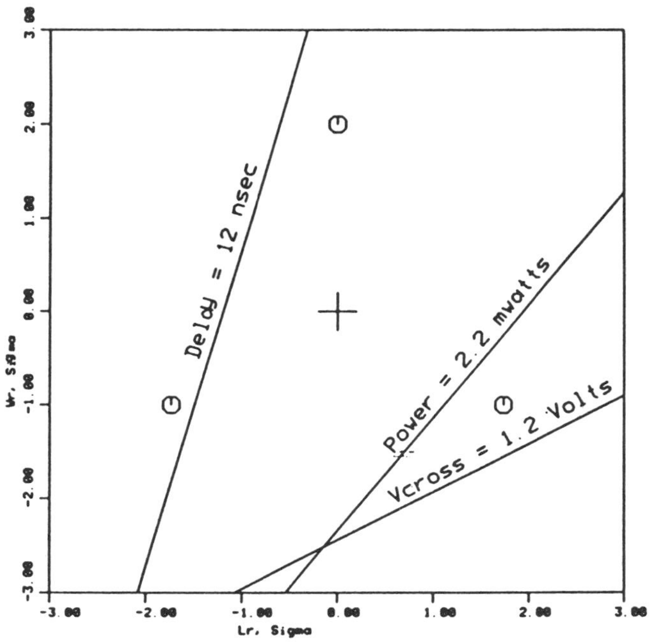

Fig. 8. Points selected for simulation runs and linear yield body boundaries for n-MOS address decoder/driver circuit.

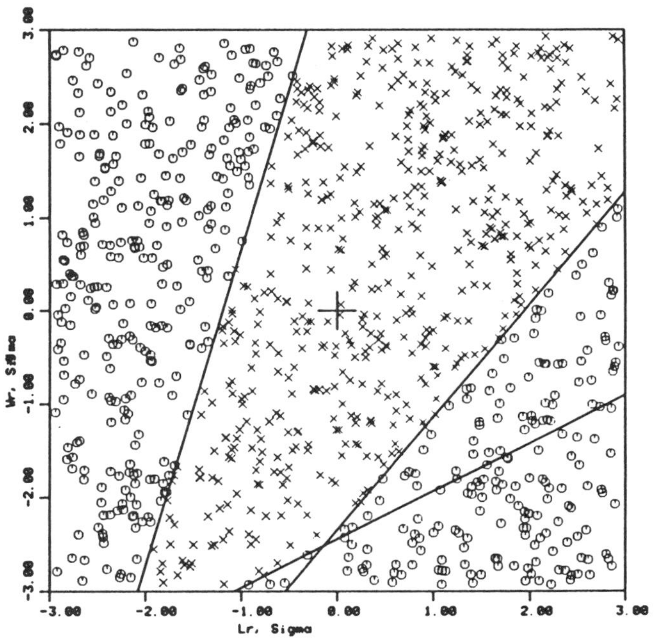

Fig. 9. A comparison of the Monte Carlo simulation results with the linear estimate of the yield body. + Simulated circuit is within the yield body. O Simulated parameters failed one or more constraints.

vertices are separated by 2 sigma from the nominal design point. From analysis of the output of these simulations, coefficients for a linear approximation of the boundaries for each of the constraints can be obtained

$$F(Lr, Wr, Cox, Vfb) = A*Lr + B*Wr + C*Cox + D*Vfb + E.$$

Fig. 8 shows the three points selected for simulation runs and the linear boundaries for the constraints in a two-dimensional parameter space with *Lr* and *Wr* as the variables. The parametric yield is estimated by numerical integration of the probability distribution function over the volume of the yield body included within these boundaries. For a large number of both static and dynamic MOS circuits investigated, this initial estimate agrees within 2 percent to that obtained from a 1000-sample Monte Carlo estimate if the closest constraint boundary is greater than 2 sigma from the nominal design point. If greater accuracy is required, additional simulations are performed at the points of closest approach of the constraint boundary to the nominal design point; because the PDF drops rapidly away from the mean, this area of the yield body boundary is most critical in determining the yield. A comparison of the Monte Carlo results with the linear estimate of the boundaries is shown in Fig. 9. The boundary between the successful samples and the failure is approximated very well by the linear estimate of the constraint limits. A 3D perspective of the yield body for the same circuit and performance constraints with *Cox* as an additional variable is shown in Fig. 10. In addition to the four principal factors (*Lr*, *Wr*, *Cox*, and *Vfb*), SPYE can also be used to determine the effect of temperature and supply voltage on the overall parametric yield.

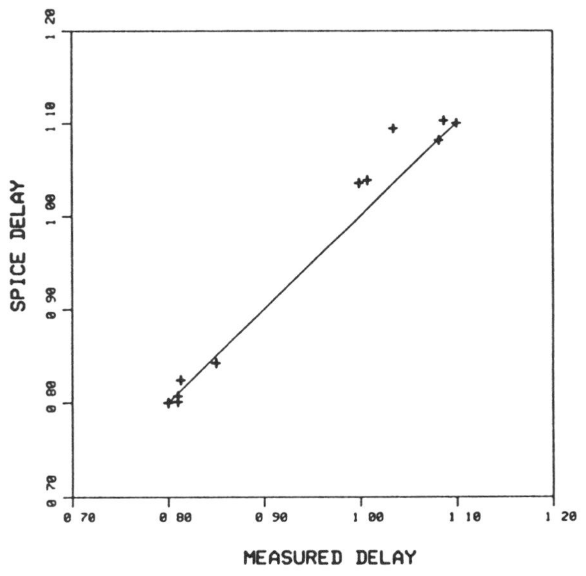

Fig. 10. 3-D perspective of yield body for n-MOS address decoder/driver with *Lr*, *Wr*, and *Cox* as the statistical variables. SPICE calculated delay versus measured delay.

VI. Experimental Results

Experimental verification of the method has been made by a comparison of the measured delay and average current for an n-MOS inverter chain with that predicted by the statistical model. For this investigation, a test bar was used which contained both a long inverter chain and individual test transistors for characterization. The inverter chain consisted of an input buffer, 7000 inverters, and an output buffer. The test transistors were designed for automatic measurement of $I-V$ characteristics and SPICE model parameter extraction. The size of the devices were

TABLE II
INVERTER CHAIN STATISTICS

	Delay, nsec	
Slice 8, measured	0.84	+/- 0.03
calculated	0.85	+/- 0.04

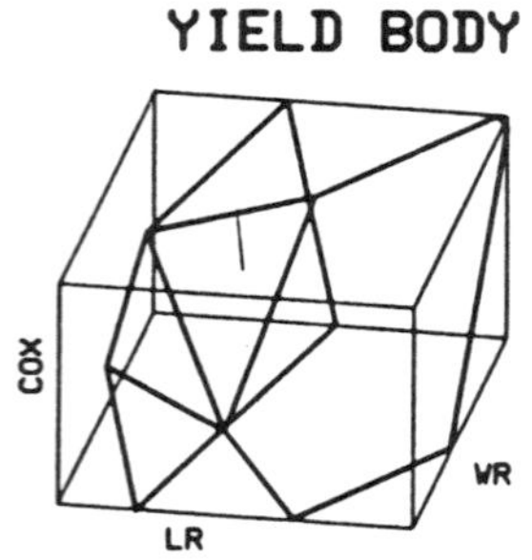

Fig. 11. Comparison of measured inverter delay per stage for 7000 inverter chain with SPICE calculated delay using measured value of principal factors.

chosen to be representative of devices in a memory circuit being designed, to permit length and width reduction measurement, and to establish the length and width dependence of the SPICE model parameters:

Width	2.0	5.5	5.5	5.5	11.0
Length	1.1	1.1	1.65	2.2	1.1.

Delay for the inverter chain was measured for all of the die from each of nine slices in the same lot. For each of the same die, BETA and threshold voltage were extracted from the linear region $I-V$ curves. From these linear BETA's, the length and width reduction factors were obtained using the expression

$$BETA = (W - Wr)/(L - Lr)*K \text{ prime}.$$

Since K prime is proportional to the oxide capitance

$$K \text{ prime} = U\text{eff} \cdot Cox$$

variations in K prime can be used to estimate the oxide capitance variation. In a similar way, flat-band voltage variation can be obtained from the residuat variation in $VT0$ after removal of length, width, and oxide capitance dependence

$$VT0 = Vfb + F(Wr, Lr, Cox).$$

In addition to the parameters extracted from the linear region, complete SPICE model parameters were obtained for the mean devices on one slice selected as representative of the lot.

A comparison of the mean delay for the inverter chains on each of the nine slices, with the delay calculated from SPICE using the extracted model parameters and the measured length and width reduction factors, is shown in Fig. 11. Excellent correlation is seen between the measured and calculated delays, which demonstrates that changes in the four principal statistical variables dominate the variation in circuit delay. Details on the statistical variation in delay per stage for Slice #8 (the slice used for SPICE model extraction) is shown in Table II.

VII. SUMMARY

This statistical modeling approach is an efficient and accurate method for parametric yield estimation for MOS digital circuits. It provides the designer with a figure of merit on the manufacturability of the design; the number of simulations required is comparable with that needed for a "worst case analysis." The method can also be used to characterize standard cells or define process tolerance required to obtain an acceptable yield for circuit designs, such as dynamic RAM's, which are produced in high volume.

An integrated approach to automated statistical parameter extraction for iterdie process variations has been developed. This method allows the joint distribution of process variations to be represented by three model vectors, which can be used to calculate the spread in transient response of the circuit. It is shown that the use of width, length, gate oxide, and flat-band voltage as principal factors enables the entire distribution to be adequately represented. Computationally efficient interpolative formulas are developed to allow the entire process statistical distribution to be simulated in SPICE2 [12]. This will enable performance optimization and design centering across the process windows.

APPENDIX

The current equations:

1) Linear Region: $V_{GS} \geqslant V_{G\,\mathrm{SAT}}$

$$I_{D,MSH} = \alpha_x \beta_n \{2(V_{GS} - V_T)V_{DS} - \alpha_x V_{DS}^2\}$$

2) Saturation Region: $V_{G\,\mathrm{SAT}} \geqslant V_{GS} > V_T$

$$I_{D,MSH} = \beta_n (V_{GS} - V_T)^2 \times \{1 + \lambda(\alpha/\alpha_x)^2 (V_{DS} - V_{D\,\mathrm{SAT}})\}$$

3) Cutoff Region: $V_{GS} \leqslant V_T$

$$I_{D,MSH} = 0.$$

In the above equations

$$\beta_n = \beta_{n0}/\{1 + \theta(V_{GS} - V_T)\}$$
$$\beta_{n0} = K_p(W - W_R)/(L - L_R)$$
$$V_T = V_{FB} + 2\phi_f + BE\sqrt{2\phi_f - V_{BS}} - DE\sqrt{V_{DS}}$$
$$\alpha_x = \alpha + \gamma(V_{GS} - V_T)$$
$$V_{D\,\mathrm{SAT}} = (V_{GS} - V_T)/\alpha_x \quad V_{G\,\mathrm{SAT}} = V_T + \alpha_x V_{DS}.$$

ACKNOWLEDGMENT

The authors are grateful to G. Giles, C. Sullivan, and V. Kie for their assistance in the above research.

REFERENCES

[1] P. Yang and P. K. Chatterjee, "Statistical modeling of small geometry MOSFET's," in *IEDM Tech. Dig.*, 1982.

[2] P. Cox, P. Yang, and P. Chatterjee, "Statistical modeling for efficient parametric yield estimation of MOS VLSI circuits," in *IEDM Tech. Dig.*, pp. 242–245, 1984.

[3] S. W. Director and G. D. Hachtel, "The simplicial approximation approach to design centering," *IEEE Trans. Circuits Syst.*, vol. CAS-24, pp. 363–371, 1977.

[4] M. R. Lightner and S. W. Director, "Multiple criterion optimization with yield maximization," *IEEE Trans. Circuits Syst.*, vol. CAS-28, pp. 781–790, Aug. 1981.

[5] H. L. Abdel-Malek and J. W. Bandler, "Yield optimization for arbitrary statistical distributions," *IEEE Trans. Circuits Syst.*, vol. CAS-27, 253–262, Apr. 1980.

[6] P. I. Sucia and R. L. Johnston, "Experimental derivation of the source and drain resistance of MOS transistors," *IEEE Trans. Electron Devices*, vol. ED-27, 1980.

[7] D. B. Scott, unpublished.

[8] J. E. Leiss and P. K. Chatterjee, "A new approach to modeling ion implanted channels in MOSFETs and applications to sensitivity analysis," presented at the 40th Annual Device Res. Conf., 1982.

[8] S. Selberherr, A. Schutz, and H. Potzl, "Investigation of parameter sensitivity of short-channel MOSFETs," *Solid-State Electron.*, vol. 25, 1982.

[9] A. Papoulis, *Probability, Random Variables, and Stochastic Process.* New York: 1965.

[10] P. K. Chatterjee and J. E. Leiss, "An analytic charge-sharing predictor model for submicron MOSFET's," in *IEDM Tech. Dig.*, 1980.

[11] T. Toyabe and S. Asai, "Analytical models of threshold voltage and breakdown voltage of short-channel MOSFET's derived from two-dimensional analysis," *IEEE Trans. Electron Devices*, vol. ED-26, 1979.

[12] K. N. Ratnakumar, J. D. Meindl, and D. L. Scharfetter, "New IGFET short-channel threshold voltage model," in *IEDM Tech. Dig.*, 1981.

[12] L. A. Akers, M. M. E. Begnwal, and F. Z. Custode, "A model of a narrow-width MOSFET including tapered oxide and doping encroachment," *IEEE Trans. Electron Devices*, vol. ED-28, 1982.

[13] P. Yang and P. K. Chatterjee, "SPICE modelling for small geometry MOSFET circuits," *IEEE Trans. Computer-Aided Design*, vol. CAD-1, Oct. 1982.

An Extrapolated Yield Approximation Technique for Use in Yield Maximization

DALE E. HOCEVAR, MEMBER, IEEE, MICHAEL R. LIGHTNER, MEMBER, IEEE, AND TIMOTHY N. TRICK, FELLOW, IEEE

***Abstract*—This paper is concerned with the computational problem of maximizing the yield of circuits. A statistical Monte Carlo based approach is taken in order to compute yield estimates directly and to decrease dimensionality dependence. The main contribution of this paper is a yield extrapolation technique which is very effective in maximizing the yield along a search direction. This technique is based upon a quadratic model of the circuit. Statistical yield estimates are computed from the model and correlated sampling is used to extrapolate along the search direction. Two simple methods for determining search directions are discussed and these are used to demonstrate the overall method through several examples.**

I. INTRODUCTION

YIELD maximization for circuits is an important problem in the area of statistical circuit design. It is of ever increasing importance in today's environment of CAD techniques; especially for integrated circuits, where, because of smaller and smaller geometries, achieving respectable yields is much more difficult. Yield maximization can be approached from either a deterministic or a statistical viewpoint. The deterministic methods employ nonlinear programming techniques and are often based on geometric concepts [1]-[13]. However, deterministic methods tend to be more difficult to implement, and suffer from the curse of dimensionality because their computational complexity depends critically on the dimension of the design parameter space [13]. Furthermore, in relation to yield analysis, no practical means exist for computing the accuracy of deterministic methods; at most only a lower bound on the yield can be computed [13]. Statistical methods have received much more attention recently [14]-[24]; mainly because they are based upon Monte Carlo analysis, where the yield accuracy depends only upon the sample size, and is independent of the dimension of the parameter space. Thus in this paper, a statistical Monte Carlo approach to the yield maximization problem will be taken.

In recent years, several reseachers have addressed yield optimization and related topics from a statistical viewpoint [14]-[24]. Many of these techniques iterate by determining a direction in parameter space and then move the design point a certain step length along that direction. There do not appear to be any techniques for performing optimization along those directions. The main purpose of this paper is to present an extrapolated yield approximation technique for solving this problem. The underlying methodology assumed in this paper for the yield maximization problem is an iterative one, consisting of determining a search direction and then maximizing along that direction to find the optimal step length to move the design point. Monte Carlo techniques are used to evaluate the yield functions.

In the remainder of this paper we first introduce a quadratic model, which is used to approximate the circuit performance functions. Quadratic approximations have been used by others in statistical circuit design [3], [7], [8], [12]; but our application is quite different, as it is used to generate less costly yield estimates. Next, the extrapolated yield approximation technique is presented, which is a very effective method for determining the optimal step length along the search direction. Then a short discussion of the techniques that were used to generate search directions for the examples is given. Finally, several yield maximization examples demonstrate the overall procedure, and clearly demonstrate the effectiveness of the approximation technique.

Manuscript received July 12, 1982; revised March 1, 1984. This work was supported by the National Science Foundation under Grant ENG 78-171815.

D. E. Hocevar was with the University of Illinois, Urbana, IL. He is now with Texas Instruments, Dallas, TX.

M. R. Lightner was with the University of Illinois, Urbana, IL. He is now with the Department of Electrical and Computer Engineering, University of Colorado, Boulder, CO.

T. N. Trick is with the Coordinated Science Laboratory and the Department of Electrical Engineering, University of Illinois, IL 61801.

II. QUADRATIC MODEL

The quadratic model to be used for the circuit is a set of quadratic approximations to the circuit's response or performance functions. The quadratic equations are denoted by G_i's and the response functions by g_i's.

$$\begin{aligned} G_1(x) &= x^T Q_1 x + b_1^T x + c_1 \cong g_1(x) \\ &\vdots \\ G_m(x) &= x^T Q_m x + b_m^T x + c_m \cong g_m(x). \end{aligned} \tag{1}$$

Here the Q_i's are $n \times n$ real symmetric matrices, the b_i's are $n \times 1$ real vectors, and the c_i's are real scalars. These quadratic approximations are to be fit from funtion and gradient evaluations of the g_i's on a set of grid points around the nominal point. This will be an exact fit in contrast to linear regression or least squares curve fitting. It is easily seen that this results in a square system of linear equations of dimension $M = (n + 1)(n + 2)/2$ which must be solved to determine the coefficients. Using standard techniques this requires about $n^6/6$ operations (multiplications or divisions), and hence is very costly. To reduce this cost and to simplity the problem, Abdel-Malek and

Reprinted from *IEEE Trans. Computer-Aided Design,* vol. CAD-3, no. 4, pp. 279–287, October 1984.

Bandler [8] have chosen the grid points in a special manner which actually results in closed form expressions for the coefficients. The quadratic is written in the form

$$G(x) = \frac{1}{2}(x - x_0)^T H(x - x_0) + a^T(x - x_0) + a_0 \quad (2)$$

and the grid points are x_0, $\{x_0 \pm \Delta_i\}_1^n$, and some elements of $\{x_0 + \Delta_i + \Delta_j\}_{i=1}^n$, $j = 1$, $i \neq j$, where $\Delta_i = (0, \cdots, \delta_i, \cdots, 0)$ and δ_i is a positive distance on the ith axis. In order to evaluate the coefficients in (2), $2n + 1$ circuit simulations are required if the gradient is available [29]. However, when no gradient information is available, then $n(n - 1)/2$ more circuit simulations are necessary. Partial gradient information can also be employed [29]. In circuits with a large number of parameters, the gradient should be computed. Furthermore, in modern circuit simulation programs, the sensitivities which form the gradient can be computed with little additional cost.

III. Extrapolated Yield Approximations

The purpose of this section is to present a very effective, and efficient, technique for extrapolated yield estimation which utilizes the quadratic model and correlated sampling. Basically, the yield extrapolation technique computes Monte Carlo yield estimates of the quadratic model itself. Then, when correlated sampling is used, one can take advantage of properties of the quadratic equations to achieve tremendous computational savings when the extrapolation is done along a line. In addition, the use of correlated sampling will greatly increase the accuracy of the differences between yield estimates along the line. The yield maximization methodology assumed in this paper, is that of using iterative techniques which successively generate search directions and perform line maximizations. Thus the extrapolated yield approximation technique provides an effective means of performing the line maximizations.

A. Correlated Sampling

In the yield maximization problem it is of critical importance to compute accurately the yield difference between two design points. This is mainly because in each iteration these yield differences are used in deciding where the next design point should be. Correlated sampling is a technique which can easily be used to greatly increase the accuracy of these yield differences.

Becker [27] and Becker and Jensen [28], have thoroughly discussed the accuracy problem in circuit yield comparisons and the application of correlated sampling. For circuits with the same topology, the sampling procedure is simply to use the same random sample set for both designs. This corresponds to a translation of the sample points from one design to the other, and tends to introduce a strong positive correlation between the two estimates, Y_1 and Y_2, since samples that passed (failed) in one design tend to pass (fail), in the second design. To see how this reduces the error, look at the variance of the yield difference which is

$$\sigma_d^2 = \sigma_1^2 + \sigma_2^2 - 2\rho\sigma_2\sigma_2 \quad (3)$$

and it is seen that any positive correlation helps reduce the variance. Note that when the correlation is zero, as for independent sampling, then the relative error of the yield difference, which is proportional to $\sigma_d / |Y_1 - Y_2|$, is much higher than the relative errors of the yield terms, which are proportional to σ_i / Y_i. When the sample size is only a few hundred samples, and one is comparing two yield estimates which differ by only a small percentage, there is a reasonably high probability that even the sign of the difference between the yield estimates could be incorrect [29]. Such an inaccuracy could be very costly in yield maximization, since it could send the algorithm in the wrong search direction.

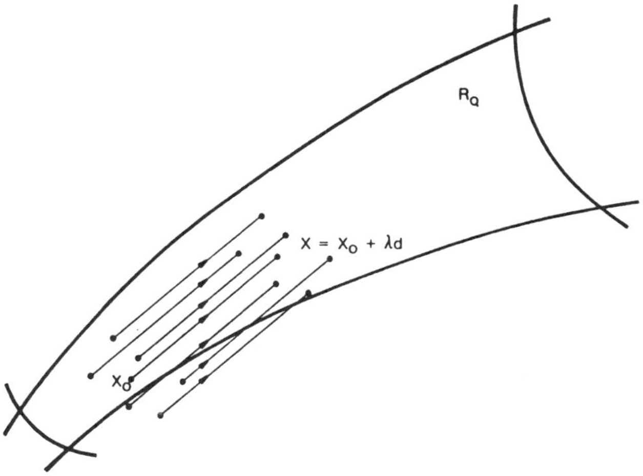

Fig. 1. Parallel translation of sample points for extrapolated yield estimation.

B. Line Searches with Invariant Parameter Statistics

An efficient technique for computing yield estimates of the quadratic model as the nominal point varies along a search direction will now be presented. Assume for now that the parameter statistics are invariant to changes in the mean. This means that the PDF at any design point is simply a translated version of the PDF at the initial nominal point. Specifically, let $p_0(x)$ be the PDF at the nominal point x_0, and let $p_a(x)$ be the pdf at the design point x_a, then $p_a(x) = p_0(x - \Delta x)$, where $\Delta x = x_a - x_0$. Now let $d \in R^n$ be an arbitrary search direction and assume that an initial yield estimate has been generated for the quadratic model using standard Monte Carlo techniques. The best way to estimate the yield of the model at various step lengths λ, a real scalar, along $x_0 + \lambda d$, is to translate the initial random sample set, along with the nominal point, and evaluate the quadratic approximations at these translated sample points. This is shown in Fig. 1. As discussed in the correlated sampling section, this tends to introduce a strong positive correlation which greatly reduces the error in the yield differences. In addition, the cost of evaluating the model at the translated points can be made insignificant by exploiting a property of the quadratic equation. Evaluating a quadratic equation along a line can be expressed as

$$\begin{aligned} G(x + \lambda d) &= (x + \lambda d)^T Q(x + \lambda d) + b^T(x + \lambda d) + c \\ &= \alpha\lambda^2 + \beta\lambda + \gamma \end{aligned} \quad (4)$$

where

$$\alpha = d^T Q d \quad (5)$$

$$\beta = 2d^T Q x + b^T d \quad (6)$$

$$\gamma = G(x) = x^T Q x + b^T x + c. \quad (7)$$

Thus the evaluation consists of evaluating a scalar quadratic equation, provided the coefficients α, β, and γ are known. The coefficient computation is *minimal,* as the γ's have already been computed in the initial model yield estimate. The β's require little effort, only an inner product for each sample point and some products dependent upon the direction, and the α's depend only upon the direction and not upon the sample points. Note that there is a set of coefficients for each performance function at each sample point. Thus once the coefficients have been computed, one can evaluate the model yield at several design points along the search direction at a trivial cost, even with large sample sizes.

The object now is to maximize the yield of the quadratic model along this direction. Because the yield of the model can only be estimated, the line maximization can only be approximately solved. Normal line search techniques, such as Fibonacci's search and quadratic or cubic interpolation, are not really applicable because of the jitter or noise in the yield estimates. Also, the model yield estimate as a function of λ, for a particular set of random samples, is actually a discontinuous, up and down staircase function. Something that can, and has been done, is to fit a quadratic regression curve to the yield estimates using several step lengths along the search direction and take the maximum given by the regression equation as the maximum along a line. However, it was found that these regression curves do not necessarily follow the data very well, and it is much easier to hand pick the maximum point (at least in the development stages of the technique). In the interactive routines that have been written to implement this technique, the *quadratic model* yield is estimated at several evenly spaced points in a user defined interval along the search direction. The user interactively adjusts this interval until the region of maximum yield is found. From this information the designer can easily find approximately where the maximum occurs, and in addition the designer can observe how the yield responds along this direction.

C. Approximation with Variant Parameter Statistics

In the last section the parameter PDF was assumed to be invariant, only the mean changed as the nominal point shifted. With respect to integrated circuits this is not usually the case, at least with some of the commonly used parameters. Many statistical circuit design procedures are restricted to the invariant situation. Recently Maly, Strojwas, and Director [30] have discussed this problem in relation to statistical circuit design. In this section the extrapolated yield estimation technique is extended for a particular situation of variant parameter statistics.

It is assumed that the parameter statistics are Gaussian, and that the standard deviations are a fixed percentage of the mean, and that the correlation factors are invariant to changes in the mean. If one thinks of the ± 3 standard deviation points as tolerance limits, then this relates to a situation where, as the mean or nominal point varies, the tolerance limits remain a fixed percentage of the mean. Let the random parameter vector X be expressed as

$$X = \operatorname{diag}(x) r = [x_1 r_1, \cdots, x_n r_n]^T \tag{8}$$

where r is normal, $(N(1, \cdots, 1), \Sigma)$, with

$$\Sigma = \begin{bmatrix} \sigma_1 & 0 & \cdots & \cdot \\ 0 & \sigma_2 & & \cdot \\ \vdots & & \ddots & \cdot \\ \cdot & & & \sigma_n \end{bmatrix} \cdot \begin{bmatrix} 1 & \rho_{12} & \cdots & \rho_{1n} \\ \rho_{21} & 1 & & \cdot \\ \vdots & & & \vdots \\ \rho_{n1} & \cdot & \cdot & 1 \end{bmatrix} \cdot \begin{bmatrix} \sigma_1 & 0 & \cdots & \cdot \\ 0 & \sigma_2 & & \cdot \\ \vdots & & \ddots & \cdot \\ \cdot & & & \sigma_n \end{bmatrix} \tag{9}$$

where the σ's and ρ's are fixed standard deviations and correlations, respectively. Then, X is $N(x, \Sigma_x)$ where

$$\Sigma_x = \begin{bmatrix} x_1\sigma_1 & 0 & \cdots & \cdot \\ 0 & \sigma_2 & & \cdot \\ \vdots & & \ddots & \cdot \\ \cdot & & & x_n\sigma_n \end{bmatrix} \cdot \begin{bmatrix} 1 & \rho_{12} & \cdots & \rho_{1n} \\ \rho_{21} & 1 & & \cdot \\ \vdots & & & \vdots \\ \rho_{n1} & \cdot & \cdot & 1 \end{bmatrix} \cdot \begin{bmatrix} x_1\sigma_1 & 0 & \cdots & \cdot \\ 0 & \sigma_2 & & \cdot \\ \vdots & & \ddots & \cdot \\ \cdot & & & x_n\sigma_n \end{bmatrix} \tag{10}$$

Thus the standard deviations of X are fixed percentages of the mean x and the correlation factors are constant.

In order to do correlated sampling for the extrapolated yield approximation, the same random samples r^i must be used again. If the nominal point x_0 is translated to $x\lambda = x_0 + \lambda d$, then the ith translated random sample is

$$X_\lambda^i = \begin{bmatrix} (x_{01} + \lambda d_1) r_1^i \\ \vdots \\ (x_{0n} + \lambda d_n) r_n^i \end{bmatrix} = X^i + \lambda d^i \tag{11}$$

where

$$d^i = [d_1 r_1^i, \cdots, d_n r_n^i]^T \tag{12}$$

is the modified translation direction for the ith sample point. Thus the new sample set for the design point x_λ is a translated version of the sample set for x_0, only each sample translates along a different direction. The quadratic approximation $G(x)$ can be evaluated along each of these directions as

$$G(X_\lambda^i) = \lambda^2 \alpha^i + \lambda \beta^i + \gamma^i \tag{13}$$

where

$$\alpha^i = d^{iT} Q d^i \tag{14}$$

$$\beta^i = 2X^{iT} Q d^i + b^T d^i \tag{15}$$

$$\gamma^i = G(X^i) = X^{iT} Q X^i + b^T X^i + c. \tag{16}$$

The difference between this situation, and the invariant statistics situation, is that more computation is required to compute the coefficients. The cost of this is dominated by the cost of computing the α^i's; the cost of obtaining the α^i's and β^i's is about the same as the cost of evaluating the yield of the quadratic model at the nominal point, which by the way, produces the γ^i's. The end result is that (13) can be used to greatly reduce the cost of the yield estimate computation along a search direction.

IV. Generation of Search Directions

In order to apply the previous techniques to perform yield maximization, one must generate search directions. Two simple techniques were chosen for this so that the extrapolated yield approximation technique could be tested in yield maximization examples. As will be seen more clearly through the examples, better search direction generation techniques are needed. The first technique was to compute statistical estimates of the yield gradient. Write the yield as

$$Y = \int_{R^n} I(x) p(x, x_0)\, dx$$

where

$$I(x) = \begin{cases} 0, & x \notin R_a \\ 1, & x \in R_a \end{cases}$$

and R_a is the region of acceptability. Analytically the gradient of the yield with respect to the nominal point x_0 can be expressed as

$$\nabla Y = \int_{R_n} I(x) \nabla \ln \{p(x, x_0)\} p(x, x_0)\, dx \tag{17}$$

but in practice the estimate

$$\nabla Y* = \frac{1}{N} \sum_{i=1}^{N} I(x^i) \nabla \ln \{p(x^i, x_0)\} \tag{18}$$

is generated by random simulations where the x^i's are samples from the probability density function (PDF), $p(x, x_0)$. The second order derivative can be treated in an identical fashion. In relation to circuit yield optimization, Aleksandrov *et al.* [19] seem to be the first to have pointed this out, although it appears that only recently this technique has received attention [15], [19]-[23].

One can envision implementing well-known unconstrained optimization techniques, such as steepest ascent and Newton's method, for the yield maximization problem. One drawback to these yield gradient estimation techniques is that the PDF, $p(x, x_0)$, must be differentiable, and one must be able to evaluate these derivatives. Another drawback is that the accuracy of these derivative estimates is very poor as discussed in [29].

Another technique for generating a search direction which does not rely upon statistical information and which is simple to compute is the midpoint method. The idea is that one searches in both directions from the nominal point along each axis, until the intersection with the boundary of the acceptable region of the quadratic model is found. This is very simple to do with the quadratic model by exploiting the same properties as were used with the extrapolated yield approximation; in fact, the coefficients of the scalar quadratic equations are found with practically no extra computation. To define a search direction one could use the midpoints of each of the line segments, for each axis, that lie in the acceptable region of the quadratic model to define an auxiliary point, as shown in Fig. 2. Then the search direction could be taken as the direction from the nominal point, to the auxiliary point, as shown in the figure. The motivation behind this idea is that in the first step or two of the optimization process it will quickly push the nominal point away from any nearby boundaries. The underlying concepts here are strongly related to some early work by Butler [25], [26] on large change sensitivities and also the radial approach of Tahim and Spence [14].

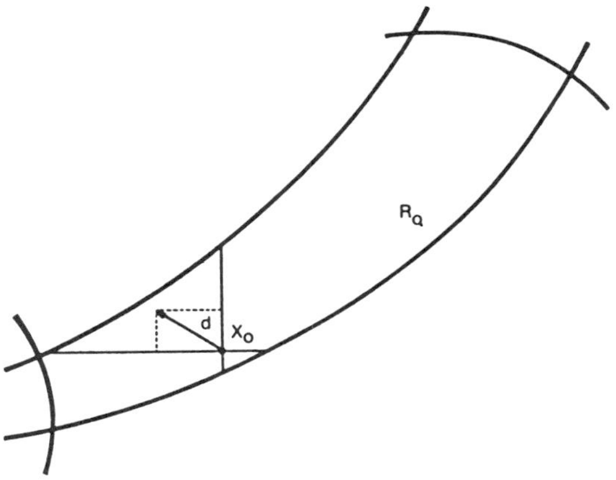

Fig. 2. Graphical depiction of midpoint direction.

V. Yield Maximization Examples

Examples which implement the previous techniques are discussed in this section. The purpose here is to demonstrate the effectiveness of the yield extrapolation technique, and its use in yield maximization. Three relatively complex filter circuits form the basis for the examples. Below is the optimization procedure that was followed.

(1) First perform a random simulation at the current nominal or design point to estimate the yield and its gradient. Set up the quadratic model using only function evaluations.

(2) Use the extrapolated yield approximation technique to estimate the yield for a fixed sample of 1000 at a set of uniformly spaced points in an interval along the scaled, gradient or midpoint direction. Interactively adjust the interval along the search direction in which these estimates are to be recomputed, if necessary, to find the region of largest increase. Choose the step length to approximately maximize the yield. (Note, the scaled gradient is the gradient estimate where each partial derivative is scaled by the square of the respective design point; this is a standard scaling technique. In addition, the length of the search direction was scaled so that it just meets the boundary of an orthotope centered about the design point that extends ±3 times the standard deviation along each coordinate. This can be thought of as a reference tolerance box; the purpose of this last scaling is to normalize the step length.)

(3) Update the current design point. If a step length was not found, because there was not an adequate increase in the estimated yield along the search direction, then the process must stop; otherwise, increase the sample size and repeat the process from step 1.

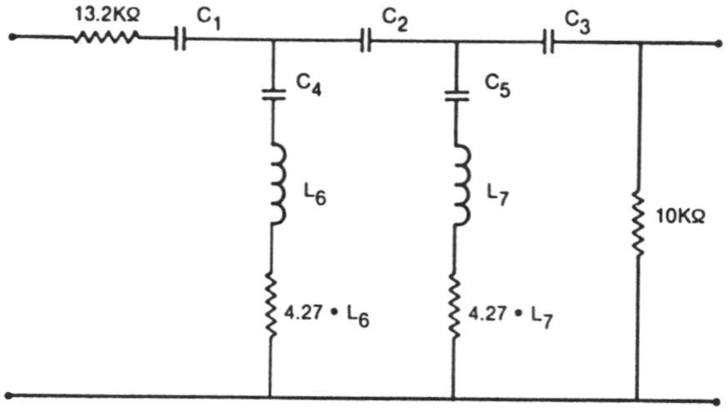

Fig. 3. Highpass *LC* ladder filter.

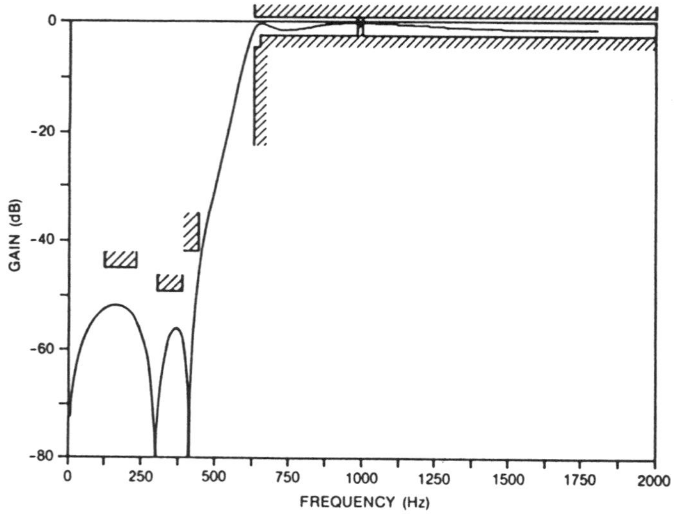

Fig. 4. Nominal response and specifications for the *LC* circuit.

A. Highpass LC Ladder Filter

Consider the highpass *LC* ladder filter shown in Fig. 3. Its response specifications are shown in Fig. 4. This filter has been used quite often in the statistical circuit design literature, as it was one of the early examples of Pinel and Roberts [12]. The specifications were set up as

$$-\infty \text{ dB} \leqslant T(170) \leqslant -45 \text{ dB}$$

$$-\infty \text{ dB} \leqslant T(350) \leqslant -49 \text{ dB}$$

$$-\infty \text{ dB} \leqslant T(440) \leqslant -42 \text{ dB}$$

$$-4 \text{ dB} \leqslant T(630) \leqslant 0.05 \text{ dB}$$

$$1.75 \text{ dB} \leqslant T(f) \leqslant 0.05 \text{ dB}$$

for $f = 650, 720, 740, 760, 940, 1040, 1800$ Hz and where T is the response magnitude referenced to the magnitude of each sample filter at 990 Hz. The 7 parameters were taken to be independent Gaussian random variables with standard deviations equal to 20/3 percent of the current design point at any iteration. This is a variant statistics example and corresponds to constant relative tolerances. The δ_i factors for the quadratic model were taken to be 1.5 times the standard deviations.

First the nominal points used by Director *et al.* [31] were selected: $C_1 = 11.1$ nF, $C_2 = 10.47$ nF, $C_3 = 12.9$ nF, $C_4 = 34.3$ nF, $C_5 = 97.3$ nF, $L_6 = 3.988$ H, and $L_7 = 2.685$ H. The midpoint method was used to determine the search directions.

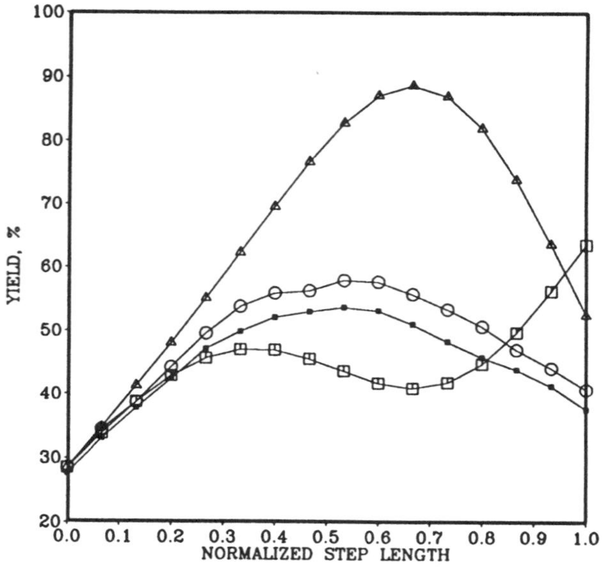

Fig. 5. Various yield estimates for *LC* filter along a search direction.

TABLE I

LC LADDER MAXIMIZATION WITH MIDPOINT DIRECTIONS FOR FIRST NORMAL POINT

Iteration	s	$Y_N(\%)$	$Y_Q(\%)$	$Y_{Q'}(\%)$	N
0	-	28.5	27.6	-	1000
1	0.52	58.5	55.3	53.9	1100
2	[illegible]	60.3	57.4	56.6	1200

Shown in Fig. 5 are various yield estimates versus the step length along the midpoint direction for the first iteration of this example. The sample size for all the correlated estimates was 1000 where the circles are actual circuit yield estimates, the solid squares are the yield estimates extrapolated from the quadratic model, and the triangles and open squares are yield estimates obtained from the approximation used by Singhal and Pinel [21]. This approximation is based upon the manipulation used in importance sampling, and here the importance sampling PDF is simply the PDF of the nominal point. The first estimate (triangles) is obtained by summing over the acceptable region, while the latter estimate (open squares) is obtained by summing over the failure region and subtracting the result from one. Notice that this technique is not very accurate except very close to the nominal point. On the other hand, the extrapolated yield approximation from the quadratic model works very well and appears to be accurate over a much larger range. More importantly, along this search direction the maximum circuit yield and the maximum yield predicted by the quadratic model occur at approximately the same step length.

Next the midpoint method was used to find the optimal yield for this circuit. The method halted after two iterations. Table I shows that the circuit yield Y_N increased from an initial value of 28.5 to 60.3 percent. The variables s and N represent the step length and the circuit sample size, respectively. The sample size for the circuit was slightly increased with each iteration in an attempt to add new statistical information. The sample size for the model remained constant. At each iteration a new quadratic model was generated. Although given the

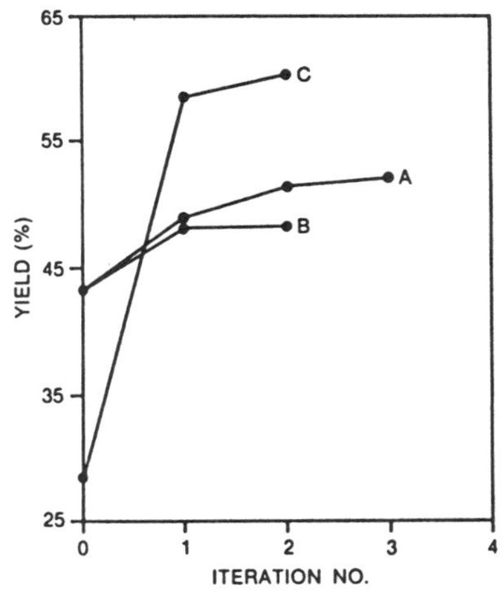

Fig. 6. Yield maximization results for the LC filter.

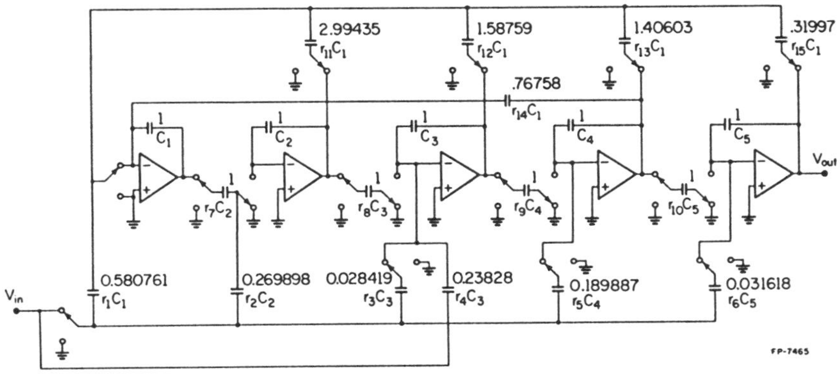

Fig. 7. Switched-capacitor filter. (The numbers are the r_i values.)

accuracy of the quadratic model, as illustrated in Fig. 5, this was probably not necessary. In Table I, Y_Q is the yield of the quadratic model gnerated at that iteration point, and Y_{Qs} is the yield extrapolated from the quadratic model generated at the previous iteration. Again note how well the extrapolated yields track the actual circuit yields.

Next the nominal point chosen by Agnew [33] was used: $C_1 = 11.8$ nF, $C_2 = 8.73$ nF, $C_3 = 10.45$ nF, $C_4 = 39.04$ nF, $C_5 = 90.13$ nF, $L_6 = 3.853$ H, and $L_7 = 3.178$ H. Again the yield extrapolated from the quadratic model compared very favorably with the actual circuit yield, even when there was considerable movement in the nominal point [29]. However, both the midpoint and the gradient methods did not perform well in optimizing the yield. In Fig. 6, curves A and B show that the initial yield was 43 percent for the nominal point, and that with the statistical gradient method (curve A) it increased to 52 percent in three iterations. The midpoint method (curve B) halted after two iterations and only resulted in a final yield of 48 percent. Whereas, curve C illustrates the improvement obtained with the midpoint method in only two iterations beginning with the first nominal point. These results indicate that there is a need for a more sophisticated method of generating search directions for yield maximization.

B. Chebyshev Low-Pass Switched-Capacitor Filter

The secon example is the switched-capacitor circuit shown in Fig. 7. This was an example used by Attaie [32]. He also gives an analytic expression for the transer function. The circuit is entirely dependent upon the 15 capacitor ratios $r_1, \cdots, r_{15}$ where the C_i's in the figure are arbitrary. A sampling frequency of 10 kHz was used. Independent Gaussian distributions were assumed for the ratios, with the standard deviations set to the specified tolerances (given later) times the current design value divided by 3; hence, variant statistics were used again. The δ_i's for setting up the quadratic model were again 1.5 times standard deviations.

The specifications are given in Fig. 8 and were checked at frequencies of 0, 600, 800, 1150, 1300, 1400, 1500, 1900, 1950, 2000, 2500 Hz. The stopband specification is -10 dB at 2500 Hz. The reference tolerances were chosen as 0.5 percent for all the ratios. The result of midpoint direction optimization are shown in Table II, and the final circuit response is shown in Fig. 8. The first step of this optimization process is extremely fruitful, and the various yield estimates along this first direction for a sample size of 1000 are given in Figs. 9 and 10. In these figures the identity of the curves is as before; that is, circles are circuit estimates, solid squares are extrapolated approximation estimates, and the triangles and open squares are, in the previous example, estimates obtained using the importance sampling concept to generate an analytic expression for the yield. These importance sampling based estimators are of no use past one step length as they look up at zero or one. Thus the figures clearly demonstrate that the extrapolated yield approximation technique can be very robust in its range of

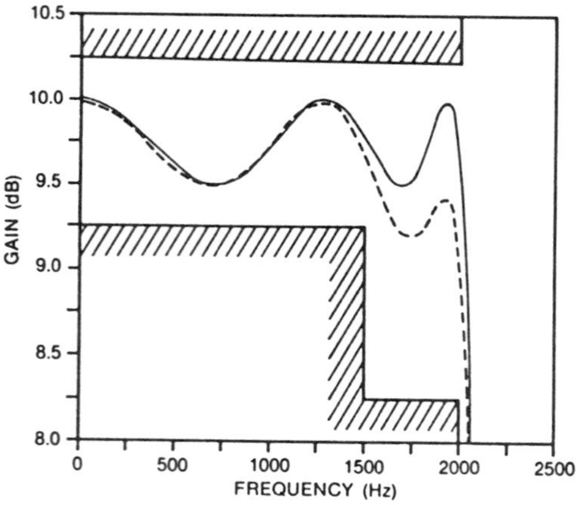

Fig. 8. Specifications and nominal passband response (solid line) for the switched-capacitor filter, and response of midpoint direction result (dashed line).

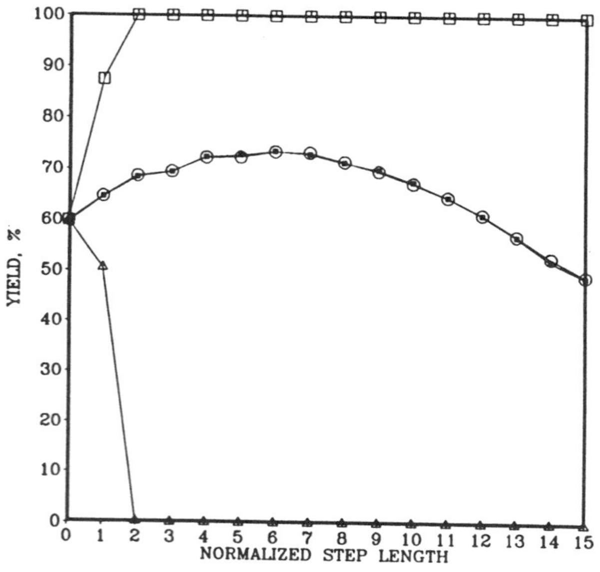

Fig. 9. Various yield estimates for the SC filter along a search direction.

TABLE II
SWITCHED CAPACITOR YIELD MAXIMIZATION

Iteration	s	Y_R(%)	Y_Q(%)	Y_{Qs}(%)	N
0	-	59.9	59.6	-	[illegible]
1	6.67	73.3	73.7	73.8	[illegible]
2	0.22	73.6	[illegible]	74.3	1200
3	1.33	73.4	[illegible]	[illegible]	[illegible]

accuracy, as it is accurate several reference tolerance boxes away.

C. A Low-pass Elliptic Switched-Capacitor Filter

As a final example the PCM fifth-order low-pass elliptic switched-capacitor ladder filter example of Antreich and Koblitz [23] was used. As in their paper the 12 parameters (capacitor ratios) were assumed to be independent random variables with normal probability density functions. The standard deviations were 2 percent of the mean values so the reference tolerance was 6 percent. The same nominal parameter values were used, and the δ_i's for setting up the quadratic model were again 1.5 times the standard deviations. The filter

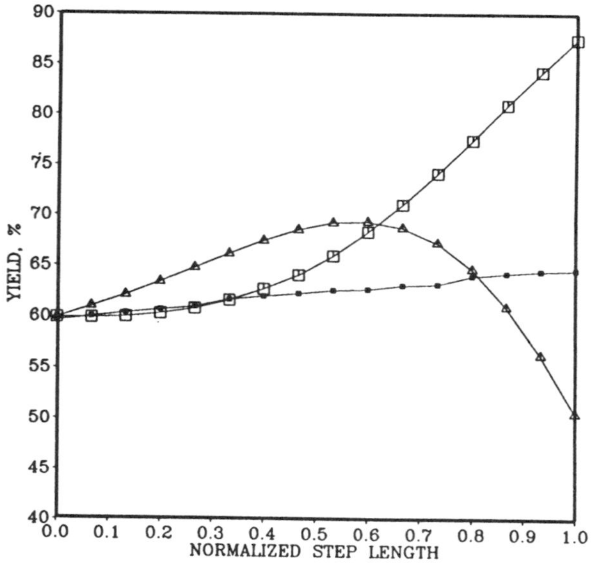

Fig. 10. Various yield estimates for the SC filter along a search direction (magnified scale).

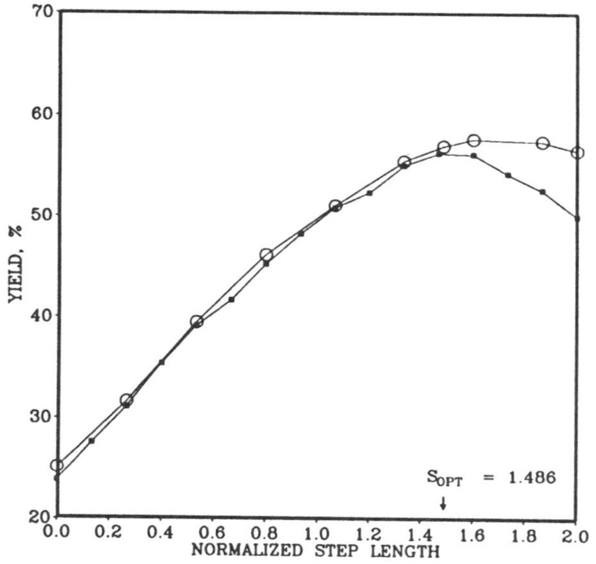

Fig. 11. Extrapolated yield and circuit yield along the optimal search direction.

specifications were the same as in [23], and these were checked at the following 22 frequencies: 1200 to 3000 Hz in 200-Hz steps, and 10^4 Hz and 10^5 Hz, all referenced to the gain at 800 Hz. In extrapolating the yield, invariant parameter statistics were used, but in optimizing the yield the standard deviations at each iteration were altered to achieve 6-percent tolerances. This was done to be consistent with the implementation of Antreich and Koblitz.

First the accuracy of the quadratic model was checked. The direction along which the yield was extrapolated was taken to be the difference between the final value of the parameters in the centered design [23] and the initial value of the parameters. As before, a normalized step length was used. The actual circuit yield and the extrapolated yield were computed at several points in the normalized step interval [0, 2]. The sample size at each point was 1000. Fig. 11 shows that there is excellent agreement between the yield extrapolated from the quadratic

TABLE III
PCM Switched-Capacitor Filter Yield Maximization with Gradients

Iteration	s	$Y_N(\%)$	$Y_Q(\%)$	$Y_{Qs}(\%)$	N
0	-	25.6	23.8	-	500
1	0.32	40.4	41.2	41.2	700
2	0.059	43.9	43.9	44.2	900

model gnerated at the initial point and the acutal circuit yield. Again the circles are circuit yield estimates and the solid squares are model yield estimates. The step length S_{opt} indicates the design center point reached by Antreich's and Koblitz's algorithms at which the yield is 59 percent.

We attempted to maximize the yield for this problem using yield extrapolation with the quadratic model and a simple gradient method for generating the search directions. The results are shown in Table III. In two iterations the circuit yield was improved from 25.6 to 43.9 percent at which point the gradient method failed to generate a fruitful search direction. This again indicates the need for a more advanced method for the generation of search directions.

VI. Conclusion

The main emphasis of this paper was the extrapolated yield approximation technique that was presented. The method consisted of a quadratic model for the circuit which was coupled with a correlated sampling technique to compute Monte Carlo yield estimates in a special manner. It was shown that this technique was very effective and efficient in determining the optimal step length when maximizing yield along a line. This approximation method has several other advantages. In contrast to many other methods, it is easy to implement. It is not restricted to any class of circuits such as linear circuits. An analytical description of the probability density function is not required. In addition, certain situations where the statistics vary as the design point is altered, can be handled.

In relation to the general approach taken here for the yield approximation problem, several comments can be made. For one, the overall approach is much less dimensionally dependent than several other deterministic techniques, and, there exists a great deal of flexibility in how the method can be applied. Specifically, one probably need not recompute the quadratic model and circuit yield estimate after each iteration, as was done in the examples. Also, the mechanism for generating search directions and for performing the line maximizations are not necessarily dependent upon one another. Thus one is free to choose any method for search directions.

It is felt that the search direction methods that were used here will probably have limited success in practice. This can be seen through the examples. Hence, much better techniques are needed for determining search directions.

References

[1] S. W. Director and G. D. Hachtel, "The simplicial approximation approach to design centering," *IEEE Trans. Circuits Syst.*, vol. CAS-24, pp. 363-372, July 1977.

[2] R. K. Brayton, S. W. Director, and G. D. Hachtel, "Yield maximization and worst-case design with arbitrary statistical distributions," *IEEE Trans. Circuits Syst.*, vol. CAS-27, Sept. 1980, pp. 756-764.

[3] L. M. Vidigal and S. W. Director, "A design centering algorithm for nonconvex regions of acceptability," *IEEE Trans. Computer-Aided Design*, vol. CAD-1, pp. 13-24, Jan. 1982.

[4] J. W. Bandler, P. C. Liu, and H. Tromp, "Nonlinear programming approach to optimal design centering, tolerancing, and tuning," *IEEE Trans. Circuits Syst.*, vol. CAS-23, Mar. 1976, pp. 155-165.

[5] L. Opalski and M. A. Styblinski, "An outer approximation algorithm for the design centering and tolerancing (DCT) in nonconvex acceptability regions," in *European Conf. Circuit Theory Design*, August 1981, pp. 665-670.

[6] E. Polak and A. Sangiovanni-Vincentelli, "Theoretical and computational aspects of the optimal design centering, tolerancing, and tuning problem," *IEEE Trans. Circuits Syst.*, vol. CAS-26, no. 9, pp. 795-813, Sept. 1979.

[7] J. W. Bandler and H. L. Abdel-Malek, "Optimal centering, tolerancing, and yield determination via updated approximations and cuts," *IEEE Trans. Circuits Syst.*, vol. CAS-25, Oct. 1978, pp. 853-871.

[8] H. L. Abdel-Malek and J. W. Bandler, "Yield optimization for arbitrary statistical distributions: Part I-Theory," *IEEE Trans-Circuits Syst.*, vol. CAS-27, pp. 245-253, Apr. 1980.

[9] H. Schjaer-Jacobsen and K. Madsen, "Algorithms for worst-case tolerance optimization," *IEEE Trans. Circuits Syst.*, vol. CAS-26, Sept. 1979, pp. 775-783.

[10] R. K. Brayton, S. W. Director, G. D. Hachtel, and L. M. Vidigal, "A new algorithm for statistical circuit design based on quasi-Newton methods and function splitting," *IEEE Trans. Circuits Syst.*, vol. CAS-26, pp. 784-794, Sept. 1979.

[11] G. D. Hachtel, T. R. Scott, and R. P. Zug, "An interactive linear programming approach to model parameter fitting and worst case circuit design," *IEEE Trans. Circuits Syst.*, vol. 27, pp. 871-881, Oct. 1980.

[12] J. F. Pinel and K. A. Roberts, "Tolerance assignment in linear networks using nonlinear programming," *IEEE Trans. Circuit Theory*, vol. CT-19, pp. 475-479, Sept. 1972.

[13] R. K. Brayton, G. D. Hachtel, and A. L. Sangiovanni-Vincentelli, "A survey of optimization techniques for integrated-circuit design," *Proc. IEEE*, vol. 69, pp. 1334-1362, Oct. 1981.

[14] K. S. Tahim and R. Spence, "A radial exploration approach to manufacturing yield estimation and design centering," *IEEE Trans. Circuits Syst.*, vol. CAS-26, pp. 768-774, Sept. 1979.

[15] M. A. Styblinski, "Estimation of yield and its derivations by Monte Carlo sampling and numerical integration in orthogonal subspaces," in *European Conf. Circuit Theory Design*, 1980, pp. 474-479.

[16] J. T. Ogrodzki and M. A. Styblinski, "Optimal tolerancing, centering and yield optimization by one-dimensional orthogonal search (ODOS) technique," *European Conf. Circuit Theory Design*, 1980, pp. 480-485.

[17] W. Strasz and M. A. Styblinski, "A second derivative Monte Carlo optimization of the production yield," *European Conf. Circuit Theory Design*, 1980, pp. 121-131.

[18] R. S. Soin and R. Spence, "Statistical exploration approach to design centering," *Proc. Inst. Elect. Eng.*, vol. 127, part G, no. 6, pp. 260-269, Dec. 1980.

[19] V. M. Aleksandrov, V. I. Sysoyev, and V. V. Shemeneva, "Stochastic optimization," *Engineering Cybernetics*, no. 5, Sept.-Oct. 1968, pp. 11-18.

[20] B. V. Batalov, Yu. N. Belyakov, and F. A. Kurmaev, "Some methods for statistical optimization of integrated microcircuits with statistical relations among the parameters of the components," *Soviet Microelectron*, (USA), vol. 7, no. 4, pp. 228-238, July-August 1978.

[21] K. Singhal and J. F. Pinel, "Statistical design centering and tolerancing using parametric sampling," *IEEE Trans. Circuits Syst.*, vol. CAS-28, pp. 692-702, July 1981.

[22] R. Koblitz, "Design centering and tolerance assignment of electrical circuits with Gaussian-distributed parameter values," *Arch. Elektro. Übertrag.* (in German), Band 34, Heft 1, Jan. 1980, pp. 30-37.

[23] K. J. Antreich and R. K. Koblitz, "Design centering by yield prediction," *IEEE trans. Circuits Syst.*, vol. CAS-29, pp. 88-96, Feb. 1982.

[24] G. Kjellström and L. Taxen, "Stochastic Optimization in System Design," *IEEE Trans. Circuits Syst.*, vol. CAS-28, no. 7, July 1981, pp. 702-715.

[25] E. M. Butler, "Realistic design using large-change sensitivities and

performance contours," *IEEE Trans. Circuits Theory*, vol. CT-18, pp. 58–66, Jan. 1971.

[26] E. M. Butler, "Large change sensitivities for statistical design," *Bell Syst. Tech. J.*, vol. 50, no. 4, pp. 1209–1224, Apr. 1971.

[27] P. W. Becker, "Finding the better of two similar designs by Monte Carlo techniques," *IEEE Trans. Reliability*, vol. R-23, pp. 242–246, Oct. 1974.

[28] P. W. Becker and F. Jensen, *Design of Systems and Circuits for Maximum Reliability or Maximum Production Yield*. New York: McGraw-Hill, 1977.

[29] D. E. Hocevar, "Automatic tuning algorithms and statistical circuit design," Rep. R-954, Coordinated Science Laboratory, (also Ph.D. dissertation), Univ. of Illinois at Urbana-Champaign, June 1982.

[30] W. Maly, A. J. Strojwas, and S. W. Director, "Fabrication based statistical design of monolith IC's," in *IEEE Int. Symp. Circuits Syst.*, Apr. 1981, pp. 135–138.

[31] S. W. Director, G. D. Hachtel, and L. M. Vidigal, "Computationally efficient yield estimation procedures based on simplicial approximation," *IEEE Trans. Circuits Syst.*, vol. CAS-25, pp. 121–130, Mar. 1978.

[32] N. Attaie, "Optimized structures for switched-capacitor filters," Report R-934, Coordinated Science Laboratory, Univ. of Illinois, Urbana-Champaign, Illinois, 1981.

[33] D. Agnew, "Design centering and tolerancing via morgin sensitivity minimization," *Proc. Inst. Elect. Eng.*, vol. 127, part G, no. 6, pp. 270–277, Dec. 1980.

Algorithms and Software Tools for IC Yield Optimization Based on Fundamental Fabrication Parameters

M. A. STYBLINSKI, MEMBER, IEEE, AND L. J. OPALSKI

Abstract—Algorithms, software tools and the relevant methodology for production yield optimization with respect to fundamental technological parameters of the IC manufacturing process (such as diffusion times and temperatures) and element layout mask dimensions are discussed. The tools developed include: STOCH-PAC—a package of new yield optimization and yield gradient estimation algorithms based on Stochastic Approximation approach and the Method of Random Perturbations; YIELD-PAC—a package of yield evaluation subroutines providing an interface to SPICE-PAC circuit simulation package; FAB-PAC—an interface to the FABRICS statistical process simulator; and IRIS (Interactive Restructurable Interface System)—a flexible user interface for efficient data manipulation, creation of different design tasks and macrotasks, restructuring the set of active subroutines, etc. These tools have been integrated into the TOY (Technological Optimization of Yield) program. Examples of yield optimization with respect to process parameters and layout dimensions using TOY are given.

I. Introduction

As INTEGRATED circuits (IC's) become increasingly complex, geometries smaller and smaller, it is becoming more and more difficult to achieve acceptable *manufacturing yields*, even if the nominal design fulfills all design constraints. The manufacturing yield is composed of two factors: the *technological yield* and the *design* (*parametric*) *yield*. The former is a result of catastrophic technological failures due, e.g., to cristal defects, and is not considered here; the latter is a result of the sensitivity of circuit performance to IC device parameter variations, caused by unavoidable variation of the manufacturing conditions from device to device and from chip to chip.

In this paper the CAD algorithms and software tools are considered, devoted specifically to the solution of the most typical problem of *statistical design centering* (SDC), where the objective is to maximize the parametric yield only with respect to the nominal values of fundamental technological IC manufacturing process parameters such as diffusion times and temperatures and element layout dimensions. Many of these parameters have very significant influence on the circuit performance and obviously also on the final value of the production yield. The choice of process parameters as designable parameters can be justified by the following approaches to IC design.

One possible design scenario is as follows: the choice of technological process parameters is a result of substantial experience and knowledge of IC process engineers and circuit designers supported by a "cut and try" approach during the initial setting of the parameters of the IC production line. Once this costly and time-consuming process is completed, the parameters of the production line are usually fixed and not allowed to be changed. Therefore, the choice of the line parameters should be optimized for the widest possible class of circuits manufactured, so that the best possible circuit performance is achieved and the largest manufacturing yield guaranteed. Therefore, the yield should be optimized with respect to technological process parameters *in advance*, for a class of circuits to be manufactured, and the results obtained used in the decision making process concerning the final setting of the technological process parameters.

Somehow different design scenario can also be envisioned, especially for (perhaps future) fully automated production lines: for each class of circuits manufactured, a "fine tuning" of process parameters is performed, in order to maximize the production yield of that particular class of circuits. In this process some additional bounds on the nominal values of process parameters should be imposed in order to fulfill some other process and design requirements.

Formally, the SDC problem was traditionally formulated as follows:

$$\max_{x \in R^n} \left\{ Y(x) = \int_{R^n} \phi(e)\, f_e(x, e)\, de \right\} \qquad (1.1)$$

where $Y(x)$ is the parametric (or design) yield, $e \in R^n$ is a vector of actual parameter values, $x \in R^n$ is the nominal point, $\phi(e) = 1$, if $e \in A$, the acceptability region in the x-space, where all circuit constraints are fulfilled, and $\phi(e)$

Manuscript received August 22, 1985; revised October 7, 1985. This work was supported in part by the Texas Engineering Experiment Station, Texas A&M University System, College Station, TX 77843. This paper is an extension of the following two conference presentations: M. A. Styblinski and L. J. Opalski, "A random perturbation method for IC yield optimization with deterministic process parameters," (IEEE Int. Symp. on Circuits and Systems, Montreal, Canada, May 7–10, 1984); and "Software tools for IC yield optimization with technological process parameters" (IEEE Int. Conf. on Computer-Aided Design, Santa Clara, CA, Nov. 12–15, 1984).

The authors are with the Department of Electrical Engineering, Texas A&M University, College Station, TX 77843, on leave from The Technical University of Warsaw, Warsaw, Poland.

IEEE Log Number 8406695.

Reprinted from *IEEE Trans. Computer-Aided Design,* vol., CAD-5, no. 1, pp. 79–89, January 1986.

$= 0$, otherwise; $f_e(x, e)$ is the probability density function (p.d.f.) of e.

As it will be shown below, formulation (1.1) is only suitable for the case of yield optimization with respect to *electrical circuit element and parameter values*; in this case the vectors x and e represent circuit elements (and/or parameters) and $f_e(x, e)$ is the p.d.f. of e. If one is able to approximate the p.d.f. of e by some differentiable density function such as the multidimensional Gaussian p.d.f. then the gradient of $Y(x)$ can be estimated from the following formula [5], [6], [16], [17]

$$\nabla Y(x) = \int_{R^n} \phi(e) \frac{\nabla_x f_e(x, e)}{f_e(x, e)} f_e(x, e)\, de$$
$$= E\{\phi(e)\, \nabla_x f_e(x, e)/f_e(x, e)\} \qquad (1.2)$$

using the same sampled points that are used for the Monte Carlo yield estimation. In (1.2) e is sampled with the p.d.f. $f_e(x, e)$ for a given x.

Several techniques for the solution of the standard SDC problem (1.1) have been proposed based on a deterministic (geometric concepts) [4], [8]–[12] or a statistical approach [5]–[7], [13]–[19]. The software tools to be described in this paper implement the statistical techniques only. The statistical methods proposed to solve (1.1), are based on the centers of gravity of "pass" and "fail" points [13], [14], asymmetry vectors [15], yield first and second derivatives [16]–[18], sampling maximizing the yield information [19], yield extrapolation techniques [7], etc. Most of these methods utilize some large sample yield and yield derivative estimators. The method proposed by Styblinski and Ruszczynski [5], [6] is based on the *Stochastic Aproximation* (SA) approach [20]–[23], in which the yield gradient estimators are only used, based on a very small number of points e^i [1] (usually just one) sampled for a current nominal point x at each algorithm step. The result of each iteration (i.e., the step direction) is a random variable, but due to a large number of iterations and specific algorithm features, the convergence to the problem solution is provided under quite mild conditions. Because of their fast initial rate of convergence and because this is the only class of statistical yield optimization methods that has very strong theoretical support [20]–[23], these methods were implemented in our yield optimization system.

In order to optimize yield w.r.t. IC process parameters and layout mask dimensions, some fast statistical simulators of the IC manufacturing process must be used. Although there exist some programs, like SUPREM [1], theoretically suitable for this purpose, only the FABRICS program [2], [3], provides fast statistical simulation due to the simplified, mostly algebraic process models used. That is why we use FABRICS as the statistical process simulator in the yield optimization package developed.

As is will be shown in Section II, formulation (1.1) cannot be directly used for yield optimization w.r.t. process parameters and layout dimensions, because some of the process parameters such as times and temperatures of various operations have very small spreads about their nominal values and therefore should be considered as *deterministic* designable parameters; the second reason is that due to the hierarchical model of generating parameter variations implemented in FABRICS, the p.d.f.'s of e parameters appearing in (1.1) are not known in analytic form, therefore the gradient formula (1.2) cannot be used.

The paper is organized as follows: technological process parameters are discussed first, then the problem formulation and new algorithms for its solution are described. The practical implementation of the yield optimization software is presented next: the packages of subroutines used are briefly presented, basic features of an interactive, user restructurable IRIS interface to the optimization software is described, and the yield optimization program TOY utilizing the software tools discussed. Finally, examples of yield optimization with respect to technological process parameters and layout mask dimensions are given.

II. Problems of Yield Optimization with IC Technological Process Parameters

Probability density functions (p.d.f.'s) of IC passive *RC* elements and active device parameters (called "electrical" IC parameters) are usually very complicated with the existence of nonlinear correlations and dependencies of the p.d.f.'s on the nominal parameter values. That is why their *direct* modeling with the use of *known* p.d.f.'s is limited to some simple cases only. In practice attempts have been made [36], [37] to combine some empirically matched nonlinear formulas with the random number generation of simple p.d.f.'s (such as normal or uniform) to account for nonlinear electrical parameter correlations and dependencies of the p.d.f.'s on nominal parameter values. In this case, however, the *resulting* p.d.f.'s of electrical parameters are not known in analytic form, which is needed to calculate the yield gradient required in the efficient gradient methods of yield optimization such as the Stochastic Approximation approach proposed in [5], [6] and mentioned in Section I.

Another approach is based on statistical modeling of the IC manufacturing process as practically implemented, e.g., in the FABRICS program [2], [3]. By carefully "tuning" the statistical simulator to a given production line, very accurate approximation to the distributions and correlations of electrical circuit parameters can be obtained. All technological process parameters are represented in FABRICS as a sum of the nominal value (denoted below by an Italic lower case letter) and a random disturbance part (denoted below by a Greek letter). The following parameter groups can be indentified [2], [3]:

$Z_1 = z_1 + \zeta_1 \in R^c$ is a vector of process control parameters such as times and temperatures of pre-deposition, drive-in, oxidation, implantation energy, gas pressures, etc. The disturbances ζ_1 are usually very small as compared to their nominal values z_1 (times and temperatures are well controlled), therefore, Z_1 can be considered as a

[1] e^i is a particular, ith realization of the vector of circuit elements, sampled with the p.d.f. $f_e(x, e)$.

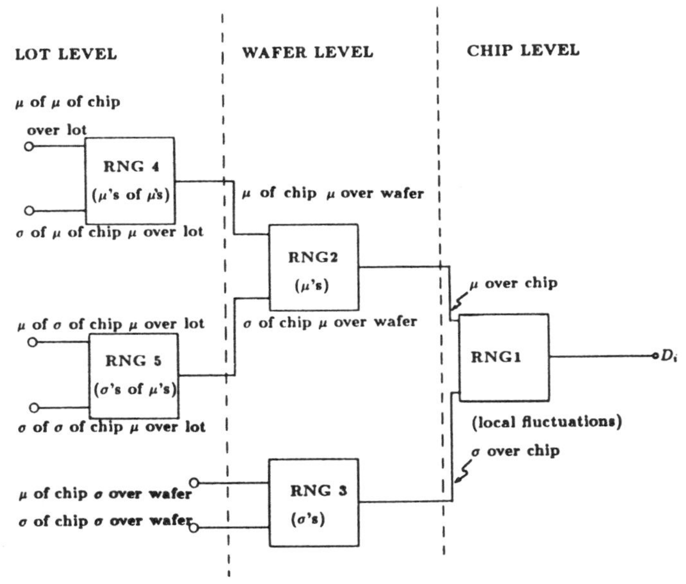

μ - mean, σ - standard deviation

Fig. 1. Hierarchical generation of statistical distribution of process disturbances D_i. RNG is a random number generator of normal or log-normal distribution.

vector of (almost) *deterministic designable* parameters. The choice of z_1 is very important for yield maximization.

$Z_2 = z_2 + \zeta_2 \in R^g$ is a vector of geometric mask (layout) dimensioins. The random disturbances ζ_2 are due to finite resolution of the photolithographic and etching processes. For large dimensions ζ_2 can be negligible w.r.t. z_2 and therefore some elements of the vector Z_2 can be also treated as (almost) deterministic designable parameters. On the other hand, for small element dimensions, typical for modern ICs, the disturbances ζ_2 cannot be neglected, since their influence on the circuit performance and production yield can be critical.

$P = p + \psi \in R^m$ is a vector of different process physical parameters such as diffusivities, enhancement factors, segregation coefficients, resistivities, impurity concentrations, etc. The nominal values p are *constant* and *not designable*.

To simplify notation, the process disturbances ζ_1, ζ_2 and the vector P (since p is fixed), will be collected into one vector $D = (\zeta_1, \zeta_2, P) \in R^d$, where $d = c + g + m$. The joint p.d.f. $f(D)$ of D is simulated in FABRICS hierarchically as shown in Fig. 1, where three levels of parameter generation are present: chip level, wafer level, and lot of wafers level. This provides correlations between circuit elements and was chosen based on the fact that some of the technological processes are performed in lots (e.g., diffusion) and some on wafers (e.g., photolithography) and that element variations within a chip are smaller than between the chips or wafers. The model shown in Fig. 1 was experimentally tested and it was demonstrated that the normal (or log-normal) p.d.f.'s at each level provide sufficient accuracy in practice. Because of the above method of generating D_i parameters, their actual (joint) p.d.f. is not known in analytic form, therefore, yield gradient estimation from (1.2) is not possible, as mentioned. In what

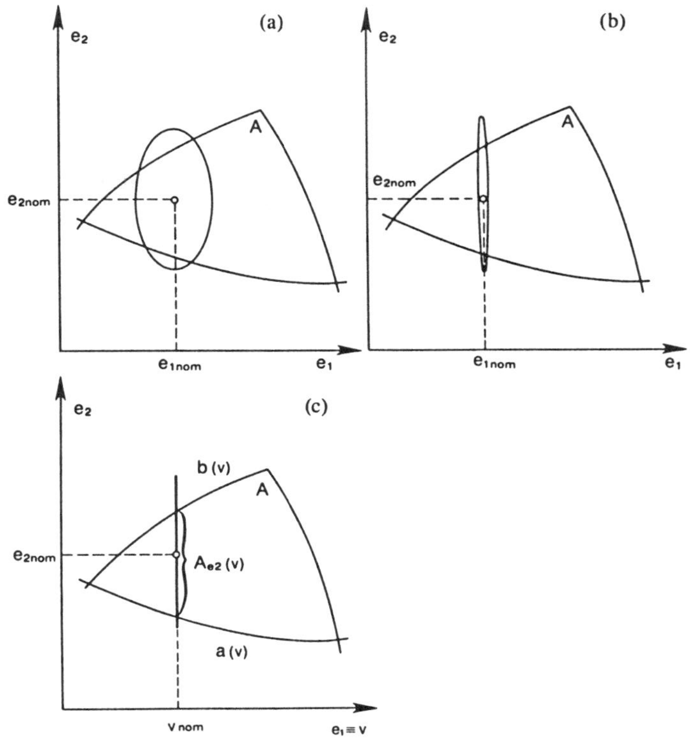

Fig. 2. Interpretation of "almost deterministic" process parameters. (a) A standard case where variations of parameters e_1 and e_2 are relatively large—this corresponds to the standard yield optimization problem (1.1). (b) Variations of e_1 are very small. (c) Approximation to case (b): parameter $e_1 \equiv v$ is strictly deterministic, e_2 is random.

follows a methodology is proposed which circumvents this problem.

Another difficulty in the application of general formulation (1) to yield optimization with respect to the process parameters discussed above, is due to the fact that some of the parameters are "almost deterministic" because of very small relative disturbances ζ_1 or ζ_2. This situation is illustrated in Fig. 2, where it is assumed that $e_1 = v$[2] parameter characterized by small variations (Fig. 2(b)) can be treated as a deterministic designable parameter (Fig. 2(c)). It is seen from Fig. 2(c) that yield can be maximized with respect to the parameter v. In the example to follow it will be demonstrated that the yield derivative w.r.t. v (or any other deterministic parameters) is not readily available in this case.

Example

Let us consider a simple two-dimensional case shown in Fig. 2(c), where $e_1 \equiv v \in R^1$ is deterministic, and e_2 is random with its nominal value fixed. Let the acceptability region be determined by two (not necessarily smooth) functions of v: $a(v)$ and $b(v)$, i.e., $a(v) \leq e_2 \leq b(v)$, as shown in Fig. 2(c). Yield can be expressed as

$$Y(v) = \int_{a(v)}^{b(v)} f(e_2)\, de_2 = \int_{-\infty}^{+\infty} \phi(v, e_2) f(e_2)\, de_2 = E\{\phi(v, e_2)\} \qquad (2.1)$$

[2]Deterministic parameters will be denoted by v in what follows.

where $f(e_2)$ is the p.d.f. of e_2, and

$$\phi(v, e_2) = \begin{cases} 1, & \text{if } a(v) \le e_2 \le b(v) \\ 0, & \text{otherwise} \end{cases} \tag{2.2}$$

defines the acceptability region A in the (v, e_2) space. Differentiating (2.1) one obtains

$$\frac{dY(v)}{dv} = f(b(v))\frac{db(v)}{dv} - f(a(v))\frac{da(v)}{dv}.$$

This leads already to some interesting observations, namely,

1) The location of the acceptability region *boundaries*, determined by $a(v)$ and $b(v)$ must be known, as well as the analytic expression for the p.d.f. $f(e_2)$ (to find $f(a(v))$ and $f(b(v))$—both of which are in general not available.

2) The derivatives of the constraining functions $a(v)$ and $b(v)$ must be known.

3) Local yield maxima can be expected.[3]

4) If $a(v)$ and $b(v)$ are not smooth or not continuous (which can be a frequent case in practice), the yield derivatives are also not continuous or might not exist at all at some points.

Maximizing yield expressed by (2.1) is equivalent to finding the maximum of the regression function $E\{\phi(v, e_2)\}$ w.r.t. v, therefore the yield optimization problem so formulated belongs in general to a class of stochastic nondifferentiable and perhaps multiextremal optimization problems.

An extension to the multidimensional case leads to very similar conclusions and yield gradient estimation algorithms so complicated that they cannot be used in practice due to the computational costs involved. Equation (2.2) defining the acceptability region A can also be interpreted as follows: $\phi(v, e_2)$ defines an acceptability region $A_{e2}(v)$ in the e_2-space of random variables (one-dimensional in our case), which is a function of the additional deterministic parameter v, as seen in Fig. 2(c). Therefore, in a general multidimensional case $v \in R^n$ and $e \in R^m$, the SDC problem should be formulated as follows:

$$\max_{v \in R^n} \left\{ Y(v) = \int_{R^n} \phi(v, e)\, f_e(e)\, de \right\} \tag{2.3}$$

where $f_e(e)$ is the p.d.f. of e. Equation (2.3) was obtained based on the intuitive reasoning described above, but it can also be derived formally applying some properties of distributions (generalized functions), and in particular the Dirac δ-functional (see Appendix I).

As seen, formulation (2.3), where v is a deterministic parameter is significantly different than formulation (1.1), since the designable deterministic parameter v appears directly in the $\phi(\cdot, \cdot)$ function, which is nondifferentiable w.r.t. v. Therefore the methods developed to solve (1.1) cannot, in general, be used to solve (2.3). Since we are interested in the gradient yield optimization methods, such as the already mentioned stochastic approximation approach proposed in [5], [6], therefore, some new yield gradient evaluation algorithms must be developed. As already shown, the direct gradient evaluation would be too costly, and in the case of process parameter modeling implemented in FABRICS, impossible at all, since the p.d.f. $f_e(e)$, corresponding in our case to the p.d.f. of the disturbance vector D (or its part) is not known in analytic form, which is required for gradient estimation. Because of that we propose to consider all nominal designable parameters z_1 and z_2 as *deterministic* designable parameters collected into the vector $v = (z_1, z_2)$, and all process disturbances $D = (\zeta_1, \zeta_2, P)$ as random variables of unknown explicit analytic form of the joint p.d.f. $f(D)$, because of the described hierarchical sampling process used in FABRICS. Therefore, in formulation (2.3) we can identify the vector e as $e \equiv D$, $v \equiv (z_1, z_2)$, $\phi(v, e) \equiv \phi(v, D)$ and $f_e(e) \equiv f(D)$, $R^n \equiv R^d$ (d is the dimension of D). This leads to the following yield optimization problem

$$\max_{v \in R^n} \left\{ Y(v) = \int_{R^d} \phi(v, D)\, f(D)\, dD \right\}. \tag{2.4}$$

We propose to solve (2.4) using a Method of *Random Perturbations* where, as it will be shown, the analytic form of $f(D)$ is not required, and it is immaterial if some of the disturbances ζ_1, ζ_2 are negligible, or not.

III. Method of Random Perturbations

3.1. Theory

The Method of Random Perturbations to be proposed originated in our observation that the speed of convergence of the Stochastic Approximation methods can be in some cases increased if the random spreads of the designable parameters of the original problem (1.1) are artificially increased, due to increased accuracy of yield derivative estimators. This can be accomplished choosing the standard deviations as $\tilde{\sigma}_i = a_i \sigma_i$, $a_i > 1$, $i = 1, \cdots, M$, where M is a number of designable parameters, with $a_i \to 1$ as $x \to \hat{x}$, the solution vector, for convergence. The proposed approach to the solution of (2.4) is, therefore, based on the following:

a) perturb the deterministic vector v with some additional random spreads $\tilde{\eta} \in R^n$ of known p.d.f. $h(\tilde{\eta}, \beta)$ (e.g., normal), with $E(\tilde{\eta}) = 0$ and $\tilde{\eta}$ independent of D; β is an additional parameter controlling the dispersion of $\tilde{\eta}$ (e.g., β can control the standard deviations of $\tilde{\eta}_1 \cdots \tilde{\eta}_n$),

b) solve the modified problem with $\beta \to 0$ as $v \to \hat{v}$, where $\hat{v}$ is the solution to the original problem. This can be formally written as

$$\max_{v \in R^n} \left\{ \tilde{Y}(v, \beta) = \int_{R^d} \int_{R^n} \phi(v + \tilde{\eta}, D)\, h(\tilde{\eta}, \beta)\, f(D)\, d\tilde{\eta} dD \right\} \tag{3.1}$$

with $\beta \to 0$ as $v \to \hat{v}$. Usually $h(\tilde{\eta}, \beta)$ is symmetrical with respect to η, i.e., $h(-\tilde{\eta}, \beta) \equiv h(\tilde{\eta}, \beta)$, hence introducing

[3] The same can obviously be expected also for the standard yield optimization problem (1.1).

a new variable $\eta \triangleq -\tilde{\eta}$, $\tilde{Y}$ can be represented as a multi-dimensional convolution of $h(\cdot, \beta)$ and $Y(\cdot)$, the original yield expression appearing in (3.1), as follows:

$$\begin{aligned}\tilde{Y}(v, \beta) &= \int_{R^n} \left\{ \int_{R^d} \phi(v - \eta, D) f(D)\, dD \right\} h(\eta, \beta)\, d\eta \\ &= \int_{R^n} Y(v - \eta)\, h(\eta, \beta)\, d\eta \\ &= \int_{R^n} h(v - \eta, \beta)\, Y(\eta)\, d\eta \qquad (3.2)\end{aligned}$$

where

$$Y(\cdot) \triangleq \int_{R^d} \phi(\cdot, D) f(D)\, dD. \qquad (3.3)$$

It turns out that the functional (3.2), called a *smoothed functional*, was first introduced in [25] and discussed in detail in [26]. Originally, the method was applied to the cases where a general function $Y(v)$ was not well behaved (e.g., nondifferentiable). It was expected that the smoothed functional (3.2) would behave better, and under some conditions the optimization of (3.2) would lead to the maximum of v. These smoothing properties are also valuable for the problem (2.4), since $Y(v)$ might be, in general, not differentiable, as mentioned.

Assuming after [26] that the p.d.f. should have the following properties:

a) $$h(\eta, \beta) = \frac{1}{\beta^n} \hat{h}\left(\frac{\eta}{\beta}\right) \qquad (3.4)$$

b) $h(\eta, \beta)$ is piecewise differentiable with respect to η;
c) $\lim_{\beta \to 0} h(\eta, \beta) = \delta(\eta)$ (the Dirac delta functional);
d) $\lim_{\beta \to 0} \tilde{Y}(v, \beta) = Y(v)$.

The gradient of (3.2) with respect to v can be found from the formula

$$\begin{aligned}\nabla_v \hat{Y}(v, \beta) &= \int_{R^n} \nabla_v h(v - \eta, \beta)\, Y(\eta)\, d\eta \\ &= \int_{R^n} \nabla_\eta h(\eta, \beta)\, Y(v - \eta)\, d\eta \\ &= \frac{1}{\beta} \int_{R^n} \nabla_\eta \hat{h}(\eta)\, Y(v - \beta\eta)\, d\eta \qquad (3.5)\end{aligned}$$

where $\hat{h}(\cdot)$ is defined by (3.4). One of the possible choices for $\hat{h}(\eta)$ is a multinormal p.d.f., [26] which leads to the following expression for $h(\eta, \beta)$:

$$h(\eta, \beta) = \frac{1}{(2\pi)^{n/2} \beta^n} \exp\left[-\frac{1}{2} \sum_{i=1}^{n} \left(\frac{\eta_i}{\beta}\right)^2\right], \qquad \beta > 0. \qquad (3.6)$$

Then the gradient of $\tilde{Y}(v, \beta)$ is expressed as

$$\begin{aligned}\nabla_v \tilde{Y}(v, \beta) &= \frac{1}{\beta} \int_{R^n} \int_{R^d} \eta \phi(v - \beta\eta, D)\, \hat{h}(\eta) f(D)\, d\eta dD \\ &= \frac{1}{\beta} E\{\tilde{\eta} \phi(v + \beta\tilde{\eta}, D)\} \qquad (3.7)\end{aligned}$$

where sampling is performed in the $R^n \times R^p$ space with the joint p.d.f. $\hat{h}(\eta) f(D)$ and the original variable $\tilde{\eta} = -\eta$ was substituted in (3.4). The unbiased gradient estimator is, therefore, expressed by

$$\hat{\nabla}_v \tilde{Y}(v, \beta) = \frac{1}{\beta} \frac{1}{N} \sum_{i=1}^{N} \tilde{\eta}^i \phi(v + \beta\tilde{\eta}^i, D^i) \qquad (3.8)$$

with N points $(\tilde{\eta}^i, D^i)$ sampled with the joint p.d.f. $\hat{h}(\hat{\eta})$ $f(D)$. Substituting $N = 1$ in (3.8) one obtains the one point gradient estimator required for the Stochastic Approximation algorithms used in [5], [6], and implemented in our yield optimization system.

Notice that to find $\hat{\nabla}_v \tilde{Y}(v, \beta)$, neither $\nabla Y(v)$ (which might not exist at all) nor the analytic form of the p.d.f. $f(D)$ have to be known. This is especially attractive in cases where $f(D)$ can only be simulated.

As shown in [24], the method is theoretically convergent to the solution of the original problem (2.4) under quite mild conditions. It has to be stressed, however, that the convergence of any statistical algorithm can be proved in a statistical sense only, i.e., most often assuming that the number of algorithm steps and the number of statistical samples used tend to infinity. In practice, the limited sample size will have effect on the algorithm convergence—there is however no theory available to study convergence properties of stochastic algorithms for the finite sample sizes. In general, which is rather obvious, the larger the sample size, the more accurate results one should expect—in practice sample sizes starting from about 100–200 up to about 1000 are typical, even if in some cases quite satisfactory results can be obtained with a much smaller number of samples, as it will be seen from the examples presented in Section V.

3.2. Algorithmic Implementation

Based on the theory presented above, a general algorithm for yield gradient estimation can be constructed as follows.

Algorithm 1 (yield gradient estimation with random perturbations)

0) Initialization (main program): select value of β, initialize various parameters.
1) Pass the vector v of nominal designable parameters from the yield optimization subroutine.
2) $i = 0$ (a subsequent sample number), $\xi = 0$ (a gradient estimator).
3) $i = i + 1$; generate $\eta^i \in R^n$ with a normalized Gaussian p.d.f.
4) Set $\theta^i = v + \beta\eta^i$, where $v = (z_1, z_2)$
   - 4a) pass θ^i to the statistical process simulator (FABRICS)
   - 4b) in FABRICS: substitute the values of θ^i parameters into z_1, z_2 vectors, hierarchially generate disturbances D^i, as shown in Fig. 1; FABRICS calculates the actual values of circuit electrical model parameters and passes them to circuit analysis subroutines (SPICE-PAC)

4c) circuit analysis: find out if the actual realization is acceptable or not, i.e., if $\phi(v + \beta\eta^i, D^i) = 0$ or 1; return the value of $\phi(\cdot, \cdot)$.

5) Set:

$$\xi = \xi + \eta^i \phi(v + \beta\eta^i, D^i) \tag{3.9}$$

(according to (3.8))

6) If $i < N$ go to 3)

7) Find the yield gradient estiamtor (3.8)

$$\xi \equiv \hat{\nabla}_v \tilde{Y}(v, \beta) = \xi/(\beta N). \tag{3.10}$$

The above algorithm was used together with the SA yield optimization subroutine proposed in [5], [6]. This subroutine is based on the following algorithm.

Algorithm 2 (Stochastic Approximation yield optimization)

1) Read x^0—the starting value of the nominal point, and the algorithm control parameters: MAXSTEPS, EPS, STEP, NMCGR.

2) Calculate the initial step size coefficient τ_0 based on yield gradient ξ^0 estimation with NMCGR samples, as follows:

$$\tau_0 = \frac{STEP}{\|\xi^0\|} \tag{3.11}$$

set $k = 0$.

3) Use the SA yield optimization algorithm to find iteratively an approximation to the solution of the yield optimization problem, according to the following iterative scheme:

$$x^{k+1} = x^k + \tau_k d^k$$

$$d^k = (1 - \rho_k)\, d^{k-1} + \rho_k \xi^k, \tag{3.13}$$

$$0 \le \rho_k \le 1$$

where ξ^k is the gradient estimator found using Algorithm 1, usually with $N = 1$ (just one sample), τ_k is step length coefficient, controlled as described below, ρ_k is the gradient averaging coefficient, changed according to the formula

$$\rho_k = \frac{\rho_{k-1}}{1 + \rho_{k-1} - R} \tag{3.14}$$

where $0 < R < 1$ is a constant controlling the rate of change of ρ_k. k assumes values: 1, 2, 3, $\cdots$ until: k = MAXSTEPS or the step length $\|\tau_k d^k\| <$ EPS.

In the above algorithm, where $\{\tau_k\} \to 0$, $\{\rho_k\} \to 0$, each new step is performed in the direction $\tau_k d^k$ of the "averaged" gradient d^k determined by (3.13). As seen from (3.13), d^k is a convex combination of the previous (old) direction d^{k-1} and a new gradient estimate ξ^k, (hence the algorithm is a stochastic analog of the deterministic conjugate gradient method); this provides the gradient averaging. The ρ_k coefficient controls the "memory" or "inertia" of the search direction d^k (large inertia for small ρ). The coefficients τ_k and ρ_k are automatically determined, based on the following heuristic reasoning: τ_k should be (on average) a maximizer of $Y(x^{k+1}) = Y(x^k + \tau_k d^k)$, hence the scalar product $E\{\langle d^k, \xi^{k+1}\rangle\} = 0$ (on average). If $E\{\langle d^k, \xi^{k+1}\rangle\} < 0$, for x^{k+1}, τ_k is too large, otherwise it is too small. Some statistical tests, based on the "learning series" (about 5 to 30 iterations, within which τ_k remains constant) are used to check the above conditions and change τ_k appropriately. The ρ_k coefficient changes within the learning series according to (3.10) in which $0 < R < 1$ stays constant within the series; $\rho_k \to R$ if $k \to \infty$ and R is responsible for the rate of change of ρ_k. From (3.13) and (3.14) it follows that R controls the search direction d^k; it should be chosen such that (on average) $E\{\langle d^{k-1}, \xi^{k+1}\rangle\} = 0$ (this condition provides a step toward the ridge of the $Y(x)$ function); if $E\{\langle d^{k-1}, \xi^{k+1}\rangle\} < 0$, R is too small, otherwise R is too large. The same learning series is used for averaging and controlling R. For examples on the application of the above algorithm see [5], [6].

IV. Practical Implementation of the IC Yield Optimization System

The practical realization of the IC yield optimization system based on the algorithms described in the previous section was a major effort, requiring interfacing several large pieces of software. As a result, a TOY (Technological Optimization of Yield) program was created, composed of the following major packages of subroutines.

a) The IRIS (Interactive Restructurable Interace System) package [27], [28], supporting the user interface to the circuit analysis and optimization software discussed below.

b) The SPICE-PAC package of circuit simulation subroutines [30], [31], based on the original SPICE2G.6 circuit simulator developed at the University of California at Berkeley. It has several additional features not available in the original SPICE program, such as a direct and dynamic access to internal circuit element values, model parameters and all output variables, which makes it suitable for efficient handling of circuit element changes.

c) The YIELD-PAC package of subroutines for Monte Carlo yield evaluation using SPICE-PAC.

d) The FAB-PAC package, for interfacing SPICE-PAC, YIELD-PAC, and other Fortran-77 routines with FABRICS [2], [3], written in the C language.

e) The STOCH-PAC subroutines for statistical optimization of the production yield, yield gradient calculation, scaling, etc.

f) Subroutines for graphical outpt (response plots), etc.

A simplified structure of the TOY program utilizing the software tools listed above is shown in Fig. 3. All program actions and the user interface to the system are coordi-

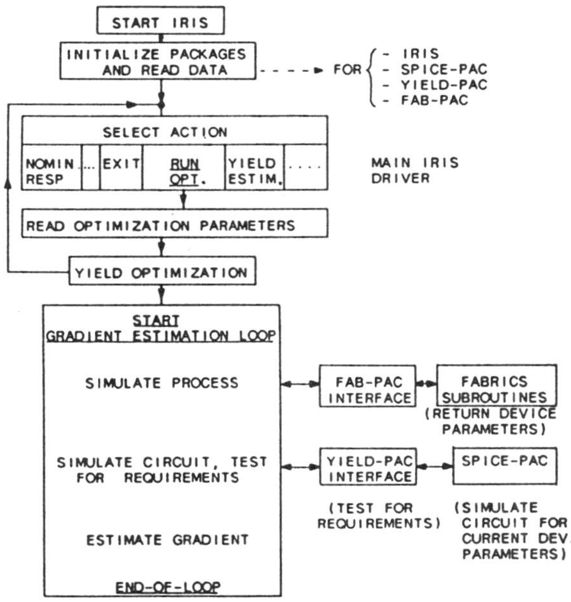

Fig. 3. Simplified structure of the TOY program for technological optimization of production yield.

nated and realized by the IRIS interface which is briefly described below.

4.1 IRIS—A General Interface to Optimization Software and its TOY Implementation

IRIS is a general purpose package supporting the user interaction with circuit analysis, optimization and statistical circuit design software. Its detailed description can be found in [27], [28].

A basic characteristic of IRIS is that it combines a menu-driven tutorial questioning (or dialog) style of interaction with a command language approach. A batch-like mode is also available through dynamically defined macro-actions composed of an arbitrary number of basic actions. Their preparation can be automated, with checking the syntax correctness, and execution is possible immediately after they were defined, or during some future sessions.

A current set of "active" subroutines can easily be restructured at several nested levels during the same optimization session, without relinking the whole system. The user can interactively manipulate (i.e., modify, list, store, print, etc.) all parameters of active subroutines as well as all circuit and technological process variables and layout dimensions.

The level of communication (prompts, menus) can be adjusted to the actual level of user's expertise (i.e., more help for a novice and less help for an expert).

Other IRIS features include: plotting nominal circuit response with zooming capabilities, estimating yield, yield optimization with user's control of the optimization process, debugging, tracing various parameter values, initialization, and others. Due to a systematic IRIS structure and the use of "templates," it is very easy to introduce new actions to the system, increasing in that way the system capabilities.

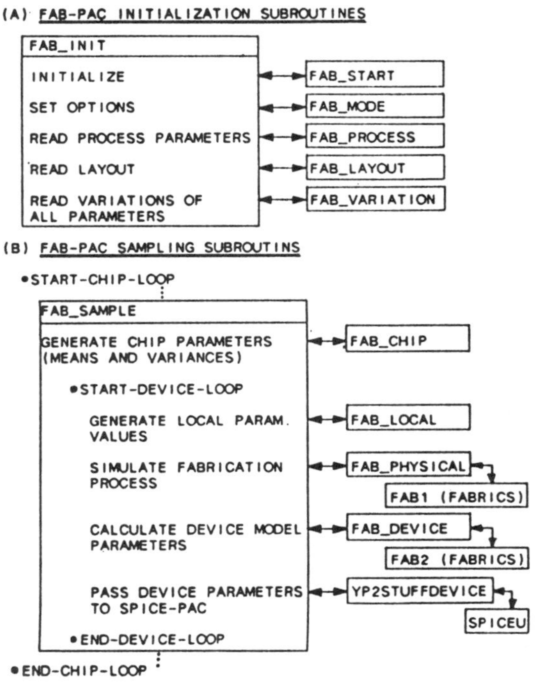

Fig. 4. Simplified structure of the FAB-PAC interface.

4.2. Internal Organization of the TOY Program

A simplified structure of TOY is shown in Fig. 3. After initialization and data reading, several actions can be chosen using the IRIS interface.

The RUN command starts the yield optimization phase, realized with the SA Algorithm 2 described in Section III. At this point a particular value of the β parameter controlling the magnitude of perturbations is selected.

The yield optimization subroutine calls the yield gradient estimation subroutine based on Algorithm 1 of Section III, and the gradient estimation loop is started. For the Perturbation Method discussed in Section III, some additional random perturbations are generated at this time, and the actual values of technological process parameters and layout dimensions are passed to FABRICS through the FAB-PAC interface.

A simplified structure of FAB-PAC is shown in Fig. 4. FAB-PAC controls the whole random sampling process (these actions are similar to the corresponding actions in the FABRICS and WAFER programs [2], [3]), and it supports the subroutine calls and data passing between the VAX-11 Fortran-77 subroutines of STOCH-PAC, IRIS and YIELD-PAC and the C language subroutines of FABRICS. The major actions of FAB-PAC are realized with its sampling routines (see (B) in Fig. 4) called from FAB__SAMPLE in the "chip-loop" (corresponding to the "gradient estimation loop" of Fig. 3). In this loop, the means and variances of chip parameters are randomly generated [2], [3], calling FAB__CHIP, for different chips within a given wafer. For a particular chip a "device-loop" is started, and the local variations of all parameters of all chip devices are sampled in the loop (FAB__LOCAL routine). Once this happens, FAB1—the major FABRICS

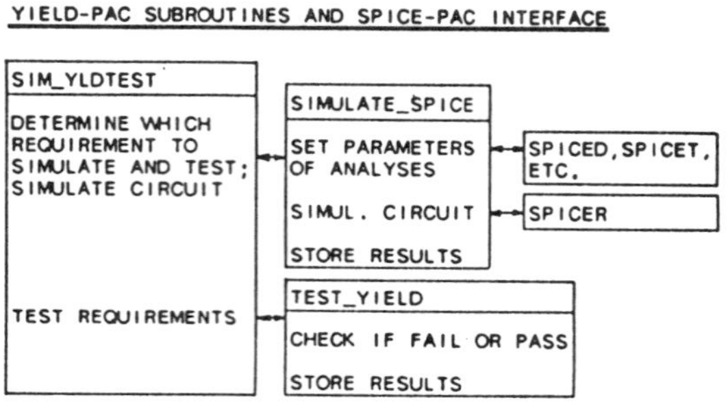

Fig. 5. Simplified structure of YIELD-PAC and its interface to SPICE-PAC.

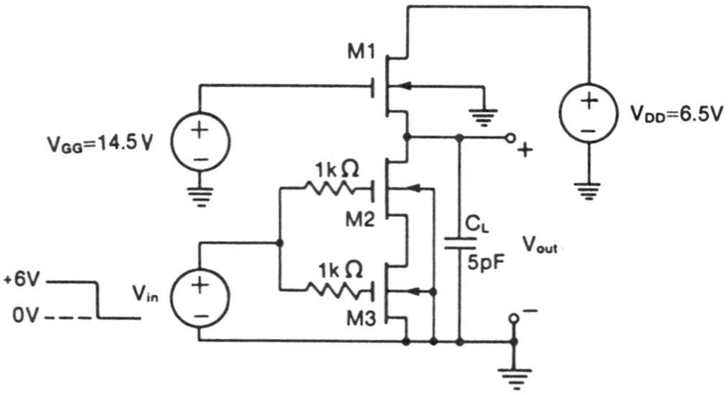

Fig. 6. NMOS NAND gate of Example 1.

simulator of the technological process is called, producing at its output all physical parameters needed for the device characterization. Then the FAB2 routine of FABRICS is called, the element and device electrical model parameters are calculated and passed to SPICE-PAC parameter updating routine SPICEU.

The next step, requiring circuit analyses and constraint checking is performed by the YIELD-PAC routines. Their simplified structure is shown in Fig. 5. All circuit requirements are written in a symbolic form in the requirement input file, and are read in advance during the initialization phase (see Fig. 3). This information is now retrieved and used to perform the required circuit simulations (e.g., ac, dc, transient, distortion, etc.). The call to SIM_YLDTEST activates the SIMULATE_SPICE routine which sets the parameters of different analyses (several calls to different SPICE-PAC routines are performed), and calls the SPICER routine which performs the actual circuit analysis. The TEST_YIELD routine then checks the outcome of each analysis against the requirements, finding the value of the $\phi(\cdot\ ,\ \cdot)$ function (see (3.8), (3.9)), and passing it back to the gradient estimation routine. After the yield gradient is estimated using (3.8), (3.9), the control is returned to the calling yield optimization routine.

When the optimization process is completed, the control is passed back to the IRIS interface. Then the β parameter is reduced and the optimization process restarted from the last obtained nominal point, as required for the convergence of the Random Perturbation Method. Based on our practical experience, it turned out that very few iterations with decreasing values of β are required (in many cases just one) for the algorithm convergence to the nominal point sufficiently close to the yield maximum (within the accuracy achievable with any statistical method).

V. Examples

Example 1

The first circuit optimized was the NMOS NAND gate (Fig. 6), used in a time-domain optimization example (in a PMOS version) by Brayton and Director [33], and Antreich and Huss [35], in an example for Simplicial Approximation by Director and Hachtel [8], as an example for multicriteria optimization design by Fraser [34] and Lightner and Director [4]. The objective of optimization was to maximize the parameteric circuit yield. The acceptability region was determined by the following requirements: $V_0 \leq 0.7$ V, $V_{out}(\text{time} = t_1) > 6.14$ V, for $t_1 = 50$ ns where V_{out} is the output voltage of the gate resulting from the application at the input a negative step function V_{in} and $V_0 = V_{out}(\text{time} = 0)$ is the ON voltage. Additionally, the circuit area was limited to 2500 μm^2 (this constraint was directly handled by the yield optimization procedure). The initial design was assumed to be the yield optimized design of Lightner and Director [4] (Table III, run 4), with the initial area of 2171 μm^2. However, our results are not comparable with those obtained in [4], since in [4] the analytical transistor models and simple statistical distribution functions were used, while in our case the Level 2 SPICE2G.6 MOSFET model was used and the IC manufacturing process simulated with FABRICS II. The designable parameters included all 39 technological process parameters available in FABRICS for the NMOS process and the widths of all transistors.[4] The yield for the initial design was equal to 20 percent (based on 50 Monte Carlo samples). The application of the Random Perturbation Method of Section III resulted in yield increase to 100 percent after 100 iterations (additional 10 samples were used for the initial yield gradient estimation and 100 to check the final yield). The area was slightly reduced to 2138 μm^2. The perturbation method used 2 percent relative perturbations. Because of the 100 percent yield, it was realized that the circuit performance could be further improved in terms of the rise time. Therefore, in the next optimization run t_1 was decreased almost twice to the value of 28 ns. As a result of the reduction of the acceptability region the yield also dropped to only about 10 percent. Then, the second optimization was performed with the relative perturbations of 1 percent. Using 20 samples for the initial gradient estimation, and after 40 iterations, i.e., using the total number of 60 additional circuit analysis, the final yield increased to 92 percent and the area slightly increased to 2188 μm^2 (checked with 50 Monte Carlo sam-

[4]All available technological process parameters were used just to show the yield optimization subroutine ability to handle a large number of parameters; in practice some of the technological parameters will have to be fixed or limited to predetermined ranges, due to other constraints and requirements normally present in IC manufacturing.

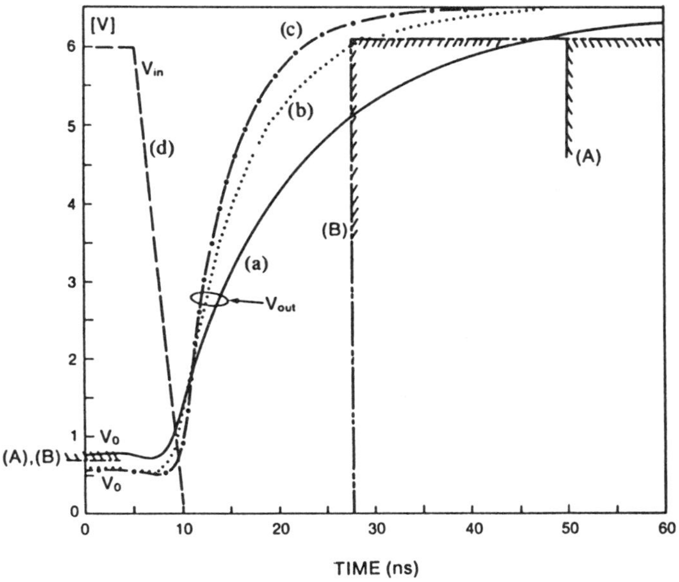

Fig. 7. Nominal transient responses and constraints for Example 1. (a) The initial response; the relevant constraints for the first optimization run are labeled as (A). (b) Optimized response after the first optimization run; the new modified constraints are labeled as (B). (c) The final response after the second optimization run. (d) The input pulse.

ples). Fig. 7 shows the change in the nominal transient response resulting from the optimization.

The results of optimization showed that the technological process parameters changed from about 0.1 percent to 17 percent. The parameters that changed by more than 10 percent were (the initial parameter value and its relative change are shown): field oxidation temp.: 1300 K (+17.1 percent), time of first gate oxidation: 600 s (−15.8 percent), time of gate anneal: 1800 s (+16 percent), temperature of first source-drain oxidation: 1200 K (+13.3 percent), time of source-drain anneal: 1200 s (−12.9 percent), time of source-drain drive-in: 2200 s (−12.3 percent), partial preasure of oxygen for preoxidation: 0.5 (−11.5 percent), time of poly doping: 800 s (+11.2 percent), time of first source–drain oxidation: 3000 s (−10.8 percent), preoxidation time: 1800 s (+10.5 percent), temperature of source–drain drive-in: 1250 K (+10 percent). The widths of M1 and M2 were reduced by −6.3 percent and −0.8 percent, respectively, and the width of M3 slightly increased by 3.3 percent.

The CPU time *per one sample* was approximately equal to 9.6 s for the first optimization, and about 5.5 s for the second run (due to a shorter transient analysis time). Out of this, about 3 s was needed to perform the statistical simulation with FABRICS. The computer used was VAX11/782 with floating point accelerator under the EUNICE operating system (UNIX emulator under the VMS operating system).

In conclusion a total number of 110 + 60 = 170 circuit analyses (with additional 200 samples used for Monte Carlo checking of yield) was sufficient not only to dramatically improve the initial yield but also to improve the nominal circuit performance and to tighten the specifications.

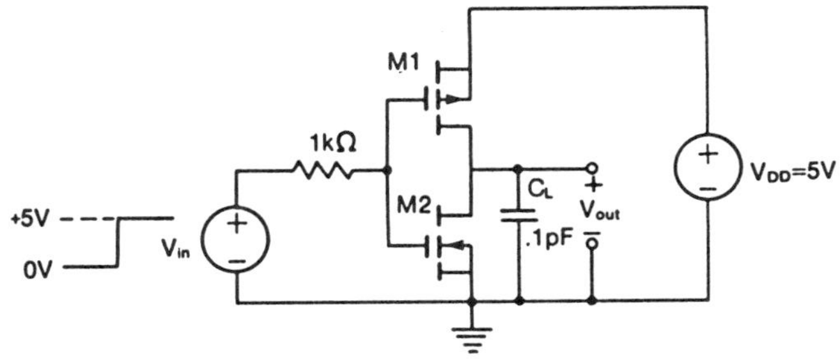

Fig. 8. CMOS inverter of Example 2.

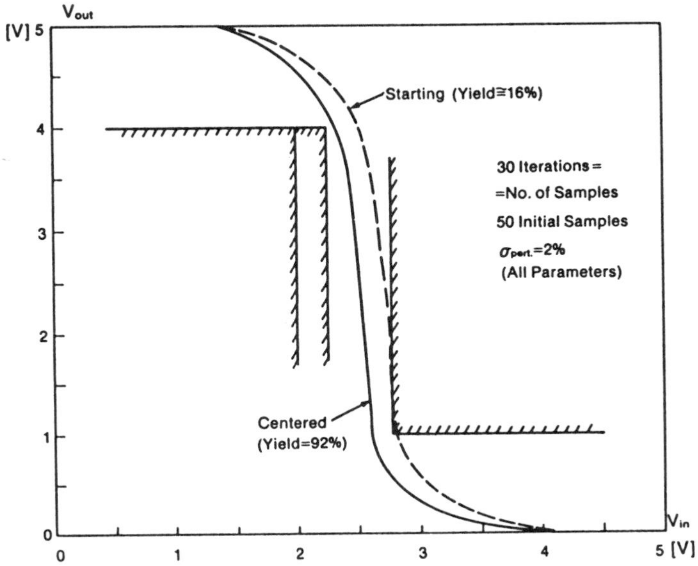

Fig. 9. Nominal dc transfer curves and constraints of the inerter of Example 2. (a) The initial transfer curve. (b) The optimized curve.

Example 2

The second circuit under consideration was the CMOS inverter shown in Fig. 8. At the beginning of optimization it was assumed that the acceptability region of the circuit is determined only by constraints imposed on the dc transfer curve of the inverter as shown in Fig. 9. All 48 process parameters available in FABRICS II for the CMOS process, and 4 layout dimensions were assumed to be designable. The initial yield was 16 percent. The optimization program using 2 percent relative perturbations increased the yield to 98 percent after only 60 iterations (additional 50 samples were used to estimate the initial gradient of the yield). Then the acceptability region was modified to further narrow down "the transition region" of the dc transfer curve as shown in Fig. 9 and a constraint on the rise/delay time was added as shown in Fig. 10. This resulted in the yield decrease to 18 percent. At this phase of optimization it was decided that only the parameters related to the gate oxidation and the length of M2 are designable, in order to improve the conditioning of the optimization problem. After 30 iterations (20 additional samples were used for the initial gradient estimation) the program found a design with yield equal to 36 percent (checked with 50 Monte Carlo samples). Figs. 9 and 10 show the changes in the nominal circuit responses after

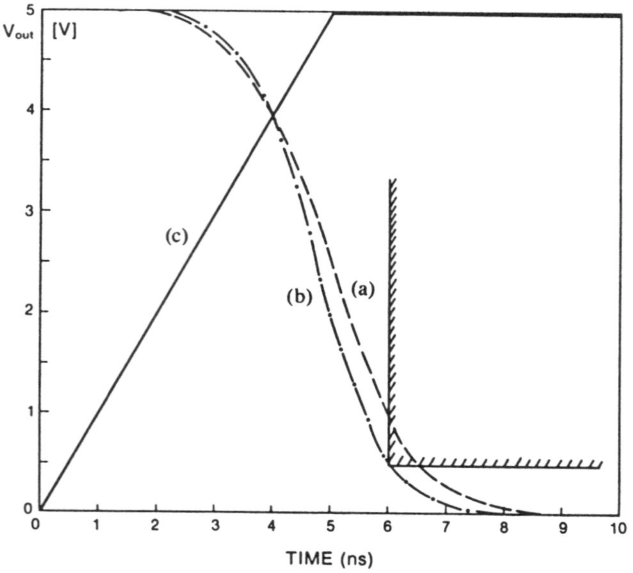

Fig. 10. Transient response for Example 2. (a) Before the second optimization run. (b) After the second optimization run. (c) Input pulse.

optimization. As seen, also in this case the yield increase was very fast.

VI. Conclusions

In this paper the algorithms and software tools for production yield optimization with respect to technological process parameters and layout dimensions were described. The FABRICS II program of Carnegie–Mellon University was interfaced with the TOY optimization program as a statistical simulator of the technological process. The Random Perturbation method used, together with the Stochastic Approximation algorithms proved to be very successful in practice, even for cases with a large number of technological parameters (about 50), leading to very fast yield increase with a relatively small total number of circuit analyses.

At the present time the size of the circuit that can be optimized is limited by the amount of time that can be devoted to optimization: a major limiting factor is the analysis time required by the SPICE circuit simulator especially in the time domain analysis; the limits of FABRICS are much less severe (less than one CPU second per active device, per simulation); the time overhead due to the optimization methods used was negligible.

Further improvements to the TOY yield optimization program and other statistical circuit design tools are planned, both on the algorithmic, as well on the user-program interaction levels.

Appendix I

Formal Derivation of (2.3)

Let $y = v + \gamma \in R^n$, independent of $e \in R^m$, be a multidimensional random variable with expectation v and a very small disturbance part $\gamma \to 0$. Therefore the joint p.d.f. of y and e, $f_y(y, e)$ can be expressed as: $f(y, e) = f_y(y) f_e(e)$, where $f_y(\cdot)$ and $f_e(\cdot)$ are the marginal p.d.f.'s of y and e, respectively. If $\gamma \to 0$, then $f(v + \gamma, e)$ becomes a singular p.d.f., which can be expressed through the Dirac delta functional as follows [32]:

$$\lim_{\gamma \to 0} f(v + \gamma, e) = \delta(y - v) f_e(e). \qquad (I.1)$$

Yield can be expressed as follows:

$$\lim_{\gamma \to 0} \left\{ Y = \int_{R^{m+n}} \phi(y, e) f_y(y) f_e(e)\, dy\, de \big|_{y = v + \gamma} \right\}$$

$$= \int_{R^m} \left\{ \int_{R^n} \phi(y, e)\, \delta(y - v)\, dy \right\} f_e(e)\, de. \qquad (I.2)$$

Taking into account the following property of the Dirac delta functional:

$$\int_{R^k} g(y)\, \delta(y - a)\, dy = g(a) \qquad (I.3)$$

one obtains from (I.2)

$$Y = Y(v) = \int_{R^m} \phi(v, e) f_e(e)\, de \qquad (I.4)$$

i.e., the yield formula appearing in (2.3) was obtained.

Acknowledgment

The authors are greately indebted to A. Strojwas for making the FABRICS-II program available together with the technological data needed for our examples; many useful discussions are also appreciated.

Fruitful discussions with W. M. Zuberek on a general philosophy of software development and SPICE-PAC in particular, are greately appreciated, as well as his readiness to make the improved versions of SPICE-PAC quickly available.

The authors also thank the two talented undergraduate students: T. L. Cole and M. D. Leonard for developing the YIELD-PAC and FAB-PAC packages; several modifications and improvements to FAB-PAC are the result of many hours of a voluntary work of T. L. Cole.

The work on the STOCH-PAC would not have been successful without discussion and help of A. Ruszczyński and W. Syski.

References

[1] D. A. Antoniadis and R. W. Dutton, "Models for computer simulation of computer IC fabrication process," *IEEE J. Solid State Circuits*, vol. SC-14, pp. 412–422, Apr. 1979.

[2] W. Maly and A. J. Strojwas, "Statistical simulation of IC manufacturing process," *IEEE Trans. Computer-Aided Design*, vol. CAD-1, July 1982.

[3] S. R. Nassif, A. J. Strojwas, and S. W. Director, "FABRICS II," *IEEE Trans. Computer-Aided Design*, vol. CAD-3, pp. 40–46, Jan. 1984.

[4] M. R. Lightner and S. W. Director, "Multiple criteria optimization with yield maximation," *IEEE Trans. Circuits Syst.*, vol. CAS-28, pp. 781–791, Aug. 1981.

[5] M. A. Styblinski and A. Ruszczynski, "Stochastic approximation ap-

proach to production yield optimization," in *Proc. 25th Midwest Symp. Circuits Syst.*, Houghton, MI, Aug. 30-31, 1982.
[6] ——, "Stochastic approximation approach to statistical circuit design," *Electron. Lett.*, vol. 19, no. 8, pp. 300-302, 1983.
[7] D. E. Hocevar, M. R. Lightner, and T. N. Trick, "An extrapolated yield approximation technique for use in yield maximization," *IEEE Trans. Computer-Aided Design*, vol. CAD-3, pp. 279-287, Oct. 1984.
[8] S. W. Director and G. D. Hachtel, "The simplicial approximation approach to design centering," *IEEE Trans. Circuits Syst.*, vol. CAS-24, pp. 363-372, July 1977.
[9] R. K. Brayton, S. W. Director, and G. D. Hachtel, "Yield maximization and worst-case design with arbitrary statistical distributions," *IEEE Trans. Circuits Syst.*, vol. CAS-27, pp. 756-764, Sept. 1980.
[10] L. M. Vidigal and S. W. Director, "A design centering algorithm for nonconvex regions of acceptability," *IEEE Trans. Computer-Aided Design*, vol. CAD-1, pp. 13-24, Jan. 1982.
[11] J. W. Bandler and H. L. Abdel-Malek, "Optimal centering, tolerancing, and yield determination via updated approximations and cuts," *IEEE Trans. Circuits Syst.*, vol. CAS-25, pp. 853-871, Oct. 1978.
[12] H. L. Abdel-Malek and J. W. Bandler, "Yield optimization for arbitrary statistical distributions: part I-Theory," *IEEE Trans. Circuits Syst.*, vol. CAS-27, pp. 245-253, Apr. 1980.
[13] W. N. Illin, "Computer aided design of electronic circuits," *Energia*, Moscow, USSR, 1972 (in Russian).
[14] R. S. Soin and R. Spence, "Statistical exploration approach to design centering," *Proc. Inst. Elect. Eng.*, vol. 127, part G, no. 6, pp. 260-262, 1980.
[15] K. S. Tahim and R. Spence, "A radial exploration approach to manufacturing yield estimation and design centering," *IEEE Trans. Circuits Syst.*, vol. CAS-26, pp. 768-774, Sept. 1979.
[16] K. J. Antreich and R. K. Koblitz, "Design centering by yield prediction," *IEEE Trans. Circuit Syst.*, vol. CAS-29, pp. 88-95, Feb. 1982.
[17] W. Strasz and M. A. Styblinski, "A second derivative Monte Carlo optimization of the production yield," in *Proc. ECCTD 1980*, Warsaw, Poland, vol. 2, pp. 121-131, Sept. 1980.
[18] K. Singhal and J. F. Pinel, "Statistical design centering and tolerancing using parameteric sampling," *IEEE Trans. Circuits Syst.*, vol. CAS-28, pp. 692-702, July 1981.
[19] G. Kjellstrom and L. Taxen, "Stochastic optimization in system design," *IEEE Trans. Circuits Syst.*, vol. CAS-28, pp. 702-715, July 1981.
[20] H. Robins and S. Monro, "A stochastic approximation method," *Annu. Math. Stat.*, vol. 22, pp. 400-407, 1951.
[21] J. Kiefer and J. Wolfowitz, "Stochastic estimation of the maximum of a regression function," *Annu. Math. Stat.*, vol. 23, pp. 462-466, 1952.
[22] H. K. Kushner and D. S. Clark, *Stochastic Approximation Methods for Constrained and Unconstrained Systems*. New York: Springer-Verlag, 1978.
[23] M. T. Wasan, *Stochastic Approximation*. New York: Cambridge Univ. Press, 1969.
[24] M. A. Styblinski and L. J. Opalski, "A random perturbation method for IC yield optimization with deterministic process parameters," in *Proc. IEEE Int. Symp. Circuits Syst.*, Montreal, Canada, May 7-10, 1984, pp. 977-980.
[25] V. Ya. Katkovnik and O. Yu. Kul'chitski, "Convergence of a class of random search algorithms," *Automat Remote Contr.*, no. 8, pp. 1321-1326, 1972.
[26] R. Y. Rubinstein, *Simulation and the Monte Carlo Method*. New York: John Wiley & Sons, 1981.
[27] L. J. Opalski and M. A. Styblinski, "A flexible interactive interface for IC minicomputer optimization software," in *Proc. Int. Symp. Mini- and Microcomputers and their Appl.*, San Francisco, CA, June 5-6, 1984, pp. 24-28.
[28] L. J. Opalski and M. A. Styblinski, "A user-reprogrammable interactive interface system for computer-aided stochastic optimization of ICs" in *Proc. 27th Midwest Symp. Circuits Syst.*, Morgantown, WV, June 11-12, 1984, pp. 587-590.
[29] M. A. Styblinski and L. J. Opalski, "Software tools for IC yield optimization with technological process parameters," in *Dig. Tech. Pap. IEEE Int. Conf. Computer-Aided Design*, Santa Clara, CA., Nov. 12-15, 1984, pp. 158-160.
[30] W. M. Zuberek, "SPICE-PAC, a package of subroutines for circuit simulation and optimization," in *Proc 26th Midwest Symp. Circuits Syst.*, Puebla, Mexico, Aug. 15-16, 1983.
[31] W. M. Zuberek and M. A. Styblinski, "SPICE-PAC Subroutines for Circuit Design," *VLSI Design*, pp. 82-83, Oct. 1983.
[32] K. S. Miller, *Multidimensional Gaussian Distributions*. New York: J. Wiley, 1964.
[33] R. K. Brayton and S. W. Director, "Computation of delay time sensitivities for use in time domain optimization," *IEEE Trans. Circuits Syst.*, vol. CAS-22, Dec. 1975.
[34] D. L. Fraser, Jr., "Modeling and optimization of MOSFET LSI circuits," Ph.D. dissertation, Univ. Florida, Gainesville, 1977.
[35] K. J. Antreich and S. A. Huss, "An interactive optimization technique for the nominal design of integrated circuits," *IEEE Trans. Circuits Syst.*, vol. CAS-31, pp. 203-212, Feb. 1984.
[36] P. Balaban and J. L. Golembeski, "Statistical analysis for practical circuit design," *IEEE Trans. Circuits Syst.*, vol. CAS-22, pp. 100-108, Feb. 1975.
[37] P. Cox, P. Yang, P. Chatterjee, "Statistical modeling for efficient parametric yield estimation," in *IEEE Int. Elect. Dev. Meet.*, 1983, pp. 242-245.

VLSI Yield Prediction and Estimation: A Unified Framework

WOJCIECH MALY, MEMBER, IEEE, ANDRZEJ J. STROJWAS, MEMBER, IEEE, AND STEPHEN W. DIRECTOR, FELLOW, IEEE

Abstract—**In this paper we present a unified framework for prediction and estimation of the manufacturing yield of VLSI circuits. We formally introduce a number of yield measures that are useful both during the design process and during the manufacturing process. This framework is general enough to bridge the gap between the traditional concepts of parametric and catastrophic yield. We provide a classification of causes of yield loss which is essential for efficient yield estimation. Finally, we relate yield to manufacturing costs which provides a common denominator for the discussion of the manufacturing process efficiency.**

I. Introduction

DUE TO inherent fluctuations in any integrated circuit (IC) manufacturing process, the yield (which is nominally viewed as the ratio of the number of chips that perform correctly to the number of chips manufactured) is always less than 100 percent. As the complexity of VLSI chips increases, and the dimensions of VLSI devices decrease, the sensitivity of performance to process fluctuations increases, thus reducing the manufacturing yield. Since profitability of a manufacturing process is directly related to yield, the search for computer-aided methods for maximizing yield through improved design methods and control of the manufacturing process has intensified dramatically.

Statistical approaches to yield modeling and optimization have been under development for a number of years. For the most part, these methods can be separated into two categories: parametric yield estimation and optimization techniques and catastrophic yield estimation and optimization techniques. In general, parametric yield optimization has been formulated as a tolerance assignment or design centering problem [3], [9], [38], [10], [4], [60], [39], [6], [19], [1], [58]. However, due to simplified assumptions about circuit element characteristics, these approaches have been proven successful only for discrete circuits with a relatively small number of designable parameters. To overcome the above drawbacks techniques for yield optimization that employ statistical process simulation techniques [17], [30], [29], [46], [34] have been developed to handle IC manufacturing process related constraints [26], [20], [56], [40], [14], [2], [8].

Methods for catastrophic yield estimation have been based upon such simplifying assumptions as yield loss being due to point defects only [33], [32], [36], [47], [48], [15]. More elaborate models that handle nonuniform defect distribution [62], [47], defect size distribution [37], [5], [25], [23], [11], [12] and the yield for chips with redundant circuitry [49], [50], [31] have also been reported.

All of the above approaches, while appropriate for the particular applications they were designed for, are not sufficient to handle all of the factors that affect yield of VLSI circuits. The purpose of this paper is to develop a unified framework from which improved methods for yield maximization can proceed. We present a general and realistic approach to predicting and estimating the manufacturing yield of VLSI circuits. All of the physical phenomena that affect manufacturing yield are taken into account. The paper is organized as follows. We begin by investigating the structure and organization of a typical VLSI fabrication process. A description of the physical causes of yield loss is then presented along with a classification scheme for possible failure modes in VLSI circuits. We then introduce formulas that can be used to estimate yield during the manufacturing process and to predict yield during the design process. Finally, we discuss the cost and profit aspects of VLSI manufacturing and relate them to yield maximization.

II. The VLSI Fabrication Process

As a first step towards developing manufacturing-based approach to CAD we formalize some terms concerning the structure and organization of a fabrication process. In this section we analyze, in general terms, the VLSI manufacturing process organization and environment, and establish relationships between what is normally called the IC design domain and IC manufacturing domain.

The IC manufacturing process involves a sequence of basic processing steps which are performed on sets of wafers called *lots*. Each wafer may contain as many as several hundred IC chips. A few of the chips on the wafer are special in that they contain *test structures* and are called *test chips*. Test structures are special purpose devices which are designed so that their performance is sensitive to the quality of specific processing steps. Examples of

Manuscript received September 19, 1985; revised October 24, 1985. This work was supported in part by the National Science Foundation under Grant ECS8203744, and by the Semiconductor Research Corporation and General Motors Research Laboratories.

The authors are with the Electrical Engineering and Computer Department, Carnegie-Mellon University, Pittsburgh, PA 15206.

IEEE Log Number 8406698.

Reprinted from *IEEE Trans. Computer-Aided Design,* vol. CAD-5, no. 1, pp. 114–130, January 1986.

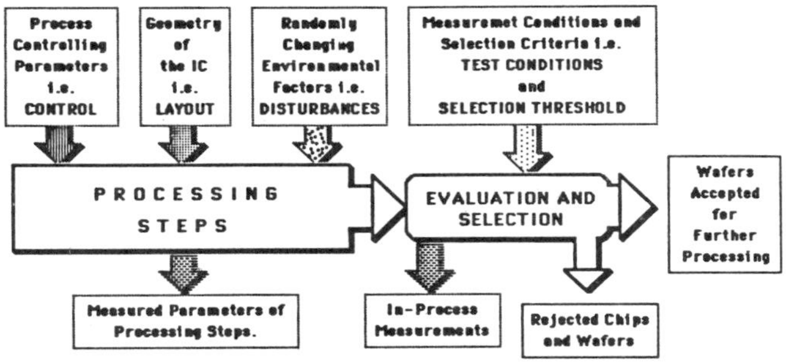

Fig. 1. Flow of information in the manufacturing operation.

test structures are long contact chains, large capacitors, and arrays of transistors of various geometries.[1]

To facilitate process observability and process control, one or more processing steps are usually separated by an evaluation or inspection, and selection step. An evaluation step is a measurement or sequence of measurements, called *in-line measurements*, performed on the wafer or, in some cases a special test wafer, that does not contain IC's. If one or more in-line measurements fall outside of some range defined by a set of *selection thresholds*, the wafers are considered defective and are discarded. The conditions under which the in-line measurements are performed, defined in terms of such parameters as voltages or currents, are called *test conditions*. We refer to a sequence of processing steps followed by an evaluation and selection step as a *manufacturing operation* as shown in Fig. 1.

The outcome of a manufacturing operation depends on three major factors: the process controlling parameters, or *control;* the geometry of the fabricated IC, or *layout*, and some randomly changing environmental factors, called *disturbances*. The *control* of a manufacturing operation is the set of parameters which can be manipulated in order to achieve some desired change in the fabricated IC structure. Examples of control parameters are processing equipment settings which determine temperatures, gas pressures, step duration, etc. The *layout* of an IC is described as a set of masks where each mask can be viewed as a collection of transparent regions within an opaque region. The *disturbances* are environmental factors that cause variations in outcome of a manufacturing operation. Such variations are inherent in any manufacturing process and can be characterized in terms of a set of random variables.

A typical IC fabrication process involves many manufacturing operations, as shown in Fig. 2. Observe that the last few manufacturing operations involve probe tests. As shown, probe tests are performed on two different types of chips: regular chips and test chips. Regular chips are classified as passing or failing, and those which pass are forwarded to the assembly and packaging operations. Since test chips are designed to provide information about

[1]In some cases, test structures are also placed on scribe lanes, i.e., on the spaces between IC chips.

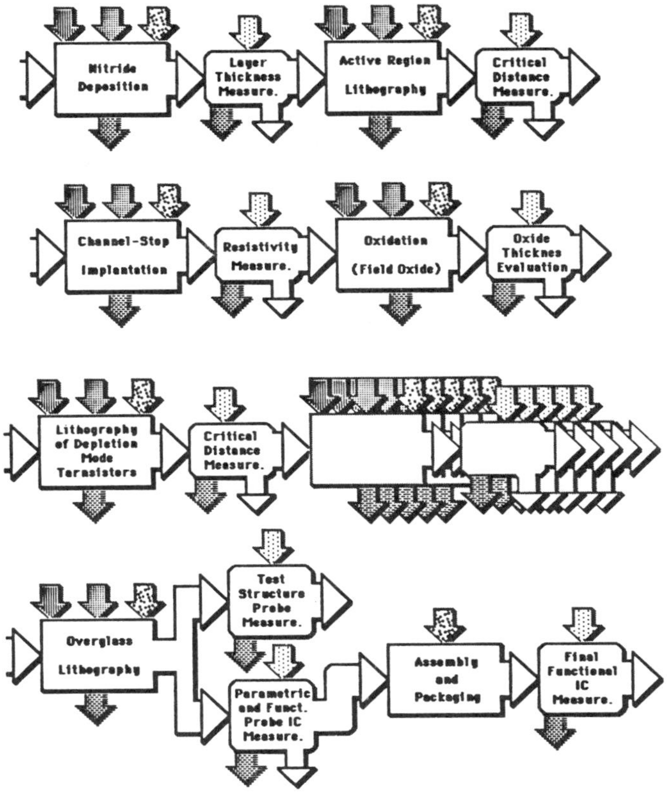

Fig. 2. Flow of information in an nMOS process.

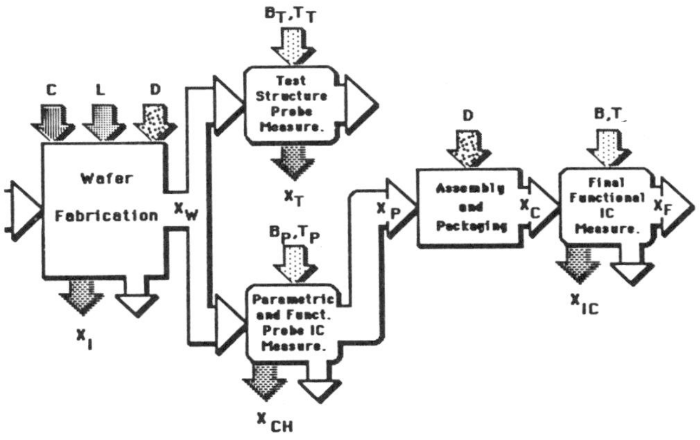

Fig. 3. Generalized structure of the IC manufacturing process.

the process performance for diagnostic purposes, and are not used to select IC chips, test chip measurements are distinguished as a separate branch in the flow diagram.

It can be concluded from Fig. 2 that the IC manufacturing process can be viewed as being composed of a wafer fabrication part, a testing part and an assembly part. The testing and assembly parts of the process involve probe measurements and testing, assembly and packaging, and final testing, as shown in Fig. 3. In the wafer fabrication part, the control parameters, layout parameters, and the disturbances are represented by the C, L, and D, respectively. The in-line measurements are denoted by X_I.

The probe and final tests are performed by testing programs that set appropriate values of the voltage and current sources and then measure voltages or currents. Two types of tests are made: parametric and functional. Parametric tests are used to detect basic discrepancies between the performance of the IC under test and desired performance and involve such quantities as power, critical signal levels and other dc parameters. Functional tests are used to detect errors in the function performed by IC chips. For digital circuits, functional tests involve the application of binary testing sequences to chip inputs and the results are compared against expected output binary vectors. For analog circuits functional evaluation depends on the type of chip, but usually consists of some frequency or dynamic response measurements. A bandwidth limitation associated with the probe capacitances is, however, an important obstacle in the implementation of some functional tests during probe measurements.

We assume that the IC probe selection is characterized by two sets of parameters, B_P and T_P. B_P is a set of parametric testing conditions and binary test vectors. T_P is a set of selection thresholds which define a *selection region*. The IC probe measurement results are contained in the set denoted by X_{CH}. We denote by B_T and X_T the test structure probe testing conditions and measurement results, respectively.

After the probe measurements, all IC chips classified as defective are rejected and the remaining chips are assembled and packaged. A full final testing of the packaged IC is then performed, and again defective IC's are rejected. The final test conditions, selection thresholds, and measurement results are denoted by B, T, and X_{IC}, respectively. Examples that illustrate selection thresholds for various selection tests are shown in Fig. 4.

In order to fully characterize the general structure of the manufacturing process, we need to introduce a set of variables, called *process state variables*, which completely describe the physical properties of the IC at any point in the fabrication process. While the process state variables are usually unknown in an actual process, they can be estimated through in-line, test structure, and final measurements. We distinguish between four sets of state variables: X_W, X_P, X_C, and X_F, which describe the "state" of an IC chip after wafer fabrication, probe selection, assembly and packaging, and final selection tests, respectively, as illustrated in Fig. 3.

We can establish a number of relationships between the variables introduced above. We define the n_i-dimensional vector X_i, $X_i = \{x_1^i, \cdots, x_{n_i}^i\}^T$, where the index i denotes the type of measurement ($i = I$ for in-line measurement, i = CH for probe measurement, etc.), and T denotes transpose. X_W and X_I which describe the IC after wafer fabrication and in-line measurements, respectively, are implicit functions of the process controls, C, the layout parameters, L, and the disturbances, D, i.e.,

$$X_W = F_W(C, L, D) \tag{1}$$

$$X_I = F_I(C, L, D). \tag{2}$$

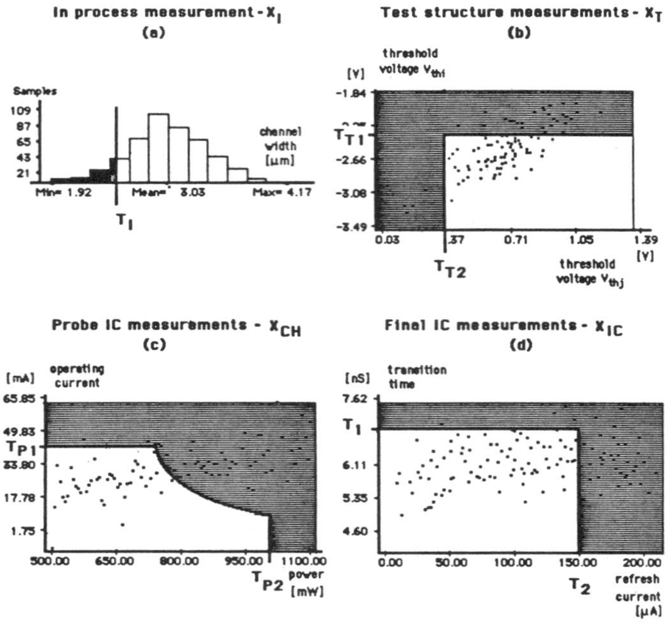

Fig. 4. Examples of the selection regions defined for: in-process measurement, shown in the histogram (a), test structure measurements, probe measurements, and final measurements shown in the scatter plots (b), (c), and (d), respectively. IC chips or wafers whose parameters are outside of the selection region, i.e., in the shaded regions, are classified as defective and are rejected from the process.

Note that F_W and F_I do not imply that explicit analytic expressions exist. Since the process disturbances, D, are random variables, X_W and X_I are random vectors and are characterized by Joint Probability Density Functions (JPDF), which will be denoted by $f_W(x_W)$ and $f_I(x_I)$, respectively.

Similarly, we can identify the following dependencies:

$$X_T = F_T\{X_W(C, L, D), B_T\} \tag{3}$$

$$X_{CH} = F_{CH}\{X_W(C, L, D), B_P\} \tag{4}$$

$$X_P = F_P\{X_W(C, L, D), B_P, T_P\} \tag{5}$$

$$X_C = F_C\{X_W(C, L, D), B_P, T_P\} \tag{6}$$

$$X_{IC} = F_{IC}\{X_W(C, L, D), B_P, T_P, B\} \tag{7}$$

$$X_F = F_F\{X_W(C, L, D), B_P, T_P, B, T\} \tag{8}$$

where the functions F_T, F_{CH}, F_P, F_C, F_{IC}, and F_F are complicated relations, that usually can only be approximated from experimental data. Of course, as in the case of X_W and X_I all of the variables on the left-hand side of (3)–(8) are random vectors which are characterized by the appropriate JPDF's.

To illustrate the physical significance of the above relations, consider a simple n-channel polysilicon-gate MOS technology. For this technology, we can identify the following variables:

- In-line measurements, X_I: field oxide thickness, gate oxide thickness, distance between edges of the field oxide in the channel of the depletion mode transistor, resistance of the polysilicon path, etc.;

- Process state vector after wafer fabrication, X_W: three-dimensional description of the geometry of all regions in the silicon substrate, the insulating and conducting layers, including distributions of the impurities in the entire IC chip, location and size of all point defects in the semiconductor substrate, parameters determining the dependence between impurity concentration and carrier mobilities, etc.;
- Test structure test conditions, B_T: set of drain-source, substrate and gate voltages used to measure I–V characteristics of a set of transistors, current of the current source used to test connectivity of a chain of contacts, etc.;
- Test structure measurements X_T: voltage drop across the polysilicon resistor connected to the output of the current source, transistor drain currents measured with the gate connected to the drain for a given supply voltage for various back-bias voltages, etc.;
- Test structure thresholds, T_T: maximal value of the voltage drop on terminals of the contact chain; maximal drain substrate current, maximal and minimal voltage drops between polysilicon resistor terminals, etc.;
- Test conditions of the IC probe measurements, B_P: voltage supply, clock frequency, current at a given output node, patterns (test vectors) of 0's and 1's on IC inputs, etc.;
- Result of IC probe measurements, X_{CH}: input current, power dissipation, voltage gain, noise margin, offset voltage, etc.;
- IC selection thresholds, T_P: maximal and minimal currents of voltage supply, minimal output voltage for a given output current, patterns of 0's and 1's on the specified IC outputs, minimal amplitude of the ac signal, etc.;
- Process state after IC selection, X_P: such as X_W excluding data that describe defective IC's. X_P could be also expressed in terms of a set of all parameters of IC element models provided that element models perfectly describe IC elements. (Examples of such parameters are SPICE MOS transistor model parameters.);
- Process state after assembly and packaging, X_C: such as X_P but expanded by the data which characterize packaging, i.e., data that describe power dissipation condition, etc.;
- Results of final functional test conditions and test sequences, B: such as B_T expanded by a full set of test vectors and all high frequency and high speed conditions such as high frequency clock rates, high frequency as signals etc.;
- Final functional test and measurements, X_{IC}: such as X_{CH} expanded by full set of results of functional test and high frequency measurements such as frequency characteristics and delays;
- Final functional test thresholds, T: a complete set of the correct results of functional tests and such quantities as maximal delay, minimal amplitude of the ac signal, and all other thresholds such as listed for T_P;
- Process state after final testing, X_F: is equivalent to X_C but can be expressed also in terms of a set of parameters of IC device models that fully characterize IC behavior under all possible test conditions.

In conclusion, we emphasize the following facts:

- Due to the process disturbances, all quantities observed in the process are random variables and have to be described by the appropriate JPDF's;
- Process state variables are essentially unknown and can only be estimated by means of some measurements;
- Relationships between process conditions and the quantities measurable in the process are usually known with a limited accuracy and in many cases this information is experimental in nature only.

We now investigate the causes of process fluctuations, and their effects on circuit performance.

III. Disturbances in the IC Manufacturing Process

The process disturbances that occur in the IC manufacturing process can be characterized in terms of the physical nature of the disturbance or in terms of the effects they have on IC performance. Effects on performance can be described in terms of changes in the electrical characteristics of the IC or in terms of changes in circuit connectivity. In this section we describe and classify process disturbances and then classify the effects that these disturbances have on performances.

3.1. Process Disturbances

Sources of the random phenomena that occur in the IC fabrication process [59], [13], [41] which can be modeled as process disturbances, can be classified as:

- **Human errors and equipment failures.**
- **Instabilities in the process conditions.** These are random fluctuations in the actual environment that surround an IC chip and arise from phenomena such as a turbulent flow of gasses used for diffusion and oxidation, inaccuracies in the control of furnace temperature, etc. Because of these instabilities, each area of wafer is exposed to slightly different environmental conditions and, hence, no two manufactured chips can possibly perform identically.
- **Material instabilities.** These are variations in the physical parameters of the chemical compounds and other materials used in the manufacturing process. Typical examples of material instabilities are: fluctuations in the purity and physical characteristics of the chemical compounds, density and viscosity of photoresist, water and gasses contamination, etc.
- **Substrate inhomogeneities.** These are local disturbances in the properties of substrate wafers which are

of three types: *point defects*, *dislocations*, and *surface imperfections*,. Point defects which are local disorders in the structure of the lattice of a semiconductor material include: *vacancies* (a lack of the atom in the lattice site), an *interstitial defect*, (an atom is located out of crystal site), and *vacancy-interstitial pairs*, called *Frenkel defects*. Point defects act as recombination centers and, therefore, affect carrier lifetime. Concentration of point defects may fluctuate and is affected by high temperature process operations. Dislocations, which are geometrical irregularities in the regular structure of the crystal lattice, are caused by crystal growth related thermal stresses and are of two different types: *edge dislocations* and *screw dislocations*. The most important characteristic of a dislocation is that it strongly interacts with point defects and impurities located in its neighborhood. As a consequence, a large increase of the impurity diffusion coefficient may occur which may lead to the formation of diffusion pipes that cause such problems as abnormal leakage currents or low breakdown voltage in p-n junctions.

- **Lithography spots.** These are lithography related disturbances of IC geometry which may be caused by the mask fabrication process as well as the lithography process itself. The mask fabrication process may produce defects of two types: transparent spots in opaque regions and opaque spots in transparent regions of the mask. The lithography process may produce spots due to dust particles on the surface of the mask and photoresist layers. Some photoresist inhomogeneities can also produce the lithography spots.

Not all of the disturbances described above have the same influence on IC performance. Thus, we are motivated to consider what electrical and geometrical effects result from these disturbances.

3.2. Process Related Deformations of IC Design

Disturbances in the process can be viewed as affecting either the geometry of an IC device and/or the electrical characteristics of an IC device. In order to characterize these effects it is useful to employ the concept of an *ideal IC*. An ideal IC is one that would result from a perfect manufacturing process, i.e., its layout is identical to that specified by the masks, and its electrical properties (e.g., impurity distributions or charge densities) are the same as those assumed by the designer during the design process. The performance of an actual IC can be viewed as a deformation of that found in an ideal IC. Actually, deformations have two components: a deterministic component, and a random component. The deterministic component of each deformation is due to simplifications used in developing the description of the ideal IC. Such simplifications are usually necessary due to the lack of the adequate process/device modeling techniques applied during the IC design process. The random component of the deformation is due to the process disturbances themselves. We are particularly interested in this random component.

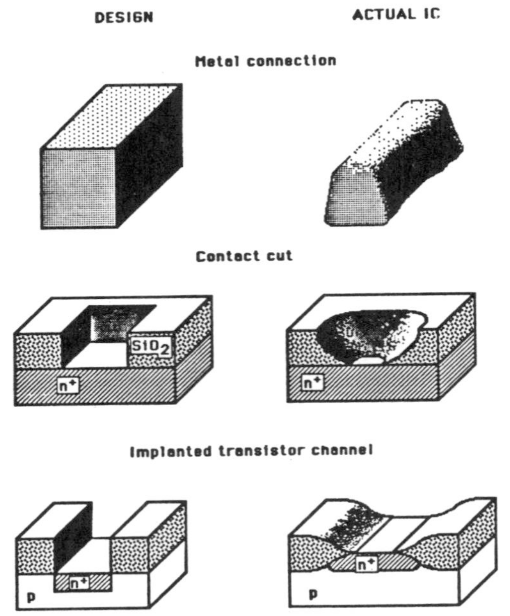

Fig. 5. Illustrations of few discrepancies between geometry of the ideal IC and geometry of the actual IC.

3.2.1. Geometrical Deformations

The geometry of an IC is determined by the two-dimensional (2-D) IC layout and the physics of the manufacturing process. In an ideal IC it is assumed that the edges of the layout define the boundaries of the various regions in an IC and that various manufacturing process steps determine the depth, or thickness, of a region. In an actual IC this is not the case, as illustrated in Fig. 5. Geometrical deformations can be of three types: *lateral effects*, *vertical effects*, and *spot defects*.

Lateral effects include all processing effects which cause the location of the boundary of the region in an actual IC to differ from the corresponding boundary in the ideal IC. A lateral effect can be composed of a *lateral edge displacement* and a *mask misalignment*.

Lateral edge displacements are caused by all processing steps with anisotropic characteristics, such as:

- Lateral diffusion;
- Lateral oxidation, i.e., bird's beak formation;
- Over- and under-etching of metal, polysilicon, oxide, and nitride layers;
- Over- and under-exposures of the photoresist and photoemulsions aplied in all lithography steps;
- UV light or e-beam diffraction on the mask edges.

An example illustrating the contribution of various lateral effects to the lateral edge displacement of the diffusion region, bounded by the edge of the field oxide, is shown in Fig. 6. Lateral edge displacements may cause drastic deformations in the topology of the actual IC by merging regions which are supposed to be disjoint. In terms of electrical behavior such deformations manifest themselves as shorts between, or breaks in, conducting paths.

Note that all edges that lie within regions of the same type are usually affected by the same set of independent disturbances, while edges that lie in different types of re-

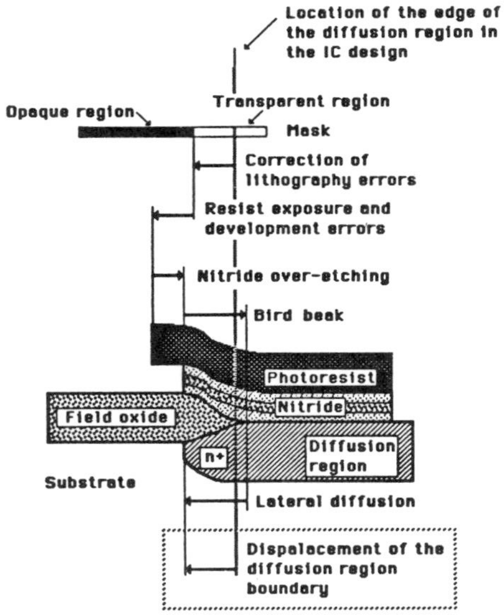

Fig. 6. Lateral displacement of the diffusion region boundary as a result of edge displacements in the mask fabrication process, resist and nitride etching, bird's beak formation and lateral diffusion.

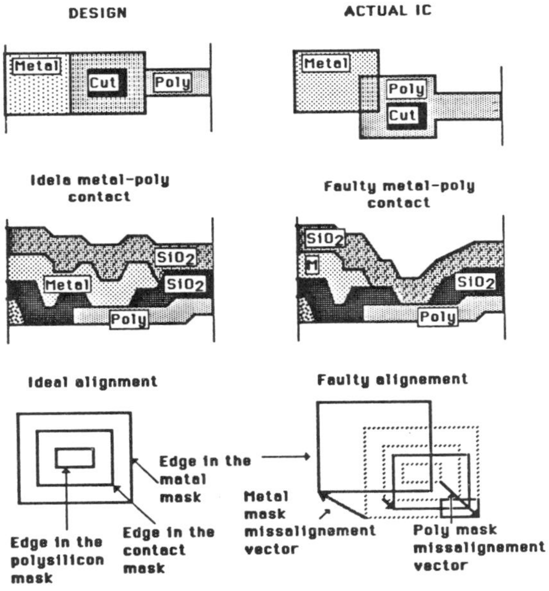

Fig. 7. Mask misalignment.

gions are affected by different sets of disturbances. There are cases, however, in which some disturbances affect different types of regions in the same way and where regions of the same type are affected by quite different sets of disturbances. For example, in the case of n-channel MOS technology, some of the disturbances that contribute to the displacement of the diffusion edge, such as depicted in Fig. 6, affect both the location of the edge of the MOS transistor drain and source as well as the location of the edges of transistor channels. On the other hand, the location of the edges of depletion mode transistors are deformed by two different sets of disturbances. Simply, the edges on the boundaries between channel and source or channel and drain are affected by the different disturbances than the edges of the active regions. Hence, in

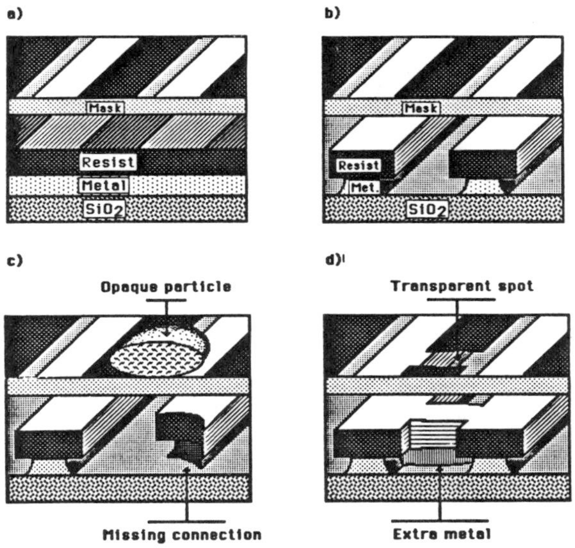

Fig. 8. Mechanism that introduce spot defects into the metal layer.

general, in an actual IC displacements of the edges of a region may either be very similar (if not identical), correlated, or totally independent.

The disturbances which significantly contribute to the lateral edge displacements are random instabilities in process conditions such as equipment vibrations, and instabilities of layers thicknesses that cause fluctuations in the amount of over- and under-etching.

Mask misalignment which is an error in the position of a lithography mask with respect to the features already engraved on the surface of a wafer, causes all edges defined by the mask to be shifted by the same amount with respect to the boundaries of the regions which already exist on the surface of the wafer. (See illustration in Fig. 7.) In fact, mask misalignment could be so significant that the geometry of an actual IC can be deformed from that of the corresponding ideal IC to the extent it results in changes of the electrical connectivity of the IC. Random errors in the mask alignment process are due to instabilities in process conditions due to limited mechanical and optical accuracy of the processing equipment, and shape variations of the wafers.

Vertical effects are deformations in the thickness of IC layers and include deformations which are due to the p-n junction depth variations and deformations in the thickness of the oxide and other deposited layers. Junction depth variations are a direct consequence of the fluctuations in the impurity concentrations while deformations in the thickness of the deposited or oxidized layers are due to process instabilities such as turbulent gas flow, temperature fluctuations, etc.

Spot defects are geometrical features that emerge during the manufacturing process that were not originally defined by the IC layout. The main source of spot defects are lithography spots. Spot defects are random in both their location and size. The size of a spot defect is determined by both the size of the actual particle that caused the spot defect and lateral deformations which disturb the

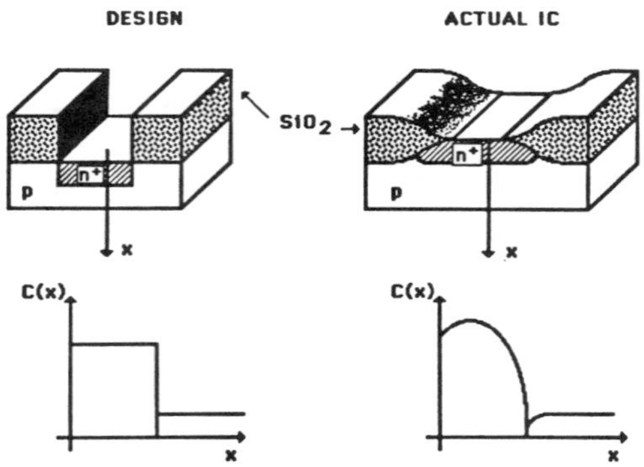

Fig. 9. Comparison of impurity distributions assumed in the channel of an ideal depletion mode transistor versus that of an actual depletion mode transistor.

location of the edges of the spot defect. The example of the mechanism which may introduce a spot defect is shown in Fig. 8.

3.2.2. Electrical Deformations

The electrical characteristics of an actual IC are determined by the three-dimensional impurity distributions in the conducting and semiconducting regions, and charge distribution in the insulating layers. In an ideal IC, a number of simplifying assumptions are typically made which allow for analysis of a design prior to manufacturing (see Fig. 9). All process models are described by simplified equations that can be solved either analytically or by simple numerical techniques. Thus the physical description of such phenomena as dependence of impurity diffusivities on concentration of vacancies and interstitials, or dependence of oxidation growth rates on the impurity concentration in the substrate is often inadequate and leads to significant discrepancies between simulation results and reality. Furthermore, most models are one-dimensional and incorporate only a few of the two-dimensional effects. The device analysis is also performed in an approximate manner and although the equations that describe device behavior are known, the boundary conditions are typically simplified. Also, a number of important physical effects are not included in the analysis (for example, heavy doping effects in bipolar transistors or thin oxide tunneling effects in MOSFET's).

The above simplifications introduce a systematic error in the analysis and cause the description of the electrical properties of an IC used in the design process to differ from physical reality. However, there also exist many phenomena that are random in nature and cause electrical deformations of IC characteristics. As was the case for geometrical deformations, we can divide these phenomena into global and local phenomena. Global electrical deformations will affect regions in different IC elements in the same way. Examples of global deformations include:

- fixed positive oxide charges in the oxide due to nonperfect (non-stechiometric) structure of the oxide;
- emitter-push effect which due to enlarged diffusivity of boron in the active base of the n-p-n transistor, caused by the emitter diffusion induced damages of the crystal lattice;
- increased junction depths under the oxide regions due to oxidation-enhanced diffusion phenomena.

These deformations are random and vary from wafer to wafer, and even across the wafer. However, within one chip their effects on IC elements are very similar. Local electrical deformations are due to

- Spikes which are tiny conducting pipes, or paths, crossing junctions or even whole regions, e.g., base in the n-p-n transistors;
- Local impurity precipitates or silicides which affect the percentage of electrically active impurities in silicon;
- Local contaminations of SiO_2-Si surface that affect surface recombination mechanisms in bipolar devices and disturb MOS transistor threshold through local concentration of surface states.

The above effects have to be treated as process related deformations of the electrical characteristics of an actual IC and have to be taken into account to obtain a realistic model for yield losses.

3.3. General Characteristics of Process Disturbances

From the above discussion we can observe that the disturbances can cause either global or local deformations. All deformations which affect all IC elements in the same way can be called *global.* Mask misalignment is an example of a disturbance that causes a global deformation of an ideal IC. We will refer to disturbances that cause global deformations as global disturbances. Similarly, disturbances that cause only local deformations (affected regions are small compared to the total area of an IC chip) will be called local disturbances. Pin-holes, spikes, spot defects are examples of local deformations.

It is also important to note that various process conditions interact very often in an indirect way. For instance, high temperature processes may cause an increase in the lithography errors due to the deformations in the shape of the wafer. Hence, in general, random fluctuations in process conditions of one processing step may result in changing process conditions of another step. Note also that the process induced deformations of any kind, local or global, geometrical or electrical, are random. For the case of global disturbances, the magnitude of process-related deformations is random and varies from one IC to another. For the case of local disturbances, both the magnitude (e.g., size of a defect) and location of the resulting deformations are random.

3.4. IC Performance Faults

We now turn to a discussion of the impact that the process disturbances have on IC device characteristics and indicate how these randomly changing characteristics affect IC performance. We then introduce a fault classification based upon the effects on performance.

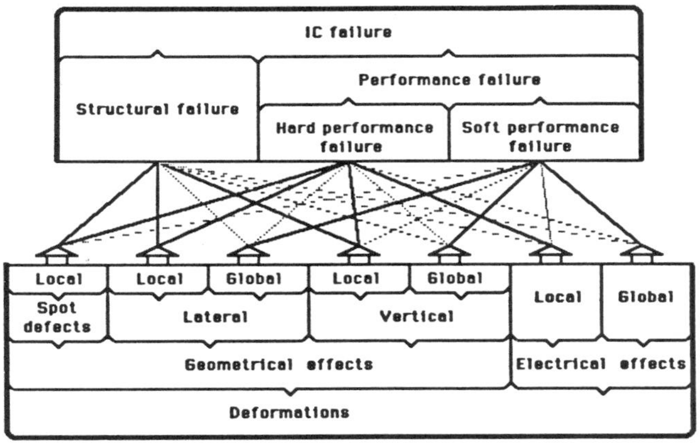

Fig. 10. Classification of IC failures and process deformations.

We begin by noting that each process disturbance may cause a number of different changes in IC performance. We will call this change in behavior a *fault*. For instance, a spot defect introduced in a lithography step may cause a short between two adjacent conducting paths, a break in the path or may only alter effective device dimensions (e.g., affect the *W/L* ratio of a MOSFET). It is therefore convenient to classify faults as either *structural faults* which are changes in the topology of the circuit, or electrical diagram, that represents an IC, or *performance faults*, in which the IC topology remains unchanged, however, circuit performance (e.g., speed or dissipated power) falls outside of some set of allowable tolerance limits. Performance faults can be further divided into two classes: *soft-performance faults* and *hard-performance faults*. If an IC is functionally correct (e.g., in a digital circuit, all state transitions occur in the proper order and the output pattern is the correct response to an input pattern vector) but some performance measure (e.g., signal delay) lies outside of the specified range, we say a soft-performance fault has occurred. If an IC does not function properly (e.g., some state transitions do not occur) or some performances are orders of magnitude from the desired values, we say a hard-performance fault has occurred.

Structural faults require some further elaboration. While some deformations will not alter the structure of an IC under no bias, or very small bias conditions, for other values of dc bias a short between two conducting paths can occur. As an example, consider the lateral diffusion in two parallel interconnect lines. Due to the expansion of depletion layers of the p-n junctions, there might be an equivalent short between these two lines under some bias conditions. On the other hand, some geometrical variations may introduce new parasitic elements (such as overlap capacitances) which do not necessarily cause an IC to be defective. Therefore, structural faults are best defined in terms of changes in the equivalent DC circuit under worst-case bias conditions.

This fault classification scheme is illustrated in Fig. 10. The bottom of the figure shows our classification of physical phenomena which cause yield loss while in the upper part of the figure, the classification is based upon effects (performance faults, structural faults). Note that each phenomena can be a cause for the variety of faults. However, some of the relations (solid lines) depicted in Fig. 10 dominate. For instance, all global effects will cause first of all soft performance failures while point defects are mainly responsible for structural failures.

IV. Measures of Process Efficiency

Because of the inherent fluctuations in an IC fabrication process discussed above, that cause variations in circuit performance, not all of the chips that are manufactured will actually meet all of the design specifications. We define the *manufacturing yield* as the ratio of the number of chips which successfully pass all of the selection steps in the process to the total number of chips that begin the fabrication process. Thus manufacturing yield is a measure of process efficiency. Since manufacturing yield is affected by both the parameters of the design (e.g., layout) as well as the parameters of the manufacturing operations themselves (e.g., control and disturbances), it is desirable to be able to *predict yield* during the design process, as well as *estimate yield* during the manufacturing process. By being able to predict yield during the design process, it may be possible to adjust the design to improve yield. Similarly during manufacturing, if the estimated yield is low, changes in the process which increase yield may be possible. Thus we are motivated to develop techniques for yield estimtion and prediction.

4.1. Yield Estimation

Before we begin our discussion of yield estimation, it is convenient to introduce the following notation (see Fig. 11):

- N maximum number of IC chips which can possibly be fabricated on W production wafers, assuming an ideal manufacturing process;
- W_W number of wafers that have reached probe selection tests;
- N_W number of IC chips on W_W wafers.
- N_P number of chips that have reached final selection step.
- N_F number of chips that have been classified as fault-free after the final selection step.
- N_I number of chips that have been rejected from the process during the in-line selection step.
- N_{CH} number of chips that have been rejected from the process in the probe selection.
- N_{IC} number of chips that have been rejected during final selection.
- N_S number of test structures measured in the probe measurements.
- N_T number of test structures classified as correctly fabricated.

We can now formally define a variety of yield estimates. As will be discussed later, each of these estimates will be useful in assessing the efficiency of different parts of the manufacturing process.

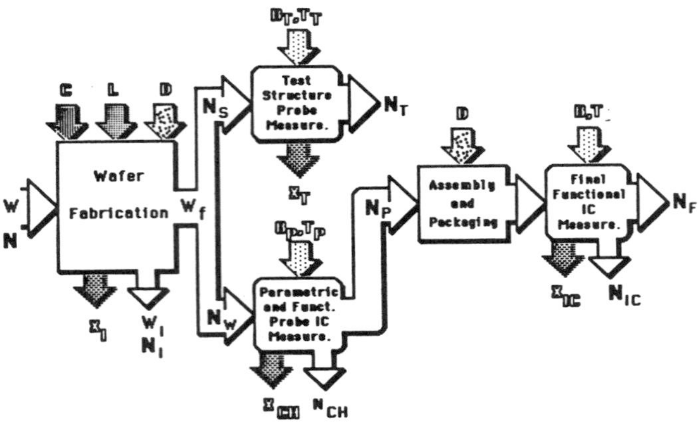

Fig. 11. Flow of IC chips in the manufacturing process.

1) The *manufacturing yield*:

$$Y_M^* = \frac{N_F}{N} \tag{9}$$

which, as indicated above, is a measure of an overall process efficiency. It indicates overall manufacturing performance and is a basic figure in evaluating manufacturing economics and planning manufacturing strategy.

2) The *probe yield*:

$$Y_P^* = \frac{N_P}{N} \tag{10}$$

which describes the efficiency of the wafer processing operations including probe measurements.

3) The *wafer yield*:

$$Y_W^* = \frac{W_W}{W} \tag{11}$$

which can also be expressed as:

$$Y_W^* = \frac{N_W}{N} \tag{12}$$

which is a measure of the quality of the wafer fabrication operations. It is the percentage of manufacturing wafers that have reached the probe tests. In industrial practice, the majority of wafer rejections is due to human errors or equipment failures therefore, wafer yield is primarily a measure of the quality of process supervision.

4) The *final testing yield*:

$$Y_{FT}^* = \frac{N_F}{N_P} \tag{13}$$

which is a measure of the percentage of IC chips that passed successfully final tests. Note that some IC chips are rejected from the process due to assembly and packaging errors and functional defects that were not detected during probe selection.

5) The *probe testing yield*:

$$Y_{PT}^* = \frac{N_P}{N_W} \tag{14}$$

which is a measure of wafer fabrication efficiency. In the case of a new product, probe testing yield shows whether a fabricated IC was designed to match the manufacturing line capabilities. In the case of correct designs, the probe testing yield is the best indicator of the process instabilities.

6) The *processing yield*:

$$Y_{PRO}^* = \frac{N_T}{N_S} \tag{15}$$

which is a measure of the discrepancy between the performance of the process and desired process parameters described in terms of model parameters of the IC elements. The processing yield is used to determine manufacturing line stability.

Finally, it has to be stressed that N_W, N_P, N_F, N_{CH}, N_{IC}, N_S, and N_T are random and, of course, have JPDF's determined by the relations characterized by (1)–(8) and therefore are related to D, L, and C as well as to the testing conditions. Thus by the careful analysis of the values of yields above one can perform the process diagnosis, which can be viewed as estimation of D, as well as evaluate quality of the process-design tuning which can lead to the corrections in L and C.

4.2. *Yield Prediction*

Note that evaluation of each of the yield estimates defined above requires the use of actual process measurements and is, therefore, inadequate for the prediction of yield during the design process prior to fabrication. In order to predict yield during the design process we need to be able to determine the probability that a single IC chip will not be rejected during fabrication. To evaluate this probability it is convenient to introduce the concept of an *acceptability region*.

Let the n_F-dimensional vector $Q_F = \{q_1^F, \cdots, q_{n_F}^F\}^T$ denote all of the functional performance measures of interest for a particular design, and the n_P-dimensional vector $Q_P = \{q_1^P, \cdots, q_{n_P}^P\}^T$ denote all of the parametric performance measures of interest for the same design. For a digital IC, functional performance is a measure of the ability to perform some desired logical operations. Parametric performance is a measure of the quality of the behavior of an IC. Quantities such as delay, power, input and output current, gains, etc. are typical measures of the parametric performance. Let the n_Q dimensional vector $Q = \{Q_F, Q_P\} = \{q_1, \cdots, q_Q\}$ $(n_Q = n_F + n_P)$ denotes all performances of interest. We now define the n_F-dimensional space S_Q^F and n_P dimensional space S_Q^P to be Euclidean spaces that contain all n_F functional performances and n_P parametric performances, respectively. Finally, we define the n_Q dimensional *performance space* as $S_Q =$

$S_Q^F \times S_Q^P$. We can now define the set of the *acceptable IC performances* as

$$A_Q = \{Q \in S_Q | \text{each } q_j \text{ is acceptable} \quad \forall j = 1, 2, \cdots, n_Q\}. \tag{16}$$

Since the performance of an IC is fully determined by either the vector of process state variables X_W, which represent all of the physical parameters of the IC after wafer processing, or the variables L, C, and D, which actually define the design, the set of acceptable performances A_Q must have its equivalent in the n_W-dimensional space of process state variables S_W. Hence, we can define the *acceptability region*

$$A_W = \{X_W \in S_W | Q(X_W) \in A_Q\} \tag{17}$$

where $Q(X_W)$ symbolizes the dependence of the performance on the process state variables. In the other words, the acceptability region is defined by a set of process state variables that assure the desired performance of the fabricated IC. Note that this region, unlike the selection regions, is not dependent on the method by which IC measurements, or measurement conditions, are defined.

Now using the concept of the acceptability region we can define *design yield*, Y, as the probability that the IC performs all functions correctly. Such a probability can be expressed as:

$$Y = \int g(x_W) f_W(x_W) \, dx_W \tag{18}$$

where $f_W(x_W)$ is JPDF associated with the random variable X_W and

$$g(x_W) = \begin{cases} 1, & \forall x_W \in A_W \\ 0, & \text{otherwise.} \end{cases} \tag{19}$$

Equations (18) and (19) can also be expressed as the integral:

$$Y = \int_{A_W} f_W(x_W) \, dx_W \tag{20}$$

where the integration is taken over all values of $X_W \in A_W$.

Design yield can also be expressed in the space of process disturbances S_D, which is an Euclidean space that contains all of the process disturbance given in the vector D. We define the acceptability region in space S_D as follows. For specific values of the variables C and L, denoted by C^0 and L^0, a disturbance vector Δ will be in the acceptability region, that is denoted by $A_D(C^0, L^0)$, if $F_W(C^0, L^0, \delta) \in A_W$ where F_W is defined by (1). In the other words, there exists, for a given L^0 and C^0, in the space of process disturbances a set of disturbances $A_D(C^0, L^0)$, which will not cause the performance of an IC to be unacceptable. This set can also be viewed as the projection of the acceptability region A_W into the space S_D. Hence, we can express design yield as

$$Y = \int_{A_D(C^0, L^0)} f_D(\delta) \, d\delta \tag{21}$$

where $f_D(\delta)$ is the JPDF which describes D.

The design yield, as defined in (18) and (19) or (21) is difficult to evaluate, because in general it is not possible to derive an analytic expression that relates IC performance to the disturbance, layout and control parameters. In other words it is not possible to obtain an explicit expression for $A_D(C^0, L^0)$. Furthermore, some of the components of D can only be vaguely defined and only a subset of state variables are actually known. However, through an analysis of the basic properties of the process disturbances, as was carried out in the previous section we can develop techniques for evaluation of design yield.

4.3. Decomposition of the Design Yield Equations

Towards this end it is convenient to view A_Q as being composed of the set of *functionally acceptable performances*

$$A_Q^F = \{Q \in S_Q^F | \text{each } q_j^F \text{ is acceptable} \quad \forall j = 1, 2, \cdots, n_F\} \tag{22}$$

and the set *parametricly acceptable performances*

$$A_Q^P = \{Q \in S_Q^P | \text{each } q_j^P \text{ is acceptable} \quad \forall j = 1, 2, \cdots, n_P\}. \tag{23}$$

Existence of two different types of performance leads us to consider two types of acceptability regions: one which assures functional correctness of the IC behavior, and another one, which assures acceptable quality of performance. Specifically, in the space of process state variables S_W, we can define the *functional acceptability region*,

$$A_W^F = \{X_W \in S_W | Q(X_W) \in A_Q^F\} \tag{24}$$

and the *parametric acceptability region*

$$A_W^P = \{X_W \in S_W | Q(X_W) \in A_Q^P\}. \tag{25}$$

Consequently, in the space of process disturbances S_D, assuming a given L^0 and C^0, we can define equivalents of functional and parametric acceptability regions as sets of disturbances which do not cause IC to have incorrect functionality or poor quality, respectively. We denote these sets by $A_D^F(C^0, L^0)$ and $A_D^P(C^0, L^0)$, respectively. Thus, we can define the *functional yield* as

$$Y_{\text{FUN}} = \int_{A_D^F(C^0, L^0)} f_D(\delta) \, d\delta \tag{26}$$

where as before $f_D(\delta)$ is the JPDF describing D, and the *parametric yield* as

$$Y_{\text{PAR}} = \int_{A_D^P(C^0, L^0)} f_D(\delta) \, d\delta \tag{27}$$

which are probabilities that the IC is correct functionally and that its performance characteristics are within some desired limits.

Decomposition of the design yield into parametric and functional components provides an opportunity for significant simplifications in evaluating design yield. This becomes evident if we note that, since $A_D^F(C^0, L^0)$ contains those process disturbances that do not disturb functional correctness of the IC, then $(1 - Y_{\mathrm{FUN}})$ can be estimated by searching for the complement of $A_D^F(C^0, L^0)$, i.e., determining the set of process disturbances that cause structural failure and/or hard performance faults only. Thus estimating $(1 - Y_{\mathrm{FUN}})$ we have to take into account such defects, as dislocations, spot and point defects, stacking faults, oxide pinholes, etc., which cause structural changes in the connectivity of an IC but are of the local nature. As a consequence, Y_{FUN}, as defined by (26), may be approximated by Y'_{FUN} given by

$$Y'_{\mathrm{FUN}} = \int_{A_D^F(C^0, L^0)} f_{D'}(\delta')\, d\delta' \tag{28}$$

where $f_{D'}(\delta')$ is the JPDF of D', which describes only the subset of D that characterizes local processing defects. By dealing with D', instead of D, we do not need to evaluate IC performance in order to determine $A_D^F(C^0, L^0)$ because process disturbances that cause structural changes in the connectivity of an IC can be found by analyzing the layout of the IC and the statistical characterization of the local defects only. In particular to determine $A_D^F(C^0, L^0)$, we need only investigate the *critical area*, of a layout, i.e., that area an IC which is vulnerable to the occurrence of a point or spot defect. Hence, Y'_{FUN} can be estimated by using one of a number existing geometrical approaches [33], [47], [48], [62], [47], [37], [5], [25], [23].

Computation of parametric yield is also simplified using this decomposition, due to the fact that a single spot or point defect, if it does not change the circuit connectivity, usually does not affect the quality of IC performance. So, local defects can be discarded from consideration when computing parametric yield. Hence, $A_D^P(C^0, L^0)$ can be estimated in the subspace of $S_{D'}$ denoted by $S_{D''}$, that contains only global deformations, such as mask misalignments and lateral edge displacements. The effect that these disturbances have on IC performance can be effectively modeled [29] and [34]. Hence, parametric yield:

$$Y'_{\mathrm{PAR}} = \int_{A_D^F(C^0, L^0)} f_{D''}(\delta'')\, d\delta'' \tag{29}$$

where $f_{D''}(\delta'')$ is the JPDF which describes D'', can be estimated. In fact, it is exactly this type of yield that is usually used as the objective function for yield maximization [9], [38], [10], [4], [60], [39], [6], [19], [1], [58].

V. Discussion

In Section 4.1, we introduced a number of yield estimates and methods for yield prediction. Experimentally based yield estimates are useful for evaluating the efficiency of different parts of the IC fabrication process and can be used to diagnose the cause of yield losses. In this section, we discuss the relationships between these estimates in order to facilitate an understanding of the physical meaning of each estimate. We then illustrate how this information may be used to analyze a manufacturing process and relate it to production cost.

5.1. Relationships Between Manufacturing and Design Yields

As we indicated above, the manufacturing yield Y_M^* is a good measure of the overall efficiency of a fabrication process. Observe that manufacturing yield can be viewed as the product of three other yield estimates: wafer yield Y_W^*, probe testing yield Y_{PT}^*, and final testing yield Y_{FT}^*:

$$Y_M^* = \frac{N_F}{N} = \frac{N_W}{N}\frac{N_P}{N_W}\frac{N_F}{N_P} = Y_W^* Y_{\mathrm{PT}}^* Y_{\mathrm{FT}}^*. \tag{30}$$

By evaluating and comparing the values of these three factors, we can identify the principal reason for a low yield in a manufacturing process. We will return to this point in the analysis of the manufacturing cost.

The manufacturing yield Y_M^* is typically close to the design yield, Y. However, in reality, IC chips cannot be tested for all possible faults (in other words, the test coverage will always be less than 100 percent) and, therefore, Y_M^* is always an overestimation of Y:

$$Y_M^* \geq Y. \tag{31}$$

In particular, both testing programs (probe and final) are typically arranged in such a way that the parametric tests are grouped together and functional tests are grouped together. If we estimate the yield components from these selective measurements, these estimates should be close to the estimates derived in the previous section. However, due to the fact that in an evaluation technique we strictly separated fault types, the values obtained from measurements will be less than the theoretical estimates. Similarly, because this strict separation of faults does not actually occur in reality, the actual yield Y will be less than the calculated product of the parametric and functional yields

$$Y \leq Y'_{\mathrm{PAR}} Y'_{\mathrm{FUN}} \tag{32}$$

Finally, if the acceptability region in terms of device parameters is properly established, the processing yield Y_{PRO} defined in terms of test structure measurements, should be approximately equal to the parametric yield Y_{PAR}:

$$Y_{\mathrm{PRO}} \approx Y_{\mathrm{PAR}}. \tag{33}$$

Hence, the design yields can be used to predict or derive bounds for some manufacturing yields. In the general case, however, design and manufacturing yields have been defined for different purposes, and therefore, accuracy of the yield prediction should not be evaluated by a simple comparison to the manufacturing yield.

5.2 *Examples of Yield Analysis*

The above discussion attempts to present a unified framework for various yield estimation and prediction purposes. However, in the past, much simpler approaches to yield estimation and prediction sufficed. For example for some types of circuits, it is well known that the manufacturing yield is strongly correlated with the density of spot defects. Therefore, estimation of functional yield has been reported to be adequate enough to predict manufacturing yield for such products as static and dynamic RAM memories. On the other hand, experience with bipolar IC's may lead to the conclusion that spot defects play an insignificant role as far as yield is concerned and that estimation of parametric yield is adequate to predict manufacturing yield. We wish to present an argument that while such simplified approaches to manufacturing yield prediction were possible in certain instances, they are unrealistic for the case of VLSI technology.

Towards this end we focus our discussion on three specific examples of IC's: a moderate size CMOS gate array, a high-precision bipolar operational amplifier, and a large (e.g., 256 kbit) NMOS dynamic RAM. We will consider primary causes of yield loss for these IC's and illustrate the importance of various yield estimates. To compare the dominating failure modes and the main causes for yield losses in these IC's, we present the following figures. According to our previous classification, we have divided failure modes into two groups: structural and performance. The causes for failure have been divided into geometrical and electrical, and each of these is divided into global and local. To provide additional insight into the yield loss mechanisms in these IC's, we also specify which yield components (for both manufacturing and design yields) are affected the most.

A CMOS gate array is an example of an IC where high manufacturing yield is quite common. Due to relaxed layout design rules and very conservative design of functional blocks, the soft performance failures are perhaps the most common. One of the main reasons for yield loss must be an excessive signal delay due to long interconnections. Hence, the estimation and prediction of parametric yield seems to be the only yield related problem which may be important in the design and manufacture of gate arrays. Fig. 12 illustrates the relation between faults and disturbances for this type of circuit.

For the case of the bipolar op amp, which is a relatively small size IC, yield must be primarily sensitive to vertical effects due to relaxed layout design rules. More specifically, in bipolar IC's vertical impurity distributions almost exclusively determine the important device parameters such as current gain coefficients and resistances, and hence op amp performance. Hence, as it is summarized in Fig. 13, the wafer yield loss is primarily due to global variations in impurity profiles, such as improper junction depths and excessive emitter dip-effect. These faults are typically detected in the junction depth measurements and may lead to rejection of wafers or even entire lots. Variations in

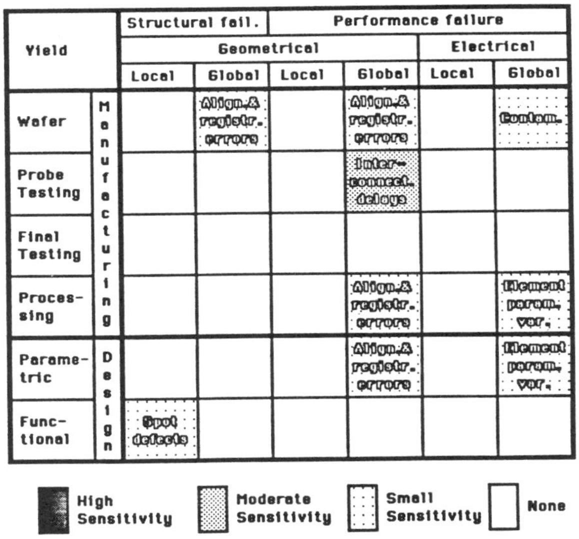

Fig. 12. Yield sensitivity of a CMOS gate array.

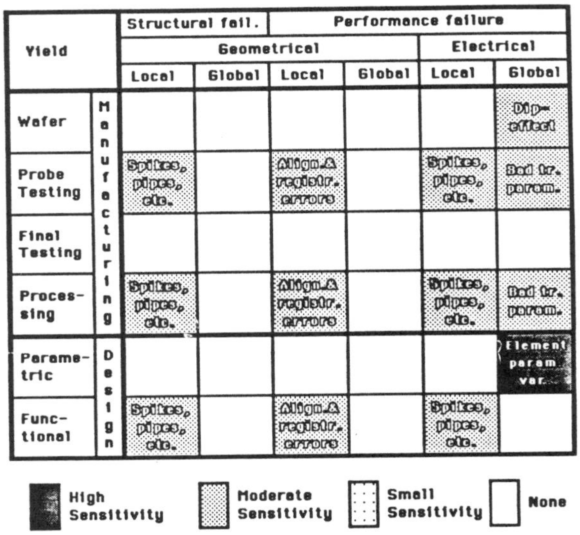

Fig. 13. Yield sensitivity of an operational amplifier.

impurity profiles and surface quality is also one of the most important reasons for a low probe testing yield. Other reasons include such local defects as diffusion spikes or pipes which may led to shorts between emitter and collector regions, or to creating regions with significantly reduced base width which are susceptible to second breakdown effects. Soft performance faults can be also caused by mismatch in device parameters caused by linewidth variations due to lithographic processes. Observe that since in this type of IC, final tests are almost identical to the probe tests, the percentage of chips rejected after final assembly should be extremely small. Also note, that for this case, test structure measurements can be used efficiently to diagnose the op amp failure modes, and therefore, the processing yield is quite similar to probe testing yield. Finally, as suggested in Fig. 13, parametric yield losses due to device parameter variations are more significant than functional yield losses. Therefore, it is critical to "cen-

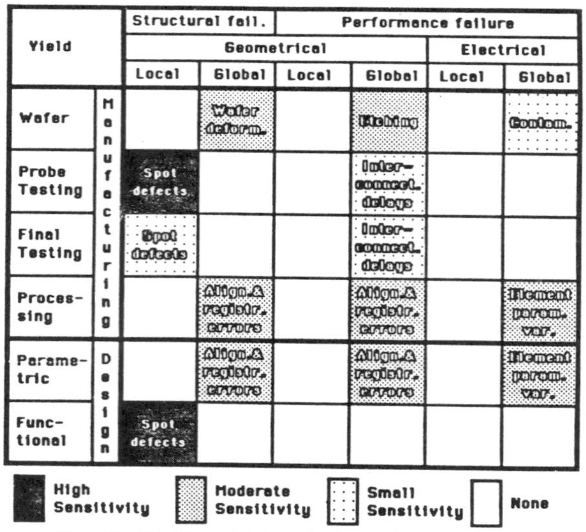
Yield
Structural fail.
Performance failure
Geometrical
Electrical
Local
Global
Local
Global
Local
Global
Wafer
Probe Testing
Final Testing
Proces-sing
Parame-tric
Func-tional
Manufacturing
Design
Wafer deform.
Etching
Contam.
Spot defects
Inter-connect delays
Spot defects
Inter-connect delays
Align.& registr. errors
Align.& registr. errors
Element param. var.
Align.& registr. errors
Align.& registr. errors
Element param. var.
Spot defects
High Sensitivity
Moderate Sensitivity
Small Sensitivity
None

Fig. 14. Yield sensitivity of a large dynamic RAM.

ter" the op amp design to make it less sensitive to the inherent process fluctuations. Hence, for the design and manufacturing of advanced bipolar circuit both parametric and functional yield estimates may be required.

The relation between yield losses and causes for a large dynamic RAM is illustrated in Fig. 14. Of course, the manufacturing yield of such an IC is much lower than yield which can be achieved for the IC's discussed above. The dominating failure mechanisms for such VLSI circuit are also different. In order to achieve the density required to implement large dynamic RAM's, minimum device sizes and very aggressive spacing rules have to be employed. This results in large yield sensitivity to spot defects. Thus structural failures that are caused by spot defects are the dominant factors which degrade yield. Usually these faults are detected by probe tests, and are either repaired, if there is sufficient redundancy on the chip, or the chips are rejected to avoid costly assembly operations. Thus probe testing yield losses have been marked in Fig. 14 as the most significant. It has to be noted, however, that an IC of such a complexity is sensitive to practically all process steps and global variations introduced in these steps. Therefore, during memory fabrication, careful in-line testing is performed to detect excessive variations in etching, water deformations or ion contamination. Hence, some wafers are rejected during the process and wafer yield is affected. Another interesting failure mode is soft parametric failure due to insufficient speed in either the write or read cycle. This failure can be caused by either excessive interconnect parasitics or incorrect device parameters (such as transconductance). In very fast dynamic memories, this failure mode seems to be very important. Hence we see that for the case of advanced memory circuits most types of process disturbances can contribute to yield loss and for effective design, as well as manufacturing, all types of failure modes and yields have to be considered.

Again, the above discussion has been presented to illustrate the point that while it has been the case that in less advanced technologies, or relaxed design rules, yield losses are dominated by a single mechanism and a simplified approach to the yield analysis can be employed, for VLSI circuits yield is determined by multiple phenomena. Therefore, an approach to yield analysis that combines various yield estimation methods seems to be unavoidable.

5.3. Yield and Production Cost

The yield measures introduced in this paper have been typically used for yield maximization purposes. The oldest and most practically sound approaches were experimentally-based and have been applied for maximization of both manufacturing and probe yields. Yield improvements have been achieved by changing process controls, IC layout or minimizing variability of the process disturbances. Although a complete discussion of these approaches is beyond the scope of this paper, it is important to note their limitations. Specifically, these approaches attempt to improve process efficiency using yield as the sole criterion and thus overlook overall manufacturing process strategy. We wish to present the argument that process efficiency has to be viewed from a wider perspective than just yield maximization. Towards this end, we present a discussion of the fabrication cost minimization problem.

The objective of an IC fabrication process is to produce a given number of working IC's within some given time period. To maximize profit, these IC's must be fabricated at minimal cost. Naturally, the higher the manufacturing yield, the lower the production cost per chip. However, since yield does not take into account the time and cost associated with manufacturing operations, nor the constraints on the number of chips that have to be manufactured within certain time period, it is not sufficient to predict costs. In fact, any cost function should include such factors as a penalty for not meetng the specified number of chips and the cost of operation and maintenance of the fabrication equipment, in addition to yield. These considerations are even more complicated if the cost of rework of selected manufacturing steps is taken into account. Such a complex cost function is necessary to perform a rigid statistical quality control and cost minimization process. However, since our goal here is to demonstrate the relationship between the production cost per chip and manufacturing yield, we will restrict our attention to much simpler expressions for the cost function.

To derive an expression for the overall cost of manufacturing W wafers (which contain N chips), we have to take into account the cost of fabrication for the chips that went through the entire fabrication process (up to the final selection tests) as well as the cost of fabricating chips (or wafers) that were rejected at the intermediate selection points. Towards this end we note that the cost of producing the $(W - W_W)$ wafers that are rejected during wafer processing, denoted by C_1, is

$$C_1 = (W - W_W)(C_P^* + C_W) \quad (34)$$

where C_W is the cost of a wafer and C_P^* is the equivalent processing cost per wafer taking into account that wafers were rejected at different stages of the wafer processing phase. Since it is more convenient to carry out an analysis in terms of chips rather than wafers, C_1 can be expressed equivalently as:

$$C_1 = (N - N_W)(c_P^* + c_W) \tag{35}$$

where the costs are adjusted per chip.

Similarly, the cost of producing the $(N_W - N_P)$ chips that were rejected in the probe selection tests, C_2, is given by

$$C_2 = (N_W - N_P)(c_{PT} + c_P + c_W) \tag{36}$$

where c_{PT} is the probe testing cost per chip and c_P is the actual cost of wafer processing operations (including material cost, and equipment operation cost and maintenance) per chip.

Finally, the cost of producing the N_P chips that reach final selection tests, C_3, is given by

$$C_3 = N_P(c_{FT} + c_A + c_{PT} + c_P + c_W) \tag{37}$$

where c_{FT} is the cost of final selection tests per chip and c_A is the overall assembly cost per chip.

Hence, the entire cost of manufacturing N chips is equal to

$$C = N_P(c_{FT} + c_A + c_{PT} + c_P + c_W) + N_W\left(1 - \frac{N_P}{N_W}\right)(c_{PT} + c_P + c_W) + N\left(1 - \frac{N_W}{N}\right)(c_P^* + c_W). \tag{38}$$

Note that only N_F out of N_P chips are fault-free and can be sold to customers. Therefore, the overall cost, C, divided by a number of good chips, N_F, is a good measure of fabrication process efficiency and should be minimized in the process.

$$\begin{aligned}\frac{C}{N_F} &= \frac{N_P}{N_F}(c_{FT} + c_A + c_{PT} + c_P + c_W) \\ &+ \frac{N_W}{N_F}\left(1 - \frac{N_P}{N_W}\right)(c_{PT} + c_P + c_W) \\ &+ \frac{N}{N_F}\left(1 - \frac{N_W}{N}\right)(c_P^* + c_W) \\ &= (c_{FT} + c_A + c_{PT} + c_P + c_W) \\ &+ \frac{N_P - N_F}{N_F}(c_{FT} + c_A + c_{PT} + c_P + c_W) \\ &+ \frac{N_W}{N_F}(1 - Y_{PT}^*)(c_{PT} + c_P + c_W) \\ &+ \frac{N}{N_F}(1 - Y_W^*)(c_P^* + c_W) \\ &= (c_{FT} + c_A + c_{PT} + c_P + c_W) \\ &+ \frac{Y_P^*}{Y_M^*}(1 - Y_{FT}^*)(c_{FT} + c_A + c_{PT} + c_P + c_W) \\ &+ \frac{Y_W^*}{Y_M^*}(1 - Y_{PT}^*)(c_{PT} - c_P + c_W) \\ &+ \frac{1}{Y_M^*}(1 - Y_W^*)(c_P^* + c_W).\end{aligned} \tag{39}$$

The above equation indicates that while the cost per good chip decreases if the manufacturing yield increases, the problems of cost minimization and yield maximization are not equivalent. For example, if the wafer yield loss is critical, better process management and supervision are necessary. If the probe yield loss is a concern, more stringent wafer rejection criteria based upon in-line measurements should be imposed (i.e., more adequate quality control system is necessary). Finally, if the yield loss during final test is significant, one of two conditions is present: either the fault coverage in the probe tests is not sufficient or the assembly operations have not been performed correctly. In the first case, the probe testing program has to be expanded, in the latter case, assembly operations have to be modified.

The transition from this cost function to profit maximization is trivial, since the profit per chip can be expressed as the difference between the market value of the IC chip and the cost of fabricating the chip. Hence, even such simple considerations can be useful in the market analysis, production planning and scheduling. However, as we stated in the beginning of this section, additional factors such as the time required to satisfy demand for a particular chip and the penalty for not meeting the demand should enter the profit considerations.

The above discussion indicates that maximization of individual yield components does not necessarily lead to an improvement in overall process efficiency. It is also important to note that to realistically solve the production cost minimization problem, a general framework for viewing process efficiency, such as presented in this paper, is necessary.

VI. Applications of the Proposed Framework and Conclusions

In this paper we have presented a framework for viewing the manufacturing yield of VLSI circuits. Specifically, we reviewed and classified the disturbances that cause yield loss as well as the effects that these disturbances have on performance. We then formally defined a number of yield measures which are useful both during the design and manufacturing processes, and established relationships between these yield measures. Finally, we related yield to manufacturing costs for the purpose of providing a common denominator for the discussion of manufacturing process efficiency. In this section we draw some general conclusions and relate a number of recently reported results through the framework we have developed.

The most important conclusions that we wish to emphasize are as follows.

1) The random phenomena that occur in an IC manufacturing process must be taken into account during

the design and testing phases of IC realization. Towards this end, the basic physical nature of such random phenomena have to be investigated and modeled in order to provide process and circuit designers with the adequate information about process instabilities.

2) Both functional and parametric yields are important when determining process efficiency. While it has been the case that one of these has dominated, in VLSI circuits we can not, in general, make such simplifications without exploring main causes of yield losses.

3) Both manufacturing process and IC design should be optimized with respect to a fabrication efficiency which can be expressed in terms of appropriately defined yields.

Let us now consider the relationship between some recently published yield related results and the framework we have developed. Specifically, we wish to consider the topics of yield maximization, process diagnosis and control, and IC testing.

As mentioned in Section IV, evaluation of parametric yield during the design process relies on the ability to predict the effects of process disturbances on device parameters. Therefore, a statistical process simulation technique was proposed [30], [20], [29]. FABRICS [29], [34] is a tool which implements this technique and takes as input a set of process controls and device layout, and generates device model parameters compatible with the device models implemented in SPICE. To obtain accurate simulation, FABRICS II [34] must be tuned a particular IC fabrication process which requires the determination of a set of process disturbances. To identify the probability density functions that describe process disturbances the automatic extraction system PROMETHEUS [46], based upon measurements performed on specially designed test structures [7], was developed. Thus a parametric yield prediction system can be realized using FABRICS II and a circuit simulator such as SPICE or SAMSON [44]. Further extensions of the methodology implemented in FABRICS are also considered [21], [43], [18], [54], [16], [24].

We have demonstrated that with such statistical process simulation technique, a number of design optimization, process optimization and process diagnosis problems can be solved. Some of the examples in the IC design area are:

- an op amp yield optimization versus selected layout dimensions and emitter diffusion parameters [29];
- minimization of power-delay product of the dynamic RAM in terms of nominal design and process controls reported in [34];
- worst-case design in terms of statistically independent process disturbances which allows for realistic prediction and precise estimation of the probability of occurrence of the worst case [53];
- statistical design of VLSI cells in the space of layout and process parameters [52].

Application of FABRICS II for process optimization can be illustrated by an example of processing yield optimization technique proposed in [14] and implemented in a system called PROMISE. In the area of process diagnosis very promising results were obtained as well. An efficient fault simulation system has been developed [54] and employed in the fault diagnostic system called PROD [35] which employs statistical pattern recognition techniques for process diagnosis [57], [51].

Prediction of functional yield requires the ability to determine the effects of global geometrical deformations and spot defects. In [25], [23], the functional yield model that includes both local and global random deformations of the layout was proposed. A simulation tool that employs Monte Carlo method to determine the effect of spot defects on IC layout was developed [61] and can be applied for yield estimation and redundancy analysis. A technique which evaluates a probability of failure caused by global deformations of the layout combined with spot defects effects is reported in [42]. With the above models, it is possible to develop design rules that can maximize functional yield [23] and [42]. Finally, in [27] and [45] an extension of the traditional spot defect oriented analysis of the manufacturing yield into the area of testing was suggested. Based upon likelihood of faults more efficient testing procedures have been proposed.

The cost minimization approaches to IC manufacturing process control have been proposed in [28] and [22]. Currently, a quality/adaptive control system is under development and is proposed in [55].

Thus we see that the framework described in this paper provides a uniform basis for the development of yield enhancement methods. Therefore, we hope that considerations presented in the paper will also help to motivate further development of an integrated approach to CAD, CAM, CAT of VLSI circuits.

Acknowledgment

This paper is the outgrowth of close collaboration between the authors and large number of individuals. The authors gratefully acknowledge these contributors without whom the work described could not have been accomplished. We list these individuals alphabetically: I. Chen, J. Deszczka, J. Ferguson, D. J. Giannopoulos, J. Gempel, T. Gutt, P. Kager, D. Korzec, W. Kuzmicz, K. K. Low, Z. Li, S. R. Nassif, P. K. Mozumder, P. Odryna, Z. Pizlo, R. Razdan, M. J. Saccamango, K. A. Sakallah, J. P. Shen, C. R. Shyamsundar, C. J. Spanos, M. Syrzycki, X. Tian, L. Vidigal, and H. Walker.

References

[1] K. J. Antreich and R. K. Koblitz, "Design centering by yield prediction," *IEEE Trans. Circuits Syst.*, vol. CAS-29, pp. 88–95, Feb. 1982.

[2] Y. Aoki, T. Toyabe, S. Asai, and T. Hagiwara, "CASTAM: A process variation analysis simulator for MOS LSI's," *IEEE Trans. Electron Devices*, vol. ED-31, pp. 1462–1467, 1984.

[3] P. Balaban and J. J. Golembeski, "Statistical analysis for practical circuit design," *IEEE Trans. Circuits Syst.*, vol. CAS-22, pp. 100–108, Feb. 1975.

[4] J. W. Bandler and H. L. Abdel-Malek, "Optimal centering, tolerancing, and yield determination via updated approximations and cuts," *IEEE Trans. Circuits Syst.*, vol. CAS-25, pp. 853–871, Oct. 1978.

[5] J. Bernard, "The IC yield problem: A tentative analysis for MOS/SOS

circuits," *IEEE Trans. Electron Devices*, vol. ED-25, pp. 939–944, Aug. 1978.
[6] R. K. Brayton, S. W. Director, and G. D. Hachtel, "Yield maximization and worst-case design with arbitrary statistical distributions," *IEEE Trans. Circuits Syst.*, vol. CAS-27, pp. 756–764, Sept. 1980.
[7] I-hao Chen and A. J. Strojwas, "A methodology for optimal test structure design," in *Proc. of CICC 85*, pp. 520–523, Portland, 1985.
[8] P. Cox, P. Yang, and P. Chatterjee, "Statistical modeling for efficient parametric yield estimation," in *Proc. Int. Electron Device Meeting*, pp. 242–245, 1984.
[9] S. W. Director and G. D. Hachtel, "The simplicial approximation approach to design centering," *IEEE Trans. Circuits Syst.*, vol. CAS-24, pp. 363–372, July 1977.
[10] S. W. Director, G. D. Hachtel, and L. M. Vidigal, "Computationally efficient yield estimation procedures based on simplicial approximation," *IEEE Trans. Circuits Syst.*, vol. CAS-25, pp. 121–130, Mar. 1978.
[11] A. V. Ferris-Prabhu, "Modeling of critical area in yield forecasts," *IEEE J. Solid-State Circuits*, vol. SC-20, pp. 874–878, Aug. 1985.
[12] —–, "Defect size variations and their effect on the critical area of VLSI devices," *IEEE J. Solid-State Circuits*, vol. SC-20, pp. 878–880, Aug. 1985.
[13] S. K. Ghandhi, *VLSI Fabrication Principles*. New York: Wiley, 1983.
[14] D. J. Giannopoulos and S. W. Director, "IC Fabrication Process Optimization," in *Proc. of ICCAD-84*, pp. 164–166. Santa Clara, CA, November 1984.
[15] A. Gupta, W. A. Porter, and J. W. Lathrop, "Defect analysis and yield degradation of integrated circuits," *IEEE J. Solid-State Circuits*, vol. SC-9, pp. 96–103, June 1974.
[16] P. Kager and A. J. Strojwas, "PI/C: process interpreter/compiler," in *Proc. of ICCAD 85*, Santa Clara, CA, 1985.
[17] D. P. Kennedy, P. C. Murley, and R. R. O'Brien, "A statistical approach to the design of diffused junction transistors," *IBM J. Res. Develop.*, vol. 8, no. 5, pp. 482–495, Nov. 1964.
[18] W. Kuzmicz, "Modeling of minority carrier current in heavily doped regions with application to bipolar IC simulation," in *Proc. of ICCAD-84*, pp. 182–184, Santa Clara, CA, 1984.
[19] M. R. Lightner and S. W. Director, "Multiple criterion optimization with yield maximization," *IEEE Trans. Circuits Syst.*, vol. CAS-28, pp. 781–790, Aug. 1981.
[20] W. Maly, A. J. Strojwas, and S. W. Director, "Fabrication based statistical design of monolithic IC's," in *Proc. ISCAS*, pp. 135–138, Chicago, IL, Apr. 1981.
[21] W. Maly, J. Gempel, D. Korzec, A. P. Piotrowski, and M. Syrzycki, "Statistical simulation of VLSI IC cell," in *Proc. ICCAD-83*, pp. 254–255. Santa Clara, CA, Sept. 1983.
[22] W. Maly and Z. Pizlo, "Tolerance assignment for IC selection tests," *IEEE Trans. Computer-Aided Design*, vol. CAD-4, pp. 156–162, Apr. 1985.
[23] W. Maly, "Modeling of lithography related yield losses for CAD of VLSI circuits," *IEEE Trans. Computer-Aided Design*, vol. CAD-4, pp. 166–177, July 1985.
[24] —–, "Modeling of random phenomena in CAD of IC—A basic theoretical consideration," in *Proc. of ISCAS 85*, pp. 427–430, Kyoto, Japan, 1985.
[25] W. Maly and J. Deszczka, "Yield estimation model for VLSI artwork evaluation," *Electron. Lett.*, vol. 19, no. 6, pp. 226–227, Mar. 1983.
[26] W. Maly and S. W. Director, "Dimension reduction procedure for simplicial approximation approach to design centering," *Proc. Inst. Elect. Eng.*, vol. 127, no. 6, pp. 255–259, Dec. 1981.
[27] W. Maly, J. Ferguson, and J. P. Shen, "Systematic characterization of physical defects for fault analysis of MOS IC cells," in *Proc. Int. Test Conf., 1984*. Philadelphia, PA, Oct. 1984.
[28] W. Maly and T. Gutt, "Simulation of quality control system applied in silicon IC technology," in *Simulation of Control Systems*, I. Troch (Ed.), The Netherlands: North Holland, 1978.
[29] W. Maly and A. J. Strojwas, "Statistical simulation of the IC manufacturing process," *IEEE Trans. Computer-Aided Design*, vol. CAD-1, pp. 120–131, July 1982.
[30] —–, "Simulation of bipolar elements for statistical circuit design," in *Proc. ISCAS*, pp. 788–791, Tokyo, Japan, July 1979.
[31] T. E. Mangir, "Sources of failures and yield improvement for VLSI and restructurable interconnects for RVLSI and WSI: Part I—Source of failures and yield improvements for VLSI," in *Proc. IEEE*, vol. 72, pp. 690–708, June 1984.
[32] G. E. Moore, "A simple method for modeling VLSI yields," *Electronics*, vol. 43, pp. 126–130, 1970.
[33] B. T. Murphy, "Cost-size optima of monolithic integrated circuits," *Proc. IEEE*, vol. 52, pp. 1537–1545, Dec. 1964.
[34] S. R. Nassif, A. J. Strojwas, and S. W. Director, "FABRICS II: A statistically based IC fabrication process simulator," *IEEE Trans. Computer-Aided Design*, vol. CAD-3, pp. 20–46, Jan. 1984.
[35] P. Odryna and A. J. Strojwas, "PROD: A VLSI fault diagnosis system," *IEEE Design & Test of Computers*, Dec. 1985.
[36] T. Okabe *et al.*, "A simple method for modeling VLSI yields," *Elec. Eng. Japan*, vol. 92: pp. 135–141, 1972.
[37] O. Paz and T. Lawson, Jr., "Modification of Poisson statistics: Modeling defects induced by diffusion," *IEEE J. Solid-State Circuits*, vol. SC-12, pp. 540–546, Oct. 1977.
[38] J. F. Pinel and K. Singhal, "Efficient Monte Carlo Computation of circuit yield using importance sampling," in *Proc. of ISCAS 77*, pp. 575–578, 1977.
[39] E. Polak and A. Sangiovanni-Vincentelli, "Theoretical and computational aspects of the optimal design centering, tolerancing, and tuning problem," *IEEE Trans. Circuits Syst.*, vol. CAS-26, pp. 795–813, Sept. 1979.
[40] P. J. Rankin, "Statistical modeling for integrated circuits," *IEE Proc., G*, vol. 129, pp. 186–191, Aug. 1982.
[41] K. V. Ravi, *Imperfections and Impurities in Semiconductor Silicon*. New York: Wiley, 1981.
[42] R. Razdan and A. J. Strojwas, "Statistical design rule developer," in *Proc. of ICCAD 85*, Santa Clara, CA, 1985.
[43] M. J. Saccamango and A. J. Strojwas, "Analytical modeling of small-geometry MOSFET's," in *ICCAD Dig. Tech. Papers*, pp. 170–172, 1984.
[44] K. A. Sakallah and S. W. Director, "SAMSON2: An event driven VLSI circuit simulator," Research Rep. CMUCAD-84-37, SRC-CMU Center for Computer-Aided Design, Dep. of Elect. and Comp. Engineering, Carnegie-Mellon Univ., Sept. 1984.
[45] J. P. Shen, W. Maly, and F. J. Ferguson, "Inductive fault analysis of MOS integrated circuits," *IEEE Design & Test*, Dec. 1985.
[46] C. J. Spanos and S. W. Director, "PROMETHEUS: A program for VLSI process parameters extraction," in *Proc. of ICCAD-83*, pp. 176–177, Santa Clara, CA, Sept. 1983.
[47] C. H. Stapper, "Defect density distribution for LSI yield calculations," *IEEE Trans. Electron Devices*, vol. ED-20, pp. 655–657, July 1973.
[48] —–, "On a composite model to the IC yield problem," *IEEE J. Solid-State Circuits*, vol. SC-10, pp. 537–539, Dec. 1975.
[49] C. H. Stapper *et al.*, "Yield modeling for productivity optimization of VLSI memory chips with redundancy and partially good product," *IBM J. Res. Develop.*, vol. 24(3), pp. 398–409, May 1980.
[50] Ch. H. Stapper, "Yield model for 256K RAMs and beyond," in *Proc. of 1982 Int. Solid-State Circuits Conf.*, pp. 12–13, 1982.
[51] A. J. Strojwas and S. W. Director, "A pattern recognition based method for IC failure analysis," *IEEE Trans. Computer-Aided Design*, vol. CAD-4, Jan. 1985.
[52] A. J. Strojws, S. R. Nassif, and S. W. Director, "Optimal design of VLSI minicells using a statistical process simulator, in *Proc. 1983 ISCAS*, Jan. 1983.
[53] —–, "A methodology for worst case design of integrated circuits," in *Proc. of ICCAD-83*, Sept. 1983.
[54] A. J. Strojwas and Z. Li, "Bipolar device models in the statistical process/device simulator FABRICS II. in *Proc. of ICCAD 84*, pp. 185–187, Santa Clara, CA, 1984.
[55] A. J. Strojwas, "CMU-CAM system," in *Proc. of 22-nd DA Conf.*, pp. 319–325, Las Vegas, NV, 1985.
[56] A. J. Strojwas, S. R. Nassif, and S. W. Director, "A methodology for worst case design of integrated circuits," in *Proc. of ICCAD-83*, pp. 152–153, Santa Clara, CA, Sept. 1983.
[57] A. J. Strojwas, "A pattern recognition based system for IC failure analysis," Ph.D. dissertation, Carnegie-Mellon Univ., Sept. 1982.
[58] M. A. Styblinski and L. J. Opalski, "Random perturbation method for IC yield optimization with deterministic process parameters," in *Proc. of ICCAD-84*, pp. 977–980, Santa Clara, CA, Nov. 1984.
[59] S. M. Sze, *VLSI Technology*. New York: McGraw-Hill, 1983.
[60] K. S. Tahim and R. Spence, "A radial exploration approach to manufacturing yield estimation and design centering," *IEEE Trans. Circuits Syst.*, vol. CAS-26, pp. 768–774, Sept. 1979.
[61] H. Walker and S. W. Director, "Yield simulation for integrated circuits," in *Proc. of ICCAD-83*, pp. 256–257, Santa Clara, CA, Sept. 1983.
[62] R. M. Warner, Jr., "Applying a composite model to the IC yield problem," *IEEE J. Solid-State Circuits*, vol. SC-9, pp. 86–95, June 1974.

Author Index

Editor's Biography

Andrzej J. Strojwas was born in Poland, on August 20, 1952. He received the M.S. degree (with honors) in electronic engineering from the Technical University of Warsaw, Poland and the Ph.D. degree in electrical engineering from Carnegie Mellon University in 1982. From October 1976 to August 1979, he was a member of the faculty of the Technical University of Warsaw. During the summer of 1981 he worked at Harris Semiconductor Co. on modeling and diagnosis of the IC manufacturing process. He is currently an Associate Professor of Electrical and Computer Engineering at Carnegie Mellon University. His research interests include statistical modeling of IC fabrication processes and semiconductor devices, computer-aided design of VLSI circuits, and control and diagnosis of the IC fabrication process. In 1975, he received first prize in the Copernicus Competition for the best student in the Technical University of Warsaw. In 1985, he received an award for the best paper published in the IEEE TRANSACTIONS ON CAD OF ICAS. At the 22nd IEEE-ACM Design Automation Conference he was given the Best Paper Award. In the same year, he also received the George Tallman Ladd Award for excellence in research from Carnegie Mellon University. In 1986, he received the Presidential Young Investigator Award. He was an Associate Editor and is currently an Editor of the IEEE TRANSACTIONS ON CAD OF ICAS and he is a member of the IEEE CANDE Committee.